The Life-Cycle of Pharmaceuticals in the Environment

Related titles

Cell Culture Models of Biological Barriers: In Vitro Test Systems for Drug Absorption and Delivery
(ISBN 978-0-415-27724-2)

Drug Absorption Studies: In Situ, In Vitro and In Silico Models
(ISBN 978-0-387-74900-6)

Woodhead Publishing Series in Biomedicine: Number 51

The Life-Cycle of Pharmaceuticals in the Environment

Barrie M. Peake
Rhiannon Braund
Alfred Y.C. Tong
Louis A. Tremblay

ELSEVIER

AMSTERDAM • BOSTON • CAMBRIDGE • HEIDELBERG
LONDON • NEW YORK • OXFORD • PARIS • SAN DIEGO
SAN FRANCISCO • SINGAPORE • SYDNEY • TOKYO
Woodhead Publishing is an imprint of Elsevier

WP WOODHEAD PUBLISHING

Woodhead Publishing is an imprint of Elsevier
80 High Street, Sawston, Cambridge, CB22 3HJ, UK
225 Wyman Street, Waltham, MA 02451, USA
Langford Lane, Kidlington, OX5 1GB, UK

ISBN: 978-1-907568-25-1 (print)
ISBN: 978-1-908818-45-4 (online)

British Library Cataloguing-in-Publication Data
A catalogue record for this book is available from the British Library

Library of Congress Control Number: 2015934369

To our families

Contents

Foreword

There is an increasing global awareness of the potential detrimental effects that pharmaceuticals can have on the environment. To fully understand how such effects might arise, it is necessary to consider the full environmental life-cycle of a pharmaceutical from its manufacture and use to disposal into the environment. This includes disposal into waste systems and the efficacy of the treatment technologies that modulate the level of the pharmaceutical discharged into the environment and ultimately the exposure and effects on the biota.

Such a life-cycle study necessitates the coordination of many scientific disciplines including pharmacy, public health, environmental chemistry, wastewater engineering, and ecotoxicology. The rationale for us writing the present text stemmed from the apparent need to discuss how all these individual disciplines are involved in understanding the various aspects of the environmental fate and transport of a pharmaceutical and the potential risk to exposed biota.

We have used our extensive professional backgrounds in teaching and research at the tertiary level in the areas of pharmacy and public health (RB & AT), analytical and environmental chemistry (BMP), and ecotoxicology (LT) to introduce aspects of the major parts of the environmental life-cycle of some common pharmaceuticals in Chapters 2–6 and 8. For completeness, we have also included in Chapter 7 an introduction to the various pieces of legislation currently enacted in different countries to control the use and discharge of pharmaceuticals even though such legal aspects are not an area of expertise of any of the authors. We have also introduced the concept of green pharmacy utilizing green chemistry principles and the related topic of life-cycle assessment in Chapter 8.

In writing this text covering such a broad range of topics and given the burgeoning number of reports and research papers that are now being published in these areas, we had to restrict our discussion to selected examples which we considered appropriately illustrated at least some of the scientific issues we had introduced in a given chapter. So although we have included some 850 references, we apologize in advance if *your* contribution to some aspect of this broad topic is not covered in this text. We also agreed to focus solely on the life-cycle of *human* pharmaceuticals although we do not discuss any environmental aspects arising from the initial step involving their manufacture. We also acknowledge that a large proportion of the pharmaceutical content detected in the global environment may arise from veterinary uses, but we do not cover this aspect in the text. This is particularly relevant in a country such as New Zealand where there is intensive agricultural activity.

We hope this text will be useful as an introduction to the topic for teachers of tertiary courses and for practitioners of aspects of pharmacy, public health, environmental chemistry, environmental engineering, ecotoxicology, and environmental law. Although each chapter should contain enough material to interest professionals working in that area, we also hope that non-experts will feel they will have at least been introduced to the relevant main concepts involved in the subject area of a given chapter. For example, a practicing pharmacist, as well as being interested in the range of disposal practices for unused medication covered in Chapter 3, may like to know more about how levels of pharmaceuticals are measured in wastewater and other types of environmental samples (Chapter 4) and how they can be minimized by the various types of wastewater treatment processes discussed in Chapter 6 before final discharge into the environment. Conversely, a wastewater engineer familiar with the treatment methods discussed in Chapter 6 may wish to know more about the medical prescription and subsequent disposal practices discussed in Chapters 2 and 3, which lead to measurable levels of specific pharmaceuticals in the raw influent water coming into their treatment plant as determined by the analytical methods summarized in Chapter 4. A member of a legislative body faced with increasing public demand to "do something about the waste polluting our waterways" may appreciate the discussion in Chapter 7 of some of the legislation currently existing in various countries to deal with this matter and the likely ecological impact on the environment (Chapter 5). Finally, somebody interested in more sustainable options in using pharmaceuticals rather than just "end-of-the-pipe" solutions will appreciate the introduction to green pharmacy and life-cycle assessment in Chapter 8.

We hope our text meets some of these diverse and widespread needs in this area of increasing global concern.

Barrie Peake, Rhiannon Braund, Alfred Tong
University of Otago, Dunedin, New Zealand
Louis Tremblay
Cawthron Institute, Nelson & University of Auckland, Auckland, New Zealand
July 23, 2015

Acknowledgements

To Lisa Bucke (Department of Chemistry, OU) for her expertise in drawing all of the figures in this text.

To Prof. Bill Cooper (Department of Civil & Environmental Engineering, University of California, Irvine) for introducing three of us (BMP, RB & AYCT) to this broad topic of environmental concern.

Introduction

1

1.1 Overview: Do pharmaceuticals pose a significant environmental risk?

The global community is becoming increasingly aware of both the environmental presence and the real or potential environmental impact of man-made chemicals (*anthropogenic* compounds) such as *pesticides*, *personal care products*, and *pharmaceuticals*. In the past, the large-scale chemical wastes associated with their industrial production were considered and known to pose significant environmental risks if their environmental discharge was not monitored. However, after appropriate measures were placed worldwide on controlling the production and disposal of industrial waste, it became apparent that new *emerging* contaminants comprising many anthropogenic compounds could also be potentially detrimental to the environment.

The first group of these emerging contaminants to receive notoriety were pesticides. In 1962, a book written by Rachel Carson entitled *Silent Spring* raised issues concerning public misinformation on the safety of pesticide use in the US, particularly dichlorodiphenyltrichloroethane (DDT) and its adverse effects on bird life. This was followed by the realization that many other chemicals were commonly present in consumer products (Halden, 2010). Perchlorinated and perfluorinated hydrocarbons used as flame retardants in clothing, carpets, and upholstery have the potential to persist and bioaccumulate in wildlife, and a major issue raised by the scientific community at that time was how best to manage these chemicals in redundant consumer goods placed in landfills (Weber et al., 2011). Public concerns about emerging contaminants escalated from the discovery that certain compounds disrupted the hormonal (endocrine) systems of a range of organisms and as such have the potential to detrimentally affect human and animal functions in trace quantities over a prolonged period of exposure. For example, detrimental effects on the reproductive health of humans have been associated with the use of bisphenol A as a plasticizer in water bottles, microwaveable containers, and other household products (Ohlson and Hardell, 2000).

Given the increased global production and use of these anthropogenic chemicals in modern-day society by governments, local bodies, corporations and private individuals and the myriad of new anthropogenic chemicals consistently available over time, it could be predicted that environmental pollution associated with these chemicals would become increasingly important in the twenty-first century (Kummerer, 2010). The recognition that private individuals and small communities can pollute the environment and lead to the release of contaminants at detectable levels was a significant milestone (Daughton, 2007). Such a concept was soon realized for pharmaceuticals and personal care products, which are often collectively referred to by the environmental science community as *PPCPs*.

Personal care products are a heterogeneous group of chemical substances that are used for various health, beauty, and hygiene formulations. Such chemicals may be

The Life-Cycle of Pharmaceuticals in the Environment. http://dx.doi.org/10.1016/B978-1-907568-25-1.00001-3

present in everyday cosmetic items such as sunscreen lotions, lip balm, perfumes, shampoos, and toiletries as diverse as toothpaste, mouthwash, and shaving cream. On the other hand, *pharmaceuticals* (or *drugs*) are largely defined in various acts and regulations around the world (such as the *Medicines Act 1981* and *Medicines Regulations 1984* in New Zealand legislation) as chemical substances or *medicines* used therapeutically for humans or animals to treat illnesses or ailments. However, there are some regional variations on what constitutes a pharmaceutical or medicine. For example, animal remedies are not considered medicines under the above act and regulations in New Zealand legislation. Because of the broad definition of PPCPs, it was clear from the very onset that it is unlikely there will ever be a single approach to effectively minimize and control their discharge into the environment.

The production of pharmaceuticals is a significant economic activity for pharmaceutical industries around the world. A total of US$954 billion was spent on prescription medications globally in 2011 alone (*see IMS Total Unaudited and Audited Global Pharmaceutical Market By Region*, http://www.imshealth.com/deployedfiles/ims/Global/Content/Corporate/Press%20Room/Top-Line%20Market%20Data%20&%20Trends/2011%20Top-line%20 Market %20Data/Regional Pharma_Market_by_Spending_2011-2016.pdf, accessed 10-6-14). In addition, much money has been spent globally on the direct-to-consumer marketing of both therapeutic and lifestyle drugs. The increasing number of new drug entities and the various dosage forms available (i.e., tablets, creams, ointments, and suspensions) to satisfy the demands of modern healthcare have made pharmaceuticals more affordable to the consumer than ever before. In 2013 alone, over 100 new formulations or chemical entities were approved by the US Food and Drug Administration (FDA) for clinical use (http://www.centerwatch.com/drug-information/fda-approvals/, accessed 10-6-14).

Given the money that is spent by many governments and consumers in funding the insatiable need for pharmaceuticals in healthcare and the wide availability of therapeutic agents, the progression of a pharmaceutical from synthesis in the laboratory to market is also a costly business. Billions of dollars are typically spent by pharmaceutical companies to develop a new chemical compound with a vision that this potential drug will be a commercial success in the market (Yamane et al., 2013). Drug development typically involves three phases: a preclinical phase which seeks to gather the pharmacological behavior of the chemical generally in laboratory animals or isolated tissues, a clinical phase where the drug is tested in both healthy and sick volunteers in order to determine its effects on humans, and finally, a postmarketing surveillance phase where the new chemical entity (which is now considered a medicine) is used in the general population (Nutt and Attridge, 2014). Here, pharmacovigilance data related to the use of the new medicine are reported through everyday clinical use, and this aspect is especially important for the determination of any adverse effects which may not have been apparent or observed during the earlier stages of clinical testing (Yamane et al., 2013; Chen et al., 2014). If any excessive adverse drug reactions (ADRs) or major concerns over public safety are reported, the new medicine could be withdrawn from market, and there have been some notable historical examples of this happening.

In 1957, thalidomide was withdrawn as an over-the-counter medication for nausea and morning sickness due to the belated reports of the development of phocomelia (limb deformations in fetuses) in pregnancy (Diggle, 2001). A lawsuit against the pharmaceutical company that developed the drug (Grunenthal) ensued (Botting, 2002). In 1999, Merck and Co. (now Merck Sharpe and Dome, MSD) marketed the drug rofecoxib (Vioxx) as a specific nonsteroidal anti-inflammatory drug (NSAID) inhibitor of cyclooxygenase isoenzyme 2 (COX-2) (McIntyre and Evans, 2014). COX-2 is responsible for the production of inflammatory mediators in the human body, whereas cyclooxygenase isoenzyme 1 (COX-1) is a constitutive enzyme responsible for the maintenance of various physiological functions (Howard and Delafontaine, 2004). It was envisioned that the use of COX-2-specific inhibitors would be associated with a lower incidence of major gastrointestinal events, such as stomach ulcers and bleeding, compared to the use of nonspecific NSAIDs which also prevented mucus formation by COX-1, which has been known to occur with heavy aspirin use and hence have the potential to aggravate peptic ulcer formation (Katz, 2013). However, MSD withdrew the rofecoxib from the market in 2004 in the light of evidence that showed that its use was associated with an increased risk of myocardial infarction compared to nonspecific NSAIDs (Collins et al., 2013).

As can be seen from these instances of postmarketing withdrawals of new pharmaceuticals, drug development ventures are not only costly but also risky for a drug development company. For a new chemical entity to be a success, profits made from the marketing and sales of the drug must cover all of the earlier development costs (Bansal et al., 2009). Usually, a 10-year patent period is given to an *innovator* product (the first drug of a new chemical entity marketed) which prevents other pharmaceutical companies from producing generic products that can compete with the innovator product on the market and drastically reduce the revenue obtained by the pharmaceutical company for the development of the original product (Knezevic and Griffiths, 2011). Drug development and marketing is a high-stakes process. As we have seen, if the risks of using a drug outweigh the clinical benefits, government health departments around the world can decide to withdraw the drug from the market. This could potentially be the worst outcome for a pharmaceutical company which has invested billions of dollars on the development of a drug but cannot recuperate the costs of the development from any subsequent sales. In short, the public perception and reputation of the drug among authorities is everything to a drug development corporate. As we will discuss later, given that the risks for many of these pharmaceuticals cannot be accurately evaluated, inaccurate information about the environmental effects of pharmaceutical contaminants can generate a frenzied media phenomenon which may vilify a drug when it does not necessarily deserve such a reputation. In such a hypothetical situation where the drug is withdrawn on such a media-frenzied reputation, not only will a pharmaceutical company suffer financial loss but patients will also lose out on what could be an effective therapy. As we will discuss in Chapter 4, given that pharmaceuticals are typically detected in low concentrations in the environment and the lack of conclusive evidence in many instances that such levels are detrimental to flora and fauna (Chapter 5), will some drugs fall prey to postmarketing withdrawal in the future in the midst of unverified environmental activism and sensationalism? In Chapter 7, we look at some of the dilemmas faced by regulatory and legislative authorities associated with setting minimum guideline values to reflect the environmental concentration of particular pharmaceuticals considered to be *persistent organic pollutants* in the light of often limited information.

Investigations into the environmental impact, fate, and behavior of pharmaceuticals have only been undertaken in the last 40 years or so (Stumm-Zollinger and Fair, 1965). In the 1970–80s, the concept of natural and synthetic estrogens as contaminants in the environment was first considered and their behavior during sewage treatment studied (Tabak and Bunch, 1970; Tabak et al., 1981). The drive to learn more about contaminants in wastewater was made greater due to the concept of reuse of urban waste water beginning in the 1980s (Rice, 1999). While many cities around the world currently still do not reuse wastewater, there are some exceptions. For example, the cities of Windhoek in Namibia and Singapore are both dependent on water recycling to support their inhabitants due to the scarcity of natural sources of freshwater in these areas (Isaacson and Sayed, 1988; Tortajada, 2006; Ilemobade et al., 2009). Richardson and Bowron (1985) raised the possibility of the presence of pharmaceutical contaminants in drinking water, as potable water supplies were commonly sourced from rivers fed by sewage effluent inputs in their country (UK). Indeed, their study established the presence of pharmaceuticals in sewage effluent, river, and tap water at concentrations in the range from <6.25 ng/L up to 1 µg/L. The significant increase in sensitivity of analytical instruments over the last 10–15 years has also resulted in routine detection in the environment of such levels of individual pharmaceutical compounds as low as 1 part in 10^9 or even 10^{12} (ng/L or pg/L, respectively[1]). As such, it is plausible that until such analytical capability was developed, the environmental presence, behavior, and ecological effects that might have been occurring through the environmental presence of pharmaceuticals had largely been ignored. In Chapter 4, we discuss some of the techniques used for trace analysis of pharmaceuticals and the use of sophisticated and sensitive instruments as a means to detect and quantify them in the environment.

It is easy to ignore the public health significance of such low concentrations of pharmaceuticals, especially for doctors, pharmacists, and other health professionals, as such levels are considered subtherapeutic. As discussed in an editorial published in the *Journal of Clinical Psychopharmacology* (Shader, 2008), while some may argue that these low levels are not likely to pose any threat to human or animal health, an assurance of safety cannot be assumed when the risk associated with these low levels cannot be accurately quantified. A case in point is the toxicity of low levels of estrogens which exhibit endocrine disruptor (EDR) behavior. A concentration of 4 ng/L ethinylestradiol (EE2) has been positively observed to induce feminization in freshwater roach (*Rutilus rutilus*) (Tyler and Jobling, 2008). However, while estrogens and other steroid hormones have shown observable changes in the reproductive success of fish in laboratory studies (Lange et al., 2001), there is a paucity of information concerning their effects in wild fish populations over time (Mills and Chichester, 2005). In addition, other confounding factors concerning field studies may affect the observed results. For example, the abundance of nutrients at sites of sewage

[1] ng/L and pg/L = nanograms per liter and picograms per liter of the contaminant in an environmental matrix, respectively, such as water. For practical purposes, 1 g of water is considered to be equivalent to 1 mL of water as the density of water is close to 1 g/mL. A substance with a concentration of 1 ng/L in an environmental matrix of water is equivalent to 1 g of the substance dissolved in one billion (999999999 or approximately 0.999×10^9) grams of water. Similarly 1 pg/L is equivalent to 1 g of substance dissolved in one trillion (0.999×10^{12}) grams of water!

effluent outfalls may actually increase the abundance and diversity of fish observed in spite of measurable levels of pharmaceuticals in the same waters (Guidetti et al., 2003). Environmental field studies undertaken on aquatic organisms such as fish may not truly reflect the reproductive capacities of their populations. It is also unknown whether field observations of gonadal changes and intersexuality of fish significantly affect reproduction of fish in the wild (Tanna et al., 2013).

The effects of pharmaceuticals at low concentrations are usually chronic and subtle (Parolini et al., 2009) and are observed at levels much higher than predicted environmental concentrations (Carlsson et al., 2013; Lamichhane et al., 2013). Nonetheless, deleterious effects on plants have been previously associated with ng/L levels of a single pharmaceutical in water (Feito et al., 2013). However, the possibility of mixture toxicity in a water body due to the presence of a number of pharmaceuticals (and as we shall see later in this book, their metabolites, degradation, and transformation products) cannot be excluded (Cleuvers, 2003, 2004) (see Chapter 5). While there has been some reports of this type of mixture toxicity (Escher et al., 2010), there must be many other instances of this type of toxicity that have not been detected and/or investigated. Given the lack of pertinent toxicological and physicochemical information on all these aspects, it is not surprising that environmental risks associated with these "complex pharmaceutical cocktails" are difficult to define and evaluate.

The difficulty in quantitating environmental risk is a hallmark of all emerging contaminants. Indeed, the official US Environmental Protection Agency (EPA) website (http://water.epa.gov/scitech/cec/, accessed 31-5-14) considers contaminants of emerging concern are those that "are *being* discovered in water at levels that may be significantly different to those expected" and "because the risk to human health and the environment associated with their presence, frequency of occurrence or source *may not be known*." How are environmental risks associated with emerging contaminants, and any contaminant in general, determined? We will discuss the concept of environmental risk as appropriate to pharmaceutical pollutants in this chapter and in Chapter 4 where we will overview the presence of pharmaceuticals in the environment. Furthermore, in Chapter 5 we will estimate their risks to humans and animals based on prescription and usage data and determine whether predicted environmental concentrations have the potential to affect ecology.

Despite the unknown environmental risks arising from the use of most pharmaceuticals, there is the much publicized case reported by Oaks et al. (2004) who attributed the death of countless Oriental white-backed vultures in Pakistan to acute renal failure. However, many of these vultures were found to have fed on dead cattle that had been previously treated with diclofenac, an NSAID commonly used by farmers. Independent analytical experiments found elevated levels of diclofenac in the tissues of the dead vultures. Additionally, many of the vultures died from concomitant gout from the deposition of uric acid crystals in their kidneys and such symptoms were consistent with acute renal toxicity in the birds. This study was the first (and seemingly the *only* reported) observation of a major ecological disaster associated with a pharmaceutical and illustrates the possible detrimental consequences and *significant risks* to humans as well as animals if no intervention is undertaken to protect the environment from pharmaceutical contaminants.

Given that pharmaceuticals are not unlike pesticides and other personal care products in their potential for ecological damage, we have a duty to further understand and characterize their environmental behavior. As health professionals and scientists, it is of utmost importance that we continue to further investigate the life-cycle of these agents and quantify the environmental risk of these compounds. In doing so, only then will we be able to determine whether some or all of these substances pose a significant risk to the environment. In order to accurately quantify this risk, we have to first consider *in what way* pharmaceuticals can be hazardous substances.

1.2 Hazardous substances

A hazardous substance is a dangerous chemical or a mixture of substances whose physiochemical properties may induce *harm* to humans and animals. This harm may arise from the ability of the substance to violently react with its surroundings under a given set of environmental conditions. For example, sodium metal is hazardous because of its potential to react violently in water. If any human or animal was adjacent to the reaction site, they would be harmed in terms of (1), the imminent ignition of the hydrogen gas formed on the surface of the water and upon exposure to the ignited gas become severely burnt and (2), the elevation in the alkalinity (and hence pH) of the affected water resulting from the formation of sodium hydroxide and the disturbance of biological systems dependent on a constant pH. While we can all appreciate that most pharmaceuticals, pesticides, and personal care products do not have the capability to detonate or violently react with the ambient environment (the most hazardous pharmaceutical of this sort being glyceryl trinitrate or nitroglycerin used for angina), they could still be considered hazardous given their ability to induce morbidity or mortality in organisms by penetrating biological membranes and disrupting physiological functions. The nature of harm will obviously be dependent on the *pharmacodynamics* and *pharmacokinetics* of the pharmaceutical (as discussed in Chapters 2 and 3).

However, no harm to humans, animals, or the environment will occur if the hazardous substance is isolated in some way from the outside world. Sodium metal is completely inert in vacuum or when stored under paraffin oil. In a similar situation, a packet of tablets of the β-blocker metoprolol will not affect the physiology of a patient if it is sitting unused on a shelf (see Chapter 2). It is very clear that the *exposure* of a human or animal to the chemical is an essential requirement if harm (and, indeed, therapeutic benefits) resulting from the supposedly hazardous chemical are to be realized. The probability of such harm resulting from the exposure of an organism to a hazardous substance is expressed by the term *risk*. Mathematically, risk is defined as the product of hazard and exposure:

$$\text{Risk} = \text{Hazard} \times \text{Exposure}$$

In other words, if an organism is not exposed to the hazardous substance, the hazardous substance cannot harm the organism. Intuitively, while the hazardous nature of a particular chemical cannot be changed and is based on the intrinsic physicochemical properties of the chemical, the risk of harm resulting from the chemical substance can be minimized if exposure to the hazardous substance is also minimized. For emerging contaminants such as pharmaceuticals, the exposure of humans and animals in the environment is directly related to the extent of their discharge into the environment. As for all environmental contaminants, the concept of exposure underpins the entire environmental lifecycle analysis of pharmaceuticals. Are the levels of pharmaceuticals that are discharged into the environment of sufficient magnitude that they are harmful? Are certain types of pharmaceuticals preferentially discharged into the environment over others? Are there ways in which scientists and health professionals can work together to reduce the exposure of the environment to pharmaceutical contaminants? A central theme of the present text is to evaluate the current state of knowledge concerning factors which affect the discharge, presence, persistence, and ecological implications of pharmaceuticals in the environment. Drawing from published research in the broad fields of health, physical, and social sciences, we will explore avenues for effective environmental management and harm minimization that might arise from the global use of pharmaceuticals. In addition, we will consider whether regulations and legislation around the world are doing enough to facilitate harm minimization from these bioactive hazardous substances given their known or at least estimated environmental risk.

An initial step in the environmental management of these potential contaminants is to acknowledge and understand the various pathways by which they enter the environment and how certain processes can affect their environmental discharge. Only then will we be able to devise effective strategies to minimize any environmental pollution that might arise from them.

1.3 Pathways to the environment: Life-cycle of pharmaceuticals

The major pathways by which pharmaceuticals for human use are discharged into the environment are summarized in the lifecycle shown in Figure 1.1.

The lifecycle begins with the manufacture of the pharmaceuticals which can lead to their direct discharge into liquid or solid waste systems. In addition, solvents and other chemicals responsible for the formulation of dosage forms can be released into wastewater along with pharmaceutical *excipients*—inactive substances such as sucrose, corn starch, or magnesium stearate—which are agents used for flavoring, binding, or lubricating tablets, respectively during formulation. The manufactured pharmaceuticals are transported to pharmaceutical wholesalers who distribute them to community or hospital pharmacies for prescription to patients pursuant to a particular course of pharmacotherapy (Chapters 2 and 3). In addition, wholesalers also distribute many pharmaceuticals that are general sales or over-the-counter medications (Chapter 2) to community pharmacies, supermarkets and grocery stores, petrol stations, and/or other types of convenience stores. Pharmaceuticals are excreted from the body in urine and feces either in their unaltered (parent) form or in their metabolized form if the pharmaceutical is taken as a tablet or suspension, and from here they enter the raw sewage waste stream. The release of the drug into the sewage system will also occur with some external

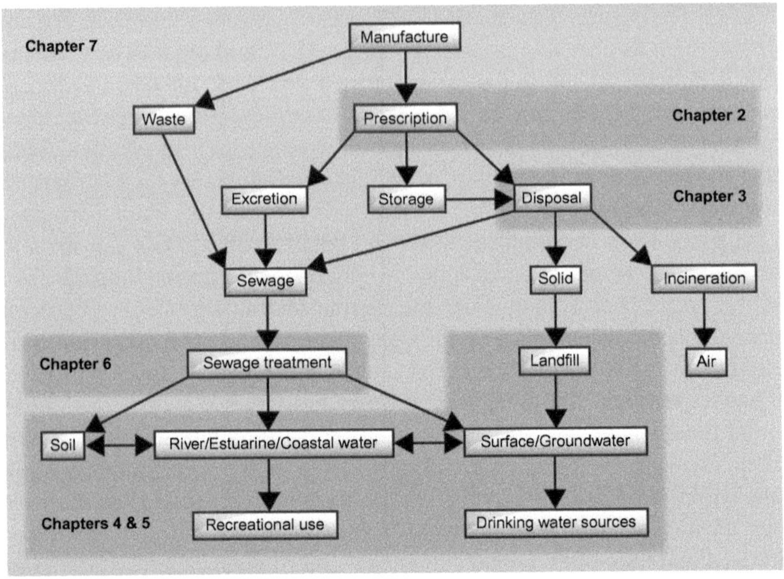

Figure 1.1 Pathways for entry into the environment of pharmaceuticals for human use. Modified from Annex in EMA (2006).

preparations, such as suppositories for rectal use. However, other external preparations, such as creams and ointments, will most likely be applied by the patient or caregiver to affected areas of skin. In such instances, bathing and swimming by a patient may contribute to the release of the external preparation into the surrounding water and the preparation is ultimately flushed unaltered into sewage or directly enters the aquatic environment in the case of people swimming in natural water bodies. Pharmaceuticals used for asthma or chronic obstructive pulmonary disorder (COPD) are usually present in metered dosage devices such as inhalers or aerosol containers. While some of the dose may be inhaled and subsequently absorbed into the systemic circulation, the remainder of the dose will be lost to the air. However, the doses contained in metered dose inhalers are small (in the range of micrograms) and the subsequent atmospheric release of active pharmaceutical ingredients associated with inhaled therapies are usually deemed to be minimal and rarely constitute a significant pathway for their entry into the environment.

In most countries, human waste material is collected in waste water (sewage) and this *influent* material either directly discharged into the environment in an untreated form or passed through one or more treatment steps before discharge as *effluent* (Chapter 6) into the natural aquatic environment such as a river, lake, estuary, or coastal or even an open ocean environment. There is also increasing use of land disposal for effluent. Solid material formed in the wastewater treatment process (*sludge*) is often buried in a landfill where, again, given time, a portion can be discharged as leachate or sometimes incinerated.

Unwanted medications, which include expired and unused drugs that are kept for indefinite periods of time, are most commonly disposed of either by the patients or caregivers through tipping them down the toilet or sink. Such disposal practices are another route by which pharmaceuticals can enter raw sewage in their parent form. Pharmaceuticals can also be disposed of into the waste bin from which they are collected and then buried in some form of landfill or incinerated directly. Over time, the leachate formed from mainly anaerobic processes in the landfill can also be collected for treatment or, if no intervention occurs, it can percolate into the surrounding soil and ultimately can become available to the groundwater. Leachate in groundwater can potentially enter drinking water reservoirs and contaminate potable water supplies. Finally, pharmaceuticals that undergo high-temperature incineration will be completely oxidized to innocuous compounds such as carbon dioxide and water that are released into the atmosphere.

From the preceding brief discussion, it should be apparent that there is a potential negative environmental impact associated with the use and disposal of any pharmaceutical. Furthermore, as we will see later, the most environmentally-acceptable disposal option may not necessarily be the one most acceptable for public health. In Chapter 3, we will examine in more detail the practices that are currently utilized by patients, caregivers, and consumers for the disposal of unused, unwanted, and expired medications. Based on current knowledge and guidelines from appropriate authorities, we will attempt to determine the environmental or public health benefits and risks of using the various disposal methods currently available.

As has already been alluded to above, one of the main reasons for the increased global awareness and resulting potential concerns about the presence and ecological impact of trace levels of pharmaceuticals in the environment has been the marked increase in the sensitivity of the analytical instrumentation and associated sampling techniques. An introduction to current sampling and analytical identification and quantification methods will be covered in Chapter 4 and illustrated with examples of reported levels of some common pharmaceuticals in various environmental systems.

Once pharmaceuticals enter the environment, their fate and persistence are largely determined by the efficiency of the various degradation mechanisms that remove them from environmental compartments. There are various points on Figure 1.1 which offer opportunities for the degradation of a pharmaceutical. In a sewage treatment system, the most basic treatment involves primary sedimentation and secondary microbial degradation using activated sludge. While the primary sedimentation processes separate the solid and liquid sewage by sedimentation of solids and the separation of greases, fats, and oils, secondary treatment further breaks down the sewage by facilitating the action of aerobic microbes. In many parts of the world, tertiary treatments such as chlorination or ozonation are also being successfully used to further remove or at least inactivate pathogenic microbes and some sewage contaminants such as pharmaceuticals that remain in the waste stream even after secondary sewage treatment. In Chapter 6, we explore the fundamentals of the main sewage treatment proceeses, the physicochemical interactions between pharmaceuticals during each of the major treatment steps, and include specific examples of the removal efficiencies of some major classes of pharmaceuticals using a range of current and potential sewage treatment techniques.

However, as will be apparent later in this book, there is a good chance that current knowledge or even future scientific investigation will not provide all the information about the behavior of pharmaceuticals to completely facilitate their removal from wastewater or assess their risk to the environment. In Chapter 5, we will explore the basic principles and current experimental practices that are used by environmental scientists to assess ecological impacts of pharmaceuticals and other contaminants in the laboratory. The types and roles of test organisms commonly used for ecotoxicological testing will be introduced and the benefits and disadvantages of the results obtained will be examined. The roles of mixture toxicity by concentration addition and independent action of pharmaceuticals on test organisms will also be explored. These ecotoxicological predictors may in turn facilitate cost-effective sewage treatment design (i.e., if the pharmaceutical in question is generally not toxic to most organisms tested at the typical levels discharged into the environment, is an expensive overhaul and upgrade of a sewage treatment facility justified in order to remove just this particular pharmaceutical from the waste stream? Should experimental ecotoxicological results be used to establish minimum guideline values for the discharge of particular pharmaceuticals into the environment (Chapter 7)?

Finally, it is important to realize that doctors (physicians), pharmacists, biopharmaceutical scientists, and environmental scientists can all have a role to play in minimizing the discharge of pharmaceuticals into the environment. The role of appropriate prescribing can substantially determine the quantity of pharmaceuticals necessitating subsequent disposal and treatment before their ultimate discharge into the environment. In Chapter 2, we discuss patient, dosing, and drug factors which may eventually determine the extent to which a pharmaceutical is consumed or otherwise by a patient. By understanding the *pharmacodynamics* and *pharmacokinetics* of particular drugs, prescribers can make more rational decisions about what and how much to prescribe. Such issues may be particularly pertinent given a range of similar therapies that can be used. For example, should a doctor prescribe a thiazide diuretic or a β-blocker for a hypertensive patient with gout symptoms? It is important to realize that standard operating procedures for the prescribing of certain medicines where alternatives exist may not take into account patient or drug factors. In such instances, side effects or the lack of efficacy may be unnecessarily manifested in the patient and, coupled with the possibility of excessive and irrational prescribing, can ultimately lead to the creation of unused medication which must be ultimately disposed of in some manner.

Many of the processes responsible for the discharge of pharmaceuticals into the environment are often regulated by the appropriate authorities. Chapter 7 introduces some of this legislation currently in practice in different countries around the globe.

Finally, a more holistic approach to minimizing the environmental discharge of pharmaceuticals is to employ the use of green pharmacy which incorporates the principles of green chemistry to minimize the amount of pharmaceutical waste before it occurs, rather than having to devise methods to treat it after it has been created. Both these approaches which embody aspects of environmental sustainability are introduced in the final Chapter 8 along with the not unrelated concept of life-cycle assessment (LCA) of a pharmaceutical.

1.4 Conclusions

The overall life-cycle of any pharmaceutical is a fascinating and complex issue. The management of pharmaceutical waste is a multidisciplinary pursuit and given the limited current state of knowledge on many environmental issues, even scientists, health practitioners, and government authorities cannot necessarily look into the future and decide on how best to proceed. Needless to say, much effort over just the past decade has been made by researchers to get closer to a definitive answer. The number of research papers that are being published concerning the presence, persistence, and ecological implications of pharmaceuticals in sewage and/or in the environment is increasing rapidly every year. For example, the total number of papers that can be retrieved using the Thomson-Reuters® Web of Knowledge™ search engine and published in the 18 years from 1996 up until the end of 2014 containing just the words "pharmaceutical*" together with "waste water" alone is approximately 3,440 (Figure 1.2).

A similar proliferation of research is observed if one uses related search terms such as "environment," "fate," "presence," "persistence" and more specific issues such as "ecotoxicology" and "removal efficiencies." In addition, there has been an explosion of literature on water remediation techniques and some academic journals have focussed on publishing literature related to the knowledge and management of pharmaceutical and other environmental contaminants. For example, Elsevier® publishes the *Journal of Hazardous Materials*, which specifically publishes research pertaining to "improve understanding of the hazards and risks certain materials pose to people and the environment." In addition, several aspects of managing risks from contaminants are published in the *Journal of Environmental Management*. All of

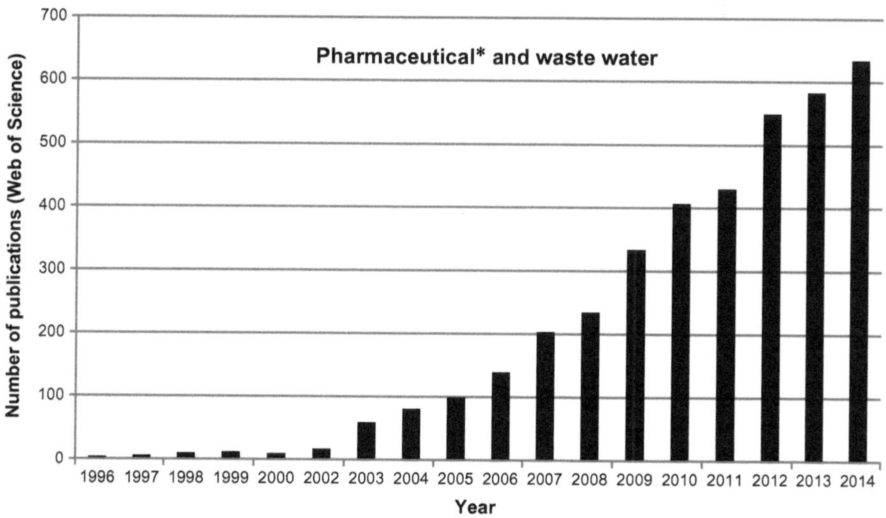

Figure 1.2 Number of publications per year for the period 1996–2014 retrieved using Web of Knowledge™ and the search words "pharmaceutical*" and "waste water."

these research and publishing initiatives pave the way to providing an objective perspective on how best to manage pharmaceutical waste on a local and global scale in the years to come.

References

Bansal, I.S., Sahu, D., et al., 2009. Evergreening—a controversial issue in Pharma Milieu. J. Intellect. Prop. Rights 14 (4), 299–306.

Botting, J., 2002. The history of thalidomide. Drug News Perspect. 15 (9), 604–611.

Carlsson, G., Patring, J., et al., 2013. Toxicity of 15 veterinary pharmaceuticals in zebrafish (Danio rerio) embryos. Aquat. Toxicol. 126, 30–41.

Chen, M.H., Ibrahim, J.G., et al., 2014. Bayesian sequential meta-analysis design in evaluating cardiovascular risk in a new antidiabetic drug development program. Stat. Med. 33 (9), 1600–1618.

Cleuvers, M., 2003. Aquatic ecotoxicity of pharmaceuticals including the assessment of combination effects. Toxicol. Lett. 142, 185–194.

Cleuvers, M., 2004. Mixture toxicity of the anti-inflammatory drugs diclofenac, ibuprofen, naproxen, and acetylsalicylic acid. Ecotoxicol. Environ. Saf. 59 (3), 309–315.

Collins, J.M., Simon, K.I., et al., 2013. Drug withdrawals and the utilization of therapeutic substitutes: the case of Vioxx. J. Econ. Behav. Organ. 86, 148–168.

Daughton, C.G., 2007. Pharmaceuticals in the environment: sources and their management. In: Petrovic, M., Barcelo, D. (Eds.), Analysis, Fate and Removal of Pharmaceuticals in the Water Cycle, Wilson and Wilson's Comprehensive Analytical Chemistry Series, vol. 50, pp. 1–58.

Diggle, G.E., 2001. Thalidomide: 40 years on. Int. J. Clin. Pract. 55 (9), 627–631.

EMA, 2006. European Medicines Agency: EMEA/CHMP/SWP/4447/00.Guideline on the environmental risk assessment of medicinal products for human use. http://www.ema.europa.eu/docs/en_GB/document_library/Scientific_guideline/2009/10/WC500003978.pdf (accessed September 9, 2015).

Escher, B.I., Bramaz, N., et al., 2010. Mixture toxicity of the antiviral drug Tamiflu (R) (oseltamivir ethylester) and its active metabolite oseltamivir acid. Aquat. Toxicol. 96 (3), 194–202.

Feito, R., Valcarcel, Y., et al., 2013. Preliminary data suggest that venlafaxine environmental concentrations could be toxic to plants. Chemosphere 90 (7), 2065–2069.

Guidetti, P., Terlizzi, A., et al., 2003. Changes in Mediterranean rocky-reef fish assemblages exposed to sewage pollution. Mar. Ecol. Prog. Ser. 253, 269–278.

Halden, R.U., 2010. Introduction to contaminants of emerging concern in the environment: ecological and human health considerations. Contaminants of emerging concern in the environment: ecological and human health considerations. Am. Chem. Soc. 1048, 1–6.

Howard, P.A., Delafontaine, P., 2004. Nonsteroidal anti-inflammatory drugs and cardiovascular risk. J. Am. Coll. Cardiol. 43 (4), 519–525.

Ilemobade, A.A., Adewumi, J.R., et al., 2009. Framework for assessing the viability of implementing dual water reticulation systems in South Africa. Water SA 35 (2), 216–227.

Isaacson, M., Sayed, A.R., 1988. Health-aspects of the use of recycled water in Windhoek, Swa/Namibia, 1974–1983—diarrheal diseases and the consumption of reclaimed water. S. Afr. Med. J. 73 (10), 596–599.

Katz, J.A., 2013. COX-2 inhibition: what we learned—a controversial update on safety data. Pain Med. 14, S29–S34.

Knezevic, I., Griffiths, E., 2011. Biosimilars—global issues, national solutions. Biologicals 39 (5), 252–255.

Kummerer, K., 2010. Pharmaceuticals in the environment. Annu. Rev. Environ. Resour. 35, 57–75.

Lamichhane, K., Garcia, S.N., et al., 2013. Chronic Effects of carbamazepine on life-history strategies of ceriodaphnia dubia in three successive generations. Arch. Environ. Contam. Toxicol. 64 (3), 427–438.

Lange, R., Hutchinson, T.H., et al., 2001. Effects of the synthetic estrogen 17-alpha-ethinylestradiol on the lifecycle of the fathead minnow (*Pimephales promelas*). Environ. Toxicol. Chem. 20, 1216–1227.

McIntyre, W.F., Evans, G., 2014. The Vioxx (R) legacy: enduring lessons from the not so distant past. Cardiol. J. 21 (2), 203–205.

Mills, L.J., Chichester, C., 2005. Review of evidence: are endocrine-disrupting chemicals in the aquatic environment impacting fish populations? Sci. Total Environ. 343 (1–3), 1–34.

Nutt, D.J., Attridge, J., 2014. CNS drug development in Europe—past progress and future challenges. Neurobiol. Dis. 61, 6–20.

Oaks, J.L., Gilbert, M., et al., 2004. Diclofenac residues as the cause of vulture population decline in Pakistan. Nature 427, 630–633.

Ohlson, C.G., Hardell, L., 2000. Testicular cancer and occupational exposures with a focus on xenoestrogens in polyvinyl chloride plastics. Chemosphere 40 (9–11), 1277–1282.

Parolini, M., Binelli, A., et al., 2009. An *in vitro* biomarker approach for the evaluation of the ecotoxicity of non-steroidal anti-inflammatory drugs (NSAIDs). Toxicol. in Vitro 23, 935–942.

Rice, R.G., 1999. Ozone in the United States of America—state-of-the-art. Ozone Sci. Eng. 21 (2), 99–118.

Richardson, M.L., Bowron, J.M., 1985. The fate of pharmaceuticals in the aquatic environment. J. Pharm. Pharmacol. 37 (1), 1–12.

Stumm-Zollinger, E., Fair, G.M., 1965. Biodegradation of steroid hormones. J. Water Pollut. Control. Fed. 37, 1506–1510.

Tabak, H.H., Bunch, R.L., 1970. Steroid hormones as water pollutants I. Metabolism of natural and synthetic ovulation-inhibiting hormones by microorganisms of activated sludge and primary settled sewage. Dev. Ind. Microbiol. 11, 367–376.

Tabak, H.H., Bloomhuff, R.N., et al., 1981. Steroid hormones as water pollutants II. Studies on the persistence and stability of natural urinary and synthetic ovulation-inhibiting hormones in untreated and treated wastewaters. Dev. Ind. Microbiol. 22, 497–519.

Tanna, R.N., Tetreault, G.R., et al., 2013. Occurrence and degree of intersex (testis-ova) in darters (Etheostoma SPP.) across an urban gradient in the Grand River, Ontario, Canada. Environ. Toxicol. Chem. 32 (9), 1981–1991.

Tortajada, C., 2006. Water management in Singapore. Int. J. Water Resour. Dev. 22 (2), 227–240.

Tyler, C.R., Jobling, S., 2008. Roach, sex, and gender-bending chemicals: the feminization of wild fish in English rivers. Bioscience 58 (11), 1051–1059.

Weber, R., Watson, A., et al., 2011. Persistent organic pollutants and landfills—a review of past experiences and future challenges. Waste Manag. Res. 29 (1), 107–121.

Yamane, N., Igarashi, A., et al., 2013. Cost-effectiveness analysis of microdose clinical trials in drug development. Drug Metab. Pharmacokin. 28 (3), 187–195.

Prescribing practices

2

2.1 Pharmaceuticals in legislative and healthcare systems

The widespread use and development of existing and new chemical entities as pharmaceuticals has meant that they have become increasingly more important as environmental contaminants of emerging concern. While pharmaceuticals can be seen as bioactive substances that can provide benefits for the clinical treatment of various health conditions in humans and animals, they can also be defined as hazardous substances that deserve appropriate handling and usage. However, it can be difficult to determine the true environmental risks arising from the clinical use of pharmaceuticals due to the current lack of information about these compounds from the time of manufacture to their destruction and/or environmental discharge. A major reason why health and environmental science researchers cannot accurately assess environmental risk from pharmaceutical contaminants is the limited information on their global and local usage patterns.

It is difficult to obtain data on the exact types and quantities of medications used worldwide because of the different healthcare and access policies specific to individual countries (see http://www.commonwealthfund.org/publications/fund-reports/2013/nov/international-profiles-of-health-care-systems, accessed 7-6-14). In the United States, private insurance schemes predominantly financed by employers cover most healthcare costs of citizens under 65 years. Medicare plans are available for some disabled and low-income citizens and citizens over this age. This is in contrast to access policies available in other countries such as New Zealand, Norway, and Sweden where a public health system funded by taxpayers administers the basic healthcare of citizens largely irrespective of age and a system of extra health benefits (such as elective surgery) can be purchased privately by individuals if they feel the need for such services. The various access policies means data on usage patterns in a given country are often not conveniently available from just one source. This difficulty is especially apparent with privately funded systems in certain countries where record keeping is often undertaken by multiple parties and not centrally monitored by the central government. Often, information on the types and amounts of medication issued to patients is held by the organization with financial responsibility, such as local governments, insurance groups, and other healthcare funding institutions, or by the consumer. This may potentially create problems for record keeping (as certain countries may be more stringent in keeping track of some classes of medicines, see below) and negate a global harmonization of efforts to effectively monitor medication usage patterns.

As we have previously mentioned in Chapter 1, the definition of "medicine" varies in the official acts and regulations of different countries. This issue further complicates the process of obtaining complete and meaningful data on pharmaceutical usage patterns. Besides the different pharmacological and therapeutic classes which can be

The Life-Cycle of Pharmaceuticals in the Environment. http://dx.doi.org/10.1016/B978-1-907568-25-1.00002-5

used to categorize medicines, they can also be classified through legislation in terms of their potential for harm through abuse or adverse effects. Incomplete data on usage patterns may also arise because classifications of medicine may have different names and be regulated differently between various countries (see Table 2.1).

Three of the countries (the United Kingdom, Australia, and New Zealand) listed in Table 2.1 differ from the United States as they have intermediate classifications of medicines apart from prescription-only and general sales medicines. Prescription-only medicines are generally defined by legislation as those that can be obtained by individuals at a pharmacy only upon presentation of an appropriate prescription from a doctor (physician). General sales medicines are those that can be sold "over the counter" (OTC) in any shop and without a prescription. Intermediate classes of OTC medicines can have

Table 2.1 Drug classification in the United States, the United Kingdom, Australia, and New Zealand

United States	United Kingdom	Australia	New Zealand
Schedule I (Controlled Substances Act 1970)	Class A controlled drug (Misuse of Drugs Act 1971)	Schedule 9 prohibited substance (SUSMP,[b] TGA[c])	Class A controlled drug (Misuse of Drugs Act 1975)
Schedule II (Controlled Substances Act 1970)	Class B controlled drug (Misuse of Drugs Act 1971)	Schedule 8 controlled drug (SUSMP,[b] TGA[c])	Class B controlled drug (Misuse of Drugs Act 1970)
Schedules III, IV and V (Controlled Substances Act 1970)	Class C controlled drug (Misuse of Drugs Act 1971)	Schedule 8 controlled drug (SUSMP,[b] TGA[c])	Class C Controlled Drug (Misuse of Drugs Act 1970)
Prescription-only medicines (incorporating new and generic drugs. US FDA)	Prescription-only medicines (Medicines Act 1968)	Schedule 4 prescription-only medicines (SUSMP,[b] TGA[c])	Prescription-only medicines (Medicines Act 1981; Medicines Regulations 1984)
No direct equivalent	No direct equivalent	Schedule 3 pharmacist-only medicines (SUSMP,[b] TGA[c])	Pharmacist-only Medicines (Medicines Act 1981; Medicines Regulations 1984)
No direct equivalent	Pharmacy-only medicines (Medicines Act 1968)	Schedule 2 pharmacy-only medicines (SUSMP,[b] TGA[c])	Pharmacy-only medicines (Medicines Act 1981, Medicines Regulations 1984)
Over-the-counter medications (OTC, US FDA[a])	General sales medicines (Medicines Act 1968)	Unscheduled substances (SUSMP,[b] TGA[c])	General sales medicines (Medicines Act 1981, Medicines Regulations 1984)

[a] Food and Drug Administration.
[b] Standard for the Uniform Scheduling of Medicines and Poisons.
[c] Therapeutic Goods Administration of Australia.
Source: Legislation available at: www.fda.gov, www.legislation.govt.nz, and www.comlaw.gov.au; accessed 14-6-14.

stricter requirements concerning their sale and include pharmacist-only medicines, which are those that can be sold OTC in a pharmacy only upon the recommendation of a pharmacist, or pharmacy-only medicines, which can only be sold OTC in pharmacies by a pharmacy staff member. The United States does not have these intermediate OTC classifications; however, there has been consideration by the FDA to establish a pharmacist-only classification (http://www.uspharmacist.com/continuing_education/ceviewtest/lessonid/1087 02/, accessed 7-6-14). In Australia, medicines classified as Class B and C Controlled Drugs are collectively placed in Schedule 8 under the *Standard for the Uniform Scheduling of Medicines and Poisons* which controls the supply and use of medicines on a national level and is enforced by the Australian Therapeutic Goods Administration. However, similarities exist between the various countries on the legislative regulation of Class B, C, or Schedule 8 medicines. For example, it is an offense to supply or possess Class B, C, or Schedule 8 medicines in any of these countries without a doctor's prescription, whereas the possession of pharmaceuticals belonging to the prescription-only classification does not constitute an offense even if the possessor does not hold a valid prescription. While the legislation in the United Kingdom and New Zealand is largely regulated at a national level, different Australian states and territories may also have their own regulations relating to the use, record keeping, and supply of medications.

While there are similarities and differences internationally between classification systems used for pharmaceuticals in different countries, collecting data on the use and supply of medicines can also be difficult because of the variability in legislative enforcement placed upon the same pharmaceutical in these different countries (Table 2.2).

It is apparent from the above information that legislative enforcement for the use, possession, and supply of one particular pharmaceutical may be entirely different from one country to another. For example, the β-lactam penicillin amoxicillin (a broad spectrum antibiotic) is consistently considered a prescription-only medicine in the United States, the United Kingdom, and New Zealand and Schedule 4 in Australia (which is equivalent to prescription-only medicine in the other three countries). However, this is not the case for omeprazole (a proton pump inhibitor for peptic ulcers) nor for pholcodine

Table 2.2 Drug classification of three pharmaceuticals in some different countries (as of 2014)

	Classification			
Medicine	**United State**	**United Kingdom**	**Australia**	**New Zealand**
Amoxicillin	Prescription-only	Prescription-only	Schedule 4	Prescription-only
Omeprazole	General sales	General sales	Prescription-only	Pharmacist-only
Pholcodine	Schedule I	Class B controlled	Schedule 8	Class C[a] controlled

[a] Not considered a "Controlled Drug" when used in low-dose cough linctuses.
Source: Legislation available at: www.fda.gov, www.legislation.govt.nz, and www.comlaw.gov.au; accessed 6-6-14.

(an opioid prescribed to treat a dry cough, see discussion about cough and cold therapies below). Omeprazole is classified as an OTC medication in the United States, the United Kingdom, and New Zealand; however, it is a prescription-only medicine (Schedule 4) in Australia. Furthermore, the extent to which omeprazole is considered an OTC medicine also varies in different countries. Omeprazole is classified as a general sales OTC medicine in the United States and the United Kingdom, whereas in New Zealand, the medication may only be sold on the recommendation of a pharmacist or through prescription. As such, there is little record keeping in the United States and the United Kingdom when omeprazole is not dispensed pursuant to a prescription in contrast to the basic information acquired by a community pharmacy in New Zealand which includes details about the customer and the quantity of omeprazole dispensed OTC.

Pholcodine is a controlled substance that may be prescribed in the United Kingdom, Australia, and New Zealand, but its use and supply is enforced at different levels of stringency by legislation in these three countries. For example, it is classified as a Class B drug in the United Kingdom, but a Class C drug in New Zealand. In addition, the drug is legally acceptable under New Zealand legislation to be sold in low-dose cough linctuses OTC in the country. In the United States, it is entirely banned for use, supply, or possession as the substance is classified as a Schedule I medicine. This would be equivalent to Class A in the United Kingdom or New Zealand and Schedule 9 classification in Australia, where its use would require special governmental or ministerial approval. In summary, the regulations for the use, supply, and possession of pharmaceuticals may not necessarily reflect the capacity of the pharmaceutical to harm patients or the environment.

The classification of medicines does not usually include vitamins and mineral supplements (Kaufman et al., 2002) although these may be regularly prescribed for a therapeutic purpose and assume the role of "medicine" in certain doses. In New Zealand, folic acid may be formulated with other vitamins and minerals at low doses in a multivitamin available at any health food store or supermarket (such as Roche® Berocca Performance™ effervescent tablets). However, if the folic acid dose exceeds 800 μg (800×10^{-6} g), it is considered to be a pharmacy-only medicine taken for the prevention of neural tube defects in the unborn fetus during pregnancy (see http://www.medsafe.govt.nz/profs/datasheet/a/Apo-Folictab.pdf, accessed 7-6-14). Iron supplements in the form of ferrous sulfate or ferrous fumarate are another example whereby they are considered to be a general sales supplement when combined in a multivitamin or a pharmacy-only medicine in a concentrated dosage form (i.e., 30 mg/mL or 325 mg ferrous sulfate tablets) and these are only available in New Zealand from pharmacies. The quantities of these types of vitamins and minerals dispensed, sold, or used by consumers or patients in this manner are very hard to reliably and accurately determine.

The above situation can also apply to substances only considered as medicines by the various acts and legislation under all circumstances. Information pertaining to the use and supply of medications in multiple classifications can be difficult to collate and accurately quantify in terms of actual medications purchased or issued. In many countries, paracetamol (acetaminophen) is prescribed as a first-line clinical treatment for osteoarthritis and other noninflammatory pain conditions and the pharmacy responsible for dispensing the drug will have a record of the quantity of paracetamol actually supplied to the patient. However, paracetamol is also widely sold OTC in many countries.

For example, in New Zealand, paracetamol is sold in supermarkets, petrol stations, and small convenience stores. While these non-pharmacy outlets are regulated to only sell packets of paracetamol containing no more than 20 tablets to prevent intentional or unintentional poisonings from the ingestion of large quantities of paracetamol, there are usually no limits imposed by these non-pharmacy outlets to restrict the number of packets that can be purchased by any individual at any one time. There is also no regulatory requirement for these non-pharmacy outlets to record the quantity purchased at the checkout. Thus, the true quantity of paracetamol that is actually purchased (if not all consumed) by a particular community as a sum of the quantity dispensed via prescription and through OTC sales is impossible to accurately quantify.

The usage patterns of certain medications are also impossible to quantify if they are purchased by consumers rather than as therapies prescribed by doctors. A good example of such a group of medications is those used to treat a cough or a cold. These medications are generally used for the relief of blocked noses, dry and wet coughs, and sore throats. In New Zealand, most of these are OTC medicines generally available from both pharmacy and non-pharmacy outlets where few records are kept of their sales. Table 2.3 lists a selection of the myriad of cough and cold products available in New Zealand without a prescription or any recommendation from a pharmacist.

Table 2.3 Cough and cold therapies available OTC in New Zealand without prescription or recommendation by a pharmacist

Pharmaceutical	Therapeutic purpose	Available formulations
Benzydamine[a]	Relief of sore throats	Throat spray, lozenges, gargle solution
Bromhexine hydrochloride	Mucolytic (thins mucus secretions)	Tablets, oral liquid
Dextromethorphan hydrobromide	Relief of irritating dry cough	Lozenges, oral liquid
Flurbiprofen sodium[a]	Relief of sore throats	Lozenges
Guaifenesin	Helps to expel mucus in wet cough	Tablets, oral liquid
Hexylresorcinol	Relief of sore throats	Lozenges
Lignocaine hydrochloride	Anesthetic for sore throats	Lozenges (in combination)
Phenylephrine hydrochloride	Clears blocked nose	Drink sachets (with paracetamol), oral liquid, tablets, capsules
Oxymetazoline hydrochloride	Clears blocked nose	Nasal spray solution
Pholcodine	Relief of irritating dry cough	Lozenges, oral liquid
Iodine (in povidone-iodine)	Relief of sore throats	Gargle solution
Xylometazoline hydrochloride	Clears blocked nose	Nasal spray solution

[a] Also, a nonsteroidal anti-inflammatory drug (NSAID). See discussion later.
Source: Data obtained from Medsafe New Zealand (www.medsafe.govt.nz; accessed 7-8-14).

Some of these cough and cold medicines are notorious for recreational abuse, such as dextromethorphan and pholcodine (see above). Recreational drug users have been known to "Robo-trip" or consume large quantities of dextromethorphan to induce hallucinogenic experiences because it is so widely available in many countries often without prescription (Bobo et al., 2005; Logan et al., 2012; Burillo-Putze et al., 2013). Problematic recreational use has rendered dextromethorphan prescription-only in Sweden, although it is still an OTC medication in many other countries where its use is not regulated, including in the United States (Thurman and Ferrer, 2012). Until authorities decide to place tighter restrictions on the sale of dextromethorphan from general non-pharmacy outlets such as supermarkets, it is unlikely that any accurate information on dextromethorphan use can be obtained from any sale figures available for this drug. Dextromethorphan and other medications sold OTC through online pharmacies (see discussion below) may also hamper any efforts to quantify their use or supply.

Some food products that are not considered medicines may also contain pharmaceuticals. The vast quantities of coffee that are drunk around the world every day will no doubt contribute to the substantial amounts of caffeine excreted into sewage (Chapter 4). Per capita, coffee consumption may vary from 0.4 to 12.0 kg/year (see http://en.wikipedia.org/wiki/List_of_countries_by_coffee_consumption_per_capita, accessed 11-6-14). It is common practice in many countries to mix Indian tonic water and similar carbonated drinks with a spirit (e.g., gin and tonic). Tonic water at least contains small quantities of the antimalarial agent quinine (Fletcher, 1976). Even although tonic water is widely available, skin eruptions and allergies have been associated with its consumption particularly in excess (Muso et al., 2007; Feas et al., 2009; Marques et al., 2013). In addition, quinine has also been associated with hypoglycemia (a sudden drop in blood glucose concentration, leading to lapses in concentration, coma, or even death) in non-diabetics (Madorell et al., 2006) and agranulocytosis (a decrease in white blood cell count, inducing immunodeficiency) (Sutherland et al., 1977; Postiglione et al., 1995; Alkadi, 2007; Andres et al., 2009; Tesfa et al., 2009). However, such adverse occurrences are rare. The quantity of tonic water sold is not monitored in most countries around the world; therefore, we cannot be completely sure how much quinine from tonic water is discharged into the environment, but we can estimate quinine use from some prescription records.

Other common OTC medications that are used by large numbers of patients for their analgesic and anti-inflammatory properties are the group of nonsteroidal anti-inflammatory drugs (NSAIDs) which include aspirin, ibuprofen, and diclofenac. It is also difficult to quantify the actual amount of these NSAIDs that are purchased and consumed, given the number of formulations and brands available (see below), the number of outlets that these can be purchased from and the lack of collated consumer data describing their purchase and consumption habits. However, some qualitative trends can be deduced for the prevalence of prescription and nonprescription use of NSAIDs around the globe. Ibuprofen has been reported to be the most popular NSAID for the treatment of pediatric pain in Italy and the United Kingdom, while diclofenac is more popular in the Netherlands (Neubert et al., 2011). In addition, aspirin has been commonly reported to be used in combination with other painkillers in Spain

(Neubert et al., 2011). In the Netherlands, 84% of the market share arising from painkillers prescribed by general medical practitioners (GPs) involved four pharmaceuticals—paracetamol, ibuprofen, diclofenac, and naproxen (Bon and Jongejan, 2001). Forty-two percent of participants in an Australian study reported purchases of OTC NSAIDs outside the pharmacy; however, fewer participants were reported to use NSAIDs according to dosage instructions (Stosic et al., 2011). In a French study, 44% of participants had been prescribed an NSAID at least once over a 2 year period (2009–2010), with users of OTC NSAIDs generally being younger and female (Duong et al., 2014). The prevalence of pain in females has also been shown in a Spanish study to be higher than in males (Bassols et al., 2002) and may account for such gender differences in NSAID usage.

These European and Australian results are also applicable to the United States situation, where ibuprofen is so widely available that many users are seemingly unaware of its potentially adverse side effects (Wilcox et al., 2005) and the recommended daily dosages as specified by manufacturers are often exceeded. In addition, over 500 million prescriptions for ibuprofen have been estimated to be written each year within the United States alone (Wilcox et al., 2005). In global terms, however, the majority of prescription and nonprescription doses of NSAIDs are often taken on a relatively short-term (acute) basis by an individual when they are suffering from pain and inflammation (Moore et al., 2004). Indeed, less than 20% of a supply of ibuprofen was previously reported to be used (Moore et al., 2004) and therefore the quantity of ibuprofen supplied to patients may not generally reflect the quantities consumed (Braund and Abbott, 2007). Such disparities between the quantity dispensed and/or purchased and that consumed may give rise to household hoarding of pharmaceuticals (see below).

An overriding issue for quantitating the usage of a particular pharmaceutical in a community is the potential difficulty of collating accurate usage information from all the products, brands, and formulations sold, which may have a particular *Active Pharmaceutical Ingredient* (API). In order to quantitate the use of a particular pharmaceutical in a community, it is clearly necessary to take into account the usage patterns of all its available formulations on the market. For example, amoxicillin is available in New Zealand in capsule and suspension form. It is also available in combination with clavulanic acid in tablets. Therefore, an accurate estimate of the overall use of amoxicillin in a community should involve estimating the sum of both the use of amoxicillin alone and in combination in all available dosage forms. Similarly, in the United States, paracetamol is often combined with opioids for oral use, such as in Vicodin™ (paracetamol/hydrocodone) (Hopkinson, 1978). Thus, any quantitation of environmental risk that might arise from an API will depend at least in part on the prescription and consumption data available for all formulations of that particular pharmaceutical. Such data is likely to be specific to a particular country, which raises the important question: how useful will risk assessments generated from country-specific usage data be for predicting environmental risks posed by a particular pharmaceutical on a global scale? Given the current lack of information on usage patterns in some countries, we are still a long way from answering this question.

While all formulations and products of some specific pharmaceuticals (such as amoxicillin) are purely categorized as prescription-only medications, in many countries the legal classification of a particular pharmaceutical may depend on its formulation. This situation is illustrated by the New Zealand legislation concerning paracetamol and codeine. While 500 mg paracetamol formulated alone in tablets and capsules can be sold OTC through any non-pharmacy outlet, the OTC sale of combination paracetamol and codeine products containing more than 8 mg codeine requires the recommendation of a pharmacist (i.e., it is a pharmacist-only medicine).

Given the legislative and social aspects which govern how medicines are supplied, it is impossible to obtain accurate data on the usage of particular pharmaceuticals after their supply. Estimating the prevalence of medication use in the general population may be more realistic for determining usage patterns as it can potentially identify actual usage (i.e., where the patient reports taking the medication) as opposed to only the identification of the amount supplied by a pharmacist. Such a study was carried out in the United States by Kaufman et al. (2002) who reported that 81% of randomly selected non-hospitalized household participants had taken at least one medication and 50% had taken prescription-only medication during the previous week. However, prescription medications were more frequently used by women and older persons. In older women participants, almost one quarter had taken five or more prescription medications in the previous week. Overall, paracetamol and ibuprofen were the most commonly taken pharmaceuticals irrespective of classification (i.e., prescription-only or general sales) while conjugated estrogens (generally for menopausal hormone replacement therapy), synthetic estrogens in the oral contraceptive pill, thyroid supplements, and diuretics were the most commonly taken medications supplied on prescription only.

Paracetamol is probably the most commonly used medication worldwide. US data suggest that almost a quarter of all adults use prescription or non-prescription paracetamol on a weekly basis (Li and Martin, 2011). Self-medication (i.e., the use of medicine without the direction of a health professional) of paracetamol is also common. In a survey of 789 health science university students in Rio Grande do Sol, Brazil, 86.4% reportedly self-medicated with paracetamol and/or other herbal therapies or illicit drugs (da Silva et al., 2012). In addition, a survey of elderly participants in Campinas, Sao Paulo, Brazil, reported that 80.4% of participants had used at least one drug including paracetamol during the three days preceding the survey (de Oliveira et al., 2012). These results are consistent with those of a Spanish study which reported paracetamol to be the most commonly used analgesic among geriatric outpatients (Dominguez et al., 1992). A survey of health plan enrollees in the United States reported that 36–84% of the various cohorts in the study used paracetamol over a two-week period (Boudreau et al., 2013).

To summarize, the access policies available and regulations enforced in a particular country are largely responsible for the use and supply of different medications. For a recent and comprehensive overview of different international access policies (including the provision of pharmaceuticals) in Australia, Canada, Denmark, France, Germany, Italy, the Netherlands, New Zealand, Norway, Sweden, Switzerland, the United Kingdom, and the United States, the reader is directed to *The Commonwealth Fund* document "International Profiles of Health Care Systems" (see http://www.commonwealthfund.org/publications/fund-reports/2013/nov/international-profiles-of-health-care-systems, accessed 7-6-14).

This particular report elaborates on the structure and role of the Medicare Scheme in Australia, the National Health Service (NHS) in the United Kingdom, and the Pharmaceutical Management Agency (PHARMAC) in New Zealand (see below). For further detailed information on the particular legislative classification(s) of pharmaceuticals in specific countries, the reader should consult the relevant acts, legislation, or authorities which enforce the use and supply of drugs and poisons in that country. Some of these acts and regulations include the *Controlled Substances Act 1970* for the United States, the *Medicines Act 1981* and *Medicines Regulations 1984* for New Zealand, and the *Medicines Act 1968* for the United Kingdom. Information on all this legislation is often available online via the Internet. For example, the New Zealand *Medicines Regulations 1984* is available from the New Zealand legislation Web site (http://www.legislation.govt.nz/regulation/public/1984/0143/latest/DLM95668.html, accessed 9-7-14) and the US Controlled Substances Act is available from the US FDA Web site (www.fda.gov/regulatoryinformation/legislation/ucm148726.htm, accessed 9-7-14).

So far in this chapter, we have discussed how legislation may affect the manner in which some common drugs are made available to a patient. However, although medication is made *available* to a patient, it does not necessarily mean that they will *use* all or even part of it. The patient's perception of medications may influence whether they *adhere* or *comply* with their therapy or not. A patient prescribed a medication that is fully funded by the state government may take for granted, or even be totally unaware, of how much the medication actually costs taxpayers. To fully appreciate the role of national pharmaceutical expenditure on patients' perceptions of their pharmacotherapy, an understanding of the functions of such an expenditure system is necessary. As funding systems around the world are different, it would be impractical to elaborate on how each scheme affects patient perceptions of their pharmacotherapy. Rather, we will illustrate some of the issues using an example with which the authors are most familiar, namely the role of the New Zealand PHARMAC administrative body in controlling pharmaceutical expenditure in that country.

2.1.1 Case study: the role of PHARMAC in New Zealand pharmaceutical expenditure

Historically, New Zealanders paid nothing for subsidized or government-funded medications (see *Te Ara, Encyclopedia of New Zealand*, http://www.teara.govt.nz/en/1966/social-security/page-6, accessed 11-6-14) and instead these were paid for out of the Social Security Fund and issued by the Department of Health under a "drug tariff." The drug tariff determined the types of medicines that were subsidized under the Social Security Fund and the quantities of medications that could be issued pursuant to a doctor's prescription. Medications were made available to the patient for free if they were dispensed from community pharmacies that were registered with the national Ministry of Health. The pharmacy was subsequently reimbursed from the Fund for the cost of the medicine. However, the increasing cost of medications meant that the drug tariff could ultimately no longer support the variety and volume of medications prescribed in New Zealand. Therefore, part charges (copayments) were introduced in 1985 and these were collected from the patient by the pharmacy at the time of dispensing and went some way to apportioning medication costs between the patient and the government. However, due to increasing prescription volumes and the number of new drug entities licensed for use in New Zealand, medication expenditure continued to grow (Chui et al., 2004) and so the New Zealand government established the PHARMAC entity in 1995 in attempt to make medications once again more affordable to the consumer and at the same time allocate pharmaceutical expenditure more rationally to meet the therapeutic needs of New Zealanders (Braae et al., 1999).

There are two important strategies that enable PHARMAC to control expenditure: (1) controlled access to subsidy, and (2) price competition and comparison.

(1) *Controlled access to subsidy*. PHARMAC determines what medications are listed on the *Pharmaceutical Schedule of New Zealand* which is a list of all the prescription medications that PHARMAC will fully or partially subsidize. This does not mean that alternative products are not available in New Zealand but rather it means that if the subsidized product is listed and if patients or prescribers prefer alternative products, then they must pay their full cost.

(2) *Price competition and comparison*. PHARMAC facilitates competition between pharmaceutical companies in the process of determining which product will be listed on the *Pharmaceutical Schedule*. In recent years, this process has become especially important in the tendering for off-patent medications. When a patent for a medicine expires, other companies can create generic versions of the medicine. As these companies usually do not have to recoup the increasingly significant research and development costs incurred for the original medication, their generic products are often less expensive. PHARMAC puts out a global tender for the supply of a particular drug to the various pharmaceutical companies that formulate it. Bidding between the competitive companies occurs and the pharmaceutical company proposing the most cost-effective bid is awarded a contract with PHARMAC for sole or preferred supply to the New Zealand market for three years, at which time the supply of the drug is again put out to tender. This competitive process not only reduces the cost to the government of the medication subsidy, but assures the drug company of sales until the next round of tenders due to the product being the only subsidized product of its type on the *Pharmaceutical Schedule*.

The increasing availability of generic medications and the reduced costs of these off-patent medications has also influenced the expenditure on pharmaceuticals in New Zealand. The volume of medication being issued in New Zealand is also amplified by the increased medical needs of an aging population (Tordoff et al., 2010; Nishtala et al., 2014), the increased use of preventative therapies (Linjakumpu et al., 2002), and the increased availability of new products on the *Pharmaceutical Schedule*. However, financial barriers which prevent patients from collecting their medications still frequently occur in New Zealand and can ultimately lead to increased psychological stress, poor health, and increased illness from multiple ailments (Jatrana et al., 2011).

A major initiative to reduce medication costs is to reduce part charges or copayments incurred by patients in collecting their medications. A survey of medication underuse in seven countries (Australia, Canada, Germany, the Netherlands, New Zealand, the United Kingdom and the United States) predicted that medication underuse was associated with high out-of-pocket copayments, a below-average income, and younger study participants (Kemp et al., 2010). Furthermore, cost-related underuse of medications was negatively associated with countries that had the lowest copayments

(Kemp et al., 2010). Younger females, patients that are in lower social economic groups, those with more health problems, and those who report high levels of psychological stress are more likely to defer collecting a prescription that had been prescribed for them (Jatrana and Crampton, 2009). While the implementation of deliberate policies to increase access to medications is still desirable in many instances, they may also inadvertently lead to increased accumulation of medications that in turn may increase the quantity of unused pharmaceutical material necessitating disposal. This possibility will be explained in further detail below.

Currently in New Zealand (as of 2015) patients enrolled in a Primary Health Care Organization (PHO) and prescribed a medication that is fully subsidized under PHARMAC policy incur a NZ $5 copayment per item dispenesd from a New Zealand community pharmacy. However, once a family receives 20 items in a calendar year, the additional prescriptions are free for the remainder of the year. This means in effect that a family could pay only a total of NZ$60 per year for all their medications with no limits on the number of items. A NZ$5 copayment charge is also applicable to holders of a Community Services card, which provides reduced healthcare and medication costs as part of certain types of social welfare provided by the New Zealand Ministry of Social Development. Holders of this type of card are generally students and/or low income earners. For hospital outpatient prescriptions, the maximum charge per subsidized item is NZ$15 unless the patient has a community services card, for which a NZ$5 copayment charge per subsidized item still applies.

Before 2012, the copayment charge for primary health or community services care was NZ$3 per item. According to the New Zealand government, the increase in the copayment charge from NZ$3 to the current $5 was intended to generate more health funding with the extra money being used by the various district health boards (DHBs) responsible for the regional healthcare of New Zealanders. However, there has been speculation from opposition political parties and lobbyists that the healthcare of New Zealanders would ultimately suffer as patients were more likely to defer collecting prescriptions (as discussed above) because they would not be able to afford copayments. This could especially be the case with patients who have more than one ailment and their conditions are managed by a multitude of medications (*polypharmacy*, see below). There were also calls to use an income-based assessment to determine how much a particular patient would pay for a copayment. Despite such calls, the copayment has been retained at NZ$5 per subsidized prescription item, with the government suggesting that those in need of financial support can seek such support from Work and Income Support Services and the Ministry of Social Development. In summary, this recent increase in the copayment amount illustrates the importance of affordable medications, especially to those in lower income brackets, and any increase in the cost of medications to patients will also have the potential to reduce the quality of their healthcare, especially for low-income earners and those lowest on the social-economic ladder.

PHARMAC is a crown agency, which is a business entity that is wholly owned by the New Zealand Government and has sole responsibility for monitoring the access to and costs of medications for the entire New Zealander population. To assist it in this role, PHARMAC keeps accurate dispensing data for prescription items in order to make clinically informed decisions on the types and costs of different medications. Such information is vital for determining the quantities of medications that are issued and the types and quantities of medications which are likely to be input into the sewage and ultimately the environment (Chapter 4). However, data for usage patterns of nonsubsidized medications paid out-of-pocket by the patient and hospital medications are generally not collated.

Because there is no income-based test to determine the level of subsidy of medications when patients collect them from pharmacies in New Zealand, a significant proportion of medications received by those who can afford the copayments are in excess of their actual or presumed therapeutic needs (see discussion on *nonadherence* below). Given the advent of PHARMAC subsidizing many medications in bulk lots (see below), such excess supplies of medicines are often stored in households, with the subsequent generation of pharmaceutical waste. In the following section, we will explore the attitudes, behaviors, and practices of patients toward the use of dispensed medicines and their potential to often lead to significant quantities of unused pharmaceuticals which ultimately must be disposed of in some manner.

2.2 The role of patient medical adherence

The medication prescribed by a healthcare provider and issued from a pharmacy may not necessarily be taken by the patient. This could be due to a lack of *medical adherence* to the particular pharmacotherapy regimen specified by their doctor.

The World Health Organization (WHO) defines *medical adherence* as "the extent to which a person's behavior taking medication, following a diet, and/or executing lifestyle changes, corresponds with agreed recommendations from a healthcare provider" (Lehmann et al., 2014). Adherence is therefore the act of both implementing and persisting with a prescribed practice. *Compliance* is another term that is used interchangeably with adherence, but this former term is generally less favored than adherence as compliance suggests that the treatment has been forced upon the patient when the patient may not necessarily have agreed to participate in this particular treatment (Lehmann et al., 2014). Neither adherence or compliance should be confused with *concordance*. Concordance is an equally fostered relationship between the prescriber and the patient (Osterberg and Blaschke, 2005; Vrijens et al., 2012), where both are equally important in making health decisions pertaining to the welfare of the patient (Ahmed and Aslani, 2014). The practice of concordance in clinical therapy can result in a course of therapy that patients agree to adhere to, because they feel their choices and decisions have been actively considered by their prescriber's management of their health. Adherence is determined by how well both the patient and their prescriber have implemented a particular regimen and how long that lasts.

The term "implementation" refers to how well a particular course of medication has been initiated and "persistence" is defined as the time interval between the initiation and discontinuation of a particular course of therapy (Lehmann et al., 2014). Both the implementation and the persistence of a particular therapy are dependent on the willingness of patients to undertake behavioral changes as their perception and outlook to the prescribed pharmacotherapy is exceptionally important for therapeutic success (Platt et al., 2014). In describing adherence, it is intuitive that the underlying principle behind any *nonadherence* to medication is that the patient deviates in some way from the intent of the prescribed instructions. It is estimated that around 50% of patients are nonadherent to long-term medications (Osterberg and Blaschke, 2005).

There are two main subsets of nonadherence, namely *primary* and *secondary nonadherence* (Storm et al., 2008). *Primary adherence* refers to the situation where a prescriber issues a prescription for medication, but the patient either collects none or only some of the prescribed items from the pharmacy (Jackevicius et al., 2008; Solomon and Majumdar, 2010). In this situation, the patient has made a conscious choice to not collect their prescribed medication and so does not initiate therapy (Osterberg and Blaschke, 2005). There are several reasons that primary nonadherence may occur, including financial barriers to paying for the items, access issues in receiving the medications, or the belief that the medications are not needed (Peyrot and Rubin, 2011). For example, the cost of medication copayments often required at the time of dispensing has been previously associated with nonadherence in chronic statin-treatment regimens (Simoens and Sinnaeve, 2014). Other factors which contribute to primary nonadherence include issues involving the effectiveness of patient–prescriber communication, whether the patient has multiple medical conditions that require treatment, individual health beliefs, and the rational or irrational belief that certain classes of medication may be better for them than others (Phatak and Thomas, 2006). As the cost of medications is the predominant driver behind primary nonadherence (Wamala et al., 2007) and the act of a patient physically collecting medications is required to produce pharmaceutical waste, it is of less importance than *secondary nonadherence* (see below) from an environmental perspective. However, primary adherence may be more environmentally important when any part charges for medication are decreased and medications become more affordable to patients (Braund et al., 2007, 2008, 2009).

In *secondary nonadherence*, the patient collects the entire quantity of medication prescribed but takes only some or even none of it (Hovstadius and Petersson, 2011). The reasons for this behaviour may include wanting to appear compliant to the prescriber (known as *white coat adherence*), experiencing undesirable side effects to the medications, believing that the medications are providing no benefit, or simply forgetfulness (Osterberg and Blaschke, 2005). Particular aspects of medication therapy increase the incidence of secondary nonadherence (Vermeire et al., 2001; Osterberg and Blaschke, 2005; Hansen et al., 2009). For example, certain personality and behavioral traits of the patient (i.e., conscientiousness and medication-information seeking behavior), increasing the dosing frequency (i.e., more than once a day), having a complex dosing regimen, treatment of asymptomatic disease, and the experience of adverse side effects in pharmacotherapy can increase the rates of secondary nonadherence (Osterberg and Blaschke, 2005).

2.2.1 Secondary nonadherence and conscientiousness

Patients who display conscientious personality traits are more likely to adhere to correctly taking their prescribed medication (Molloy et al., 2014). In psychology, *conscientiousness* is one of the "big five" personality traits that characterize human behavior. People with conscientious personalities are generally goal-orientated and organized.

In addition, they often display self-control, are tenacious and persistent in their pursuits, tend to be meticulous in planning, and are often self-reflective of health behaviors or actions. As such, they are less likely to deviate from normal routines and tend to follow any recommended guidelines of prescribers (Molloy et al., 2014). In addition, people with conscientious personalities are less likely to take medications erratically by under- or overdosing because they are more likely to keep to routines. However, there are no defined practical measures that can be currently used by prescribers to monitor such conscientiousness apart from not detecting erratic behavior in the clinic or pharmacy. Warning signs may include the patient not collecting refills or repeat dispensing of prescriptions on time, or running out of medication prematurely. This behavior may also be confounded by other behavioral aspects, such as forgetting to take medications and therefore not picking up refills or repeats in a planned and timely manner rather than intentionally or deliberately not collecting them. However, forgetfulness could also mean that a particular patient is not conscientious enough about addressing their health issues that involve taking their prescribed medications (Molloy et al., 2014).

2.2.2 Secondary nonadherence and medication-information seeking behavior

While conscientiousness may lead to increased adherence, medication-information seeking behavior of patients may contribute to secondary nonadherence (Carter et al., 2013). Patients generally question and seek information about their medication when they are concerned or worried about treatment outcomes, develop adverse drug reactions as a side effect of the pharmacotherapy, or question the efficaciousness of the treatment. In addition, the wide availability of media available in modern-day society, such as television, newspapers, lifestyle magazines, and most recently the Internet, means patients can use these as sources of information when uncertainty about their pharmacotherapy exists. However, many of these resources fail to provide completely accurate and evidence-based information about a patient's condition or medications. Consultation of these resources may lead patients to "self-diagnose" their own condition and create negative perceptions about their pharmacotherapy both leading to nonadherence (Carter et al., 2013). However, nonadherence to pharmacotherapy has only been reported to be positively correlated with autonomous medication-information seeking and not medication-information seeking from prescribers. In other words, patients who were more likely to obtain medication information concerning their pharmacotherapy from other sources are also more likely to shun or terminate any pharmacotherapy given to them by the prescriber, simply because they may not trust their advice and practice (Carter et al., 2013). A prescriber who encourages open and honest conversation with their patient can therefore minimize this form of secondary nonadherence. Prescribers should acknowledge any benefits and limitations of the pharmacotherapy they prescribe so that patients know what to expect and understand likely treatment outcomes.

2.2.3 "White coat adherence" syndrome

Patients may be more likely to take their medication immediately preceding prescriber visits than at other times. The reasons for such behavior can include that the patients may want to create the favourable impression that they are regularly taking medications when they are in reality not. Such behavior is known as the *white coat adherence* syndrome and tends to occur early on in chronic or long-term treatment therapy (Modi et al., 2012). While this effect can be positive for adherence to a particular pharmacotherapy in the short term, nonadherence to medication may start to occur once its effects "wear out" over time and if the patient is not conscientious about completing the pharmacotherapy.

A study of glaucoma patients and their use of intraocular medications (Schwartz and Quigley, 2008) suggested that white coat adherence can fool prescribers into thinking that intraocular pressure was adequately controlled and that a particular patient's glaucoma had therefore not worsened. When intraocular pressure was not adequately controlled, it could also be difficult for prescribers to assess whether this was due to nonadherence or a true lack of therapeutic efficacy. The study observed that adherence to intraocular medications was highest approximately one week before a return visit to the prescriber, with adherence rates dropping immediately afterwards (Schwartz and Quigley, 2008). Therefore, awareness and understanding of why a particular patient may not, or cannot, adhere to a particular medication is necessary for efficacious pharmacotherapy when white coat adherence is suspected. In addition, physiological and medical examinations of the patient may be a more accurate assessment of the status of any underlying disease than any claimed self-admitted adherence by patients. For example, optic tests showing visual field deterioration may determine that glaucoma has rapidly progressed in a patient when the intraocular pressure has been determined at the visit to be normal. While the latter observation can arise from a short-term adherence to medications that lower intraocular pressure, any visual field deterioration cannot be masked by any similar practice.

2.2.4 Secondary nonadherence and adverse drug reactions, side effects, or efficacy

There are many patients who are known to collect medications and then subsequently fail to take them because they are afraid of the possible side effects and/or adverse drug reactions. Side effects resulting from pharmacotherapy have been found to be one of the six key contributing factors to nonadherence in older adults (Mukhtar et al., 2014). It has also been reported that many patients do not want to read official information leaflets (accompanying the prescription or provided independently by the pharmaceutical company) or leaflets given to them by their prescriber because they were afraid of the potential side effects they would discover about their pharmacotherapy (Tsiantou et al., 2010). To them, any knowledge of side effects may present as a rational or irrational fear about their medicines. Indeed, it can be a very difficult task for the prescriber to confidently convince a patient to take their pharmacotherapy before their fears and paranoia are addressed. Many patients who refuse to take prescribed medications because of purported side effects will nonadhere because of a previous unpleasant experience with another medicine. Such previous negative emotional experiences may affect their outlook on the current course of pharmacotherapy. Patients may not be able to weigh up the benefits and risks of taking medication prescribed to treat a particular medical condition. For example, symptoms of discomfort rarely appear in patients experiencing hypertension other than knowing that their blood pressure is abnormally raised. However, it is common knowledge that palpable side effects occur when using many antihypertensive drugs. Patients may feel lethargic from taking β-blockers, or have an annoying dry cough from taking Angiotensin II Converting Enzyme (ACE) inhibitors. The emotional and physical pain or discomfort that can arise from these side effects generally overrides the determination to treat the underlying asymptomatic hypertension (Marshall et al., 2012). Having side effects could also be very confusing for the patients, especially if they have started therapy, as they may not know whether the side effects experienced are an adverse drug reaction from therapy or an underlying symptom of their medical condition under treatment.

Some courses of medication are never finished because patients incorrectly believe they are already cured by the medication they have taken so far in the course of treatment. For example, finishing a seven- or 14-day course of antibiotics is usually necessary to effectively treat bacterial infections. However, patients may subjectively decide to terminate antibiotic therapy prematurely simply because they observe that the purulent discharge has gone, the pain has disappeared, or some other change has occurred for them to believe that their condition has already been adequately treated (Gould, 2009). Such practices have led to the development of antibiotic resistance (see below). Prescribers and pharmacists should therefore encourage patients to always finish any antimicrobial therapy irrespective of any subjective feelings of enhanced wellness by the patient before the end of medication.

The time has never been more important for concordance between the patient and the prescriber. Effective therapy can only occur if the patient is completely honest about any worries, issues, or side effects manifested with a particular pharmacotherapy. In addition, prescribers need to be completely open-minded about any such worries or side effects and not lightly dismiss a patient's worries about their medicine even if they deem any reports of side effects or adverse reactions to be unusual. The root cause of any concerns, whether pharmacological or psychological, needs to be addressed and consideration be given to alternative evidence-based therapies that are likely to be more acceptable to the patient (Tsiantou et al., 2010).

An interesting perception among nonadherers is the belief that some brands of medication may be better than others. As noted above, government agencies which manage pharmaceutical expenditure in various countries cannot fund all of the available brands of a particular medication on the market. For example, under the New Zealand PHARMAC system, a patient may choose for a nonsubsidized brand to be dispensed if they pay entirely out-of-pocket for the medicine. Purchasing a nonsubsidized brand can be costly and unaffordable. For many off-patent medicines, the default will be for pharmacies to dispense a subsidized and inexpensive generic version of the medication in order to avoid the patient paying entirely out-of-pocket. However, many patients will claim that the generic brand dispensed is not as effective as the original brand. Such a situation usually happens when a new pharmaceutical company wins a supply tender (as with sole or preferred supply status in New Zealand, see discussion on PHARMAC above) leading to a change in the subsidized brand. The belief that certain brands are more efficacious than others is known as the *nocebo* effect (Weissenfeld et al., 2010). In a German study, 36.7% of participants believed that generic products were inferior in quality to innovator products and 43% of the participants reported real or perceived problems with side effects after switching to a generic brand drug (Weissenfeld et al., 2010). Some of the side effects experienced by users of generic products can be explained by the different pharmacokinetics of the innovator and generic products (i.e., different drug release and absorption profiles). There is also the possibility that some brands are not completely *bioequivalent* (i.e., the rate and extent of absorption of both products is not equal). However, many of the side effects reported were subjective, such as nausea, the urge to vomit or general malaise and headache. While two-thirds of participants in one particular study were reported to be content with generic substitutions at a pharmacy, 34% of participants using psychological medications reported additional side effects from the change (Weissenfeld et al., 2010). Prescribers should therefore be aware of funding issues surrounding medications and be attentive to patients who may still prefer the original innovator brand over any generic products. While some innovator brands that are the preferred choice of the patient may not be cost-effective for them, it is important for a prescriber to be completely open and honest about the situation and disclose to their patient the likely out-of-pocket costs they will have to pay at the pharmacy for their choice of innovator product.

2.2.5 Secondary nonadherence and complex dosing regimens

The complexity of treatment regimes can be associated with adherence. In a survey exploring the relationship of treatment complexity and intentional and unintentional nonadherence (de Vries et al., 2014), a significant positive correlation was reported between treatment complexity and nonadherence in patients with type 2 diabetes. Having to take multiple doses (i.e., a high frequency of medication) during the day was also observed to significantly increase nonadherence in adults older than 40 years (Remondi et al., 2014). More doses have been shown to be missed by patients on a twice-daily regimen compared to once-daily (Matthes and Albus, 2014), as twice-daily regimens posed more interruptions for the patient (Vervloet et al., 2013). There is also some evidence that nonadherence is more frequent for certain days of the week or time of the day. The study by Vervloet et al. (2013) reported that out of 104 nonadherent type 2 diabetics, patients were most negligent about taking their medications on Saturday evenings while multiple doses were usually more poorly timed on Sunday evenings.

However, a high medication-taking frequency may not necessarily be the only factor that influences adherence rates. Vrijens et al. (2014) explored the difference in doses missed between two different antiplateleting regimens—a once-daily dose of clopidogrel compared to twice-daily doses of ticagrelor. Over a period of 30 days, 25.7% of the patients whose histories were selected for review, missed at least one dose when twice-daily ticagrelor was prescribed, in contrast to 46.8% of the patients prescribed once-daily clopidogrel. Inhibition of platelet aggregation (prevention of blood clotting) was also controlled better with twice-daily ticagrelor compared to once-daily clopidogrel. This study emphasizes that while a general increase in dosing frequency may be negatively associated with adherence, an increase in dose frequency may be better for patients who may have otherwise forgotten to take their medications on a once-daily dosage regimen. In addition, as the period between doses is shorter (every 12h for twice-daily doses), the steady-state concentration of the drug is more likely to be maintained compared to a longer dosing period such as 24h for a once-daily dose. Therefore, it is more clinically "forgiving" if *single doses* on a twice-daily regimen are missed compared to missing single doses on a once-daily regimen (Levy, 2001).

The correct timing of doses to achieve medication adherence has been hypothesized to fit into six potential categories, the so-called "rule of sixes" (Urquhart, 1998, 2002; Osterberg and Blaschke, 2005). It is estimated that nearly one sixth of patients adhere almost perfectly to their medication regimens; one sixth take nearly all doses, but with some irregularity in timing; one sixth miss an occasional single day's dose with some timing irregularity; one sixth take drug holidays three to four times a year, with occasional omissions of doses; one sixth have a drug holiday monthly or more often, with frequent omissions of doses; and the remaining sixth take few or no doses while giving the impression of good adherence (Urquhart, 2002). While it might be inevitable that the predominant number of patients will forget to take a dose of medications at some time during a course of pharmacotherapy, a prescriber needs to be aware of a patient's lifestyle and plan an individualized pharmacotherapy around the patient's routines so that any effects from a missed dose are minimized while still achieving the overall therapeutic goal.

2.3 Strategies to minimize medication nonadherence

Patients who are nonadherent to medications often have reduced health outcomes which subsequently lead to increased health-related costs (Vermeire et al., 2001; Osterberg and Blaschke, 2005). For this reason alone, concordance between patient and prescriber is important for effective therapy.

A retrospective cohort[1] study in the US in an ambulatory care setting investigated the admission of over 400,000 cases due to nonadherence over a span of two years (Heaton et al., 2013). The visits of patients with type 2 diabetes, hypertension, or mental illness were significantly associated with nonadherence. Many of the prescribed medications usually require chronic long-term management and any withdrawal of medications usually means a loss of adequate management of these conditions. Heaton et al. (2013) concluded that targeting these high-risk groups may benefit overall adherence rates.

Many countries have established consultation programmes which aim to improve health literacy (patient's understanding of their disease(s) and their pharmacotherapy) and address problems with pharmacotherapy (such as adverse drug effects). At the same time, technical issues relating to medication administration, such as difficulties opening child-resistant caps for elderly sufferers of osteoarthritis, are addressed. A pharmacist and health promoter team which facilitates medication management through consistent communication with patients can also improve medication usage (Gerber et al., 2010).

[1] A retrospective cohort study investigates two demographically similar groups of study participants differing only by the presence or absence of one health characteristic (e.g., hypertensive vs nonhypertensive patients) by examination of past medical records over a particular timeframe. Quantitatively, such a study determines the odds or risk that the presence of a characteristic is associated with an outcome (i.e., nonadherence) relative to the absence of the characteristic.

A lack of health literacy may be a potential barrier to effective patient self-care and management of medications (Gazmararian et al., 2009). For example, a lack of health literacy and inappropriate metered-dose inhaler administration techniques have been attributed to the poor management of asthma (Williams et al., 1998). Improvement of health literacy seems to be a promising way of addressing adherence issues. A technology-based "Meducation" programme addressing antihypertensive medication adherence in veterans has been trialed in the United States (Zullig et al., 2014). Forty-percent of programme participants had a literacy level of less than 9th grade. However, after using "Meducation" calendars which communicated exactly when and what anti-hypertensives should be taken specific to a patient's needs, there were significant decreases in diastolic and systolic blood pressure and body weight after six months of intervention (Zullig et al., 2014). Another study concerning glaucoma therapy emphasizes the importance of adequate prescriber education for those with low health literacy (Sleath et al., 2014). While new patients to glaucoma therapy were usually given full instructions on how, why, and when glaucoma therapies should be used, such information was not given to low health literacy or African-American patients. An Irish study assessing the accessibility of the patient information leaflet given to patients treated with clozapine found that 95% of study participants understood a special pharmacist-designed leaflet compared to 73% for the company leaflet. Approximately a third of the study participants had low or marginal health literacy (Brosnan et al., 2012). These results suggest that prescribers may not be effectively disseminating health information to patients that need it most in order to maximize the likely benefits of their pharmacotherapy. The information provided to a patient by a pharmacist and prescriber can only be effective if it is matched to their level of health literacy.

One potential effective way to tailor information to the needs of patients with low health literacy is to use illustrative as well as verbal instructions as to how and when medications should be used. Pharmacists have a pivotal role in translating written information on package inserts and information leaflets into an accessible and interactive form. Hu et al. (2012) demonstrated that a pharmacist's verbal face-to-face counseling of patients and the presentation of photographic material on medication administration made administration of pediatric antibiotics easier to understand for caregivers. In addition, caregivers reported that face-to-face consultations saved them time deciphering written instructions (Hu et al., 2012). A lack of understanding among Latino communities in the United States as to how medications should be used has been reported to create barriers to adherence (Mohan et al., 2013). However, the use of illustrated material to demonstrate medication administration was highly preferred and study participants commented that such information could address barriers and challenges to their appropriate use of medications (Mohan et al., 2013). Illustrated information for the dissemination of written and verbal instructions has also been favorably received by pharmacists and many other healthcare providers and it has been successfully trialed in underserved US communities (Mohan et al., 2012).

The use of pictograms to communicate medication administration principles was determined to be the most successful way of disseminating health information to three distinct low-level literacy groups in Canada (Braich et al., 2011). Pictograms increased medication adherence to postoperative cataract regimens by 28 days or more compared to the study control group, which was only given verbal instructions. In addition, the intervention group who were allowed to take the pictograms home to study were more adherent to long-term treatment regimens than those who were presented with the pictogram instructions only at the clinic (Braich et al., 2011). Cross-cultural communication can also be improved with visual aids. A study of 220 patients in an anticoagulation clinic in the United States identified that nonnative English speakers were more likely to receive discordant medication administration information if prescriber instructions were disseminated verbally (Schillinger et al., 2006). However, the effectiveness of visual communication and written materials was not affected by language barriers or health literacy (Schillinger et al., 2006).

As discussed above, one of the major barriers to medication adherence is the effectiveness of the prescriber–patient communication. Often, the patient does not understand the way in which the prescriber is planning the treatment regimen. Also, the prescriber may not fully understand the behavior, lifestyle, and habits of the patient enough to prescribe the most appropriate pharmacotherapy for the patient. Such situations benefit from increased concordance between the prescriber and the patient. New electronic technologies to assess and record medication use and create medication schedules that are effective for the patient have been trialed with some success (Morrow et al., 2012).

However, improving health literacy only improves knowledge and not a patient's medication-taking behavior. A US study of newly admitted patients in three heart clinics assessed their level of health literacy, heart failure knowledge, self-efficacy (the belief in an individual's ability to achieve a desired goal), and self-care (Chen et al., 2014). While health literacy was significantly associated with knowledge concerning heart failure, it was not associated with self-care. Instead, self-efficacy was significantly associated with self-care. Another study assessing the clinical outcomes of health literacy in veterans taking five medications or more, found that patients with poorer health literacy had poorer knowledge of their medications (Mosher et al., 2012). However, these groups did not report a lower adherence rate or a higher incidence of adverse events from medication use. A further study on 125 type 2 diabetics in a primary care setting showed that health literacy was not significantly associated with self-care or medication adherence (Bains and Egede, 2011). Health literacy and knowledge of their disease states are important for patient understanding of pharmacotherapy; however, the pharmacist and other health professionals must empower patients to believe that their pharmacotherapy can be easily managed around their lifestyle before adherence is likely to be achieved in these situations (Bailey et al., 2014). So health knowledge and literacy improve compliance.

Some ways to increase such self-efficacy in medication management include increasing self-care (Bennett et al., 2009) and reducing the *pill burden*, which is defined by the quantity of medication doses that a patient needs to take on a daily basis. The pill burden may be influenced by the number of dosage forms of a particular medication taken (i.e., twice- or three-times daily dosing vs. once-daily dosing) or the different types of medication that are taken at one time by a patient (e.g., concomitant β-blocker and ACE inhibitor for antihypertensive therapy). A recent meta-analyses reported that pill burden significantly promoted nonadherence in HIV antiretroviral therapy, whereas once-daily regimens promoted adherence (Nachega et al., 2014). A similar result was observed with a sample of elderly (older than 65 years) patients prescribed multiple medications by medical staff before their discharge from an internal medicine ward in Italy (Pasina et al., 2014). Follow-up interviews for three months after discharge revealed that medication nonadherence was positively associated with the number of medications prescribed. In addition, less than 30% of the participants surveyed in the two instances of follow-up did not understand the therapeutic purpose of their medications.

Patients with disease conditions (e.g., HIV infection) that are difficult to manage with a single medication, often report adherence problems due to pill burden and complex dosing frequencies (Schaecher, 2013). Such reports of *polypharmacy*, which is defined as the prescription of concomitant multiple drugs in a rational or irrational manner, can drastically reduce the manageability of a patient's pharmacotherapy and contribute to nonadherence. However, polypharmacy is not always avoidable given that some patients can have complex and multiple medical conditions. Many schizophrenic patients are on a regimen involving multiple antipsychotic drugs because their state of psychiatric health cannot be stabilized with a single type of medication alone (Correll et al., 2011; Peuskens et al., 2014). Other reasons that prescribers do not use monotherapy include evidence-based reasons which demonstrate increased clinical benefit of polypharmacy over monotherapy and the lack of patient efficacy or tolerability to the current drug of last resort for treatment of schizophrenia, namely clozapine (Correll et al., 2011). Therefore, prescribers should attempt to reduce *unnecessary* polypharmacy; however, in certain circumstances, polypharmacy may be the best option for the patient, given that the benefits can outweigh the risks of the pharmacotherapy (Zigman and Blier, 2012).

The use of *deprescribing*—a holistic approach where inappropriate medications are identified and withdrawn to be replaced with an appropriate regimen involving thorough follow-up and support—has been recently reported by Reeve and Wiese (2014). Deprescribing is based on understanding the factors that enable or prevent a patient from stopping a particular medication regimen and allowing the safe and effective withdrawal of the medication from a patient's pharmacotherapy. In a deprescribing process, potentially inappropriate medications (i.e., whereby taking the medication is suspected to increase the risks of adverse drug effects rather than confer clinical benefits) are identified for a particular patient. The benefits and risks are then evaluated to determine whether the withdrawal of a medication is necessary or not. Subsequently, potentially inappropriate medications are prioritized for withdrawal. If it is recognized that a withdrawal is necessary, the patient is weaned off the medication. Any adverse effects arising from the withdrawal regimen are monitored and further decisions are then made to address such issues. For example, a patient may develop severe agitation from abrupt withdrawal of an opioid previously prescribed for relief of pain (analgesia). Even though the medication may not be necessary any longer, the prescriber may decide to resume the opioid and withdraw it at a slower rate to avoid the agitation experienced by the patient (Reeve et al., 2014).

Another aspect of making drugs more effective and easier to take for patients is the repackaging of dosage forms to increase the self-efficacy of patients and allow them to manage their medications more independently (Hawthorne et al., 2008; Laufs, 2011). For example, the use of calendar packs, unit-dose packaging (i.e., blister packs), and reminders have been observed to increase medication adherence to a certain extent (Heneghan et al., 2006; Yamanouchi et al., 2010). In addition, some experimental blister packs containing electronic sensors to monitor whether a unit-dose blister has been opened or not, may provide healthcare providers with more information about patient adherence (van Onzenoort et al., 2012). However, the evidence for their efficacy is equivocal (Orton and Barnish, 2005). An earlier study concerning the use of blister packs and pill organizers for adherence to vitamin tablets found that while there was some benefit from using blister packs, there was no significant evidence to show pill organizers increase adherence (Huang et al., 2000). While calendar packs and unit doses are a fool- and memory-proof way to help those taking a large quantity of medications at one time, other issues, such as the patient's inability to open the adhesive-heat sealed foil packs, may prevent medication adherence (Yamanouchi et al., 2010). Furthermore, unit-dose dispensing can jeopardize medication quality. Unit-dose dispensing usually requires the repackaging of medications outside of their original containers. Therefore, it is vital that the dispensing pharmacy uses good manufacturing practices (GMP) when repackaging medications to avoid microbial contamination of repackaged dosage forms (Abu Shaqra et al., 2014). In addition, medications in repackaged blister packs may discolor or degrade faster (Yamazaki et al., 2010). The use of organizers and repackaging medications may therefore increase adherence, but it is important for the prescriber to investigate the type of repackaging to best suit the needs of the patient, as no single method is appropriate for all situations (Cramer, 1998).

The taste of dosage forms may increase medication adherence, particularly in children (Lipchock et al., 2012). Usually intensely sweet and flavored dosage forms, as in pediatric paracetamol and ibuprofen suspensions, have a higher acceptability and tolerability profile compared to traditional tablets and capsules that do not have any flavor or are bitter (Lipchock et al., 2012). Preference to sweet-tasting formulations has also been observed to be genetically controlled and children are generally more sensitive to bitter tastes than adults (Mennella et al., 2013). In addition to sugar, the addition of acids and salts in many pediatric formulations increases acceptability of the formulation to children; however, they are usually not completely successful at suppressing the intrinsic bitterness of many medications (Mennella et al., 2013). Adding volatile-tasting flavors or edible oils, such as bubble-gum, to formulations may also increase dosage form acceptability. Nevertheless, it is difficult to find flavors which are universally acceptable to all children (Mennella and Beauchamp, 2008). Furthermore, the stability of the API must be taken into question when formulated with large concentrations of sugar and other potential excipients as the presence of sucrose may lead to decreased API stability in the dosage form (Bundgaard and Larsen, 1978). The use of intensely sweet formulations may also be discouraged for diabetics and the need for artificial sweetener replacements may therefore be necessary in their situation (Jana et al., 1994). Sweet-tasting formulations may increase medication adherence; however, whether a sweet-tasting formulation is used will need to be evaluated on a case-by-case basis.

The use of computer-related technology (*cyberhealth*) to promote medication adherence is potentially cost-effective when patients are in doubt about treatment (Alvarez-Jimenez et al., 2014). Using the Internet, patient and health records can now be streamlined and patients can often more easily manage their medications

(Stellefson et al., 2014). In the future, patients may be able to easily communicate any issues concerning their pharmacotherapy to their prescriber, who may then be able to optimize their therapy without a physical visit to the clinic for a consultation (Laffer and Feldman, 2014). The use of such technology to promote medication adherence may particularly benefit young adults and adolescents because of their general proficiency in the use of electronic devices such as smartphones as ready sources of information (Hailey and Arscott, 2014). However, further research needs to be undertaken before the long-term impact of cyberhealth on medication adherence can be assessed. Furthermore, the use of electronic medical records and medicines management software may seem quite foreign to older prescribers, who may find difficulty in mastering new e-health skills for managing a patient's medications. Other aspects, such as the ability to reimburse prescribers for their advice, the ownership of uploaded information, and the ability of online software and applets to maintain acceptable security and confidentiality of health information, also needs to be addressed before cyberhealth can be considered feasible for medicines management (Schneider et al., 2014).

Minimizing medication nonadherence is a challenging and complex area of clinical therapy. The prescribed pharmacotherapy must be tolerable and effective while at the same time manageable, accessible, and empowering to patients. Medication Use Review (MUR) programmes may be of benefit in achieving such aims (Hugtenburg et al., 2009). These programmes, such as the Medication Use Review and Adherence Support (MUR) in New Zealand (Lee et al., 2009), Home Medication Reviews (HMR) in Australia (Ghosh et al., 2014), Medication Therapy Management (MTM) in the United States (Touchette et al., 2006; Lounsbery et al., 2009), and the MUR clinical medication reviews in the United Kingdom (Latif et al., 2011), provide the opportunity for patients and pharmacists to identify areas of medication usage that can be improved. However, it is important to realize that individualized pharmacotherapy is resource- and cost-intensive and services funded by health services may not necessarily provide the level of clinical support required by patients. For example, basic health literacy to improve patient adherence is funded in New Zealand by certain DHBs through the MUR programme; however, clinical issues affecting nonadherence in the realm of primary care prescribers, such as medication changes and dose titrations (see discussion below), are not funded (Hatah et al., 2014). In addition, tackling adherence problems from both clinical and nonclinical perspectives needs to be a coordinated effort between different health professionals whereby continuity in the level of patient care is maintained (Hatah et al., 2012). Only then will MUR recommendations be useful to patients and prescribers. However, any comprehensive data on the efficacy of the many adherence interventions proposed or trialed are still lacking. Further studies to determine intervention efficacies are necessary if any of these measures used to combat nonadherence are to become evidence-based recommendations (Laufs, 2011).

Medication nonadherence is only one of the processes that create unused pharmaceuticals. Clinical problems potentially producing pharmaceutical waste may also stem from medicine or dosage inadequacies to meet the pharmacotherapy needs of the patient. The following section discusses drug and patient factors which may lead to the production of pharmaceutical waste.

2.4 "Doctor, it just doesn't work!"—Pharmacodynamics, pharmacokinetics, and pharmaceutical waste

A lack of patient adherence to medications is not the only reason for wastage of pharmaceuticals at the prescribing level. The clinical termination of therapy usually arises from the following situations—"wrong drug, wrong dosage," "wrong drug, right dosage," and "right drug, wrong dosage"—and such termination can lead to wastage of medications. Termination of therapy usually results from the lack of consideration for: (1) drug action and behavior in the particular patient; (2) the patient's medical history; (3) other concomitant prescription or OTC medications the patient is taking; and (4) adjusting the dose of a medication until optimal results are reached (i.e. the process of *dose titration*—see Section 2.4.2 below). While the consideration of these issues will not completely prevent any medication wastage arising from a clinical decision, there is no doubt that wastage can be minimized if prescribers are attentive to these issues, due to the increased possibility that a patient will be stabilized on a particular pharmacotherapy without the need to correct problems later. These concepts are explained in the following sections of this chapter.

2.4.1 *Action and behavior of drugs*

Rational prescribing is based on making an educated decision as to the most effective (efficacious) drug for the patient that will minimize any adverse effects. Drug choice should be determined by the physiological characteristics of the patient and the underlying disease or symptom to be treated. A clear example is the treatment of osteoarthritis. Paracetamol and ibuprofen are both analgesic drugs. While paracetamol has only analgesic (painkilling) and antipyretic (fever-reducing) abilities, ibuprofen is an NSAID that has analgesic, antipyretic, and anti-inflammatory effects. As osteoarthritis results from the mechanical wear and tear of the synovial joints and cartilage, the affected joints are usually painful and stiff upon movement, but not usually inflamed. While it might initially appear that both drugs seem to be perfectly acceptable for the treatment of uncomplicated osteoarthritis, paracetamol is considered to be first-line therapy over ibuprofen. This is because osteoarthritis usually does not involve inflammation and so there is no extra benefit in using a drug with additional anti-inflammatory properties, such as ibuprofen. There is also the increased risk of gastrointestinal bleeding from the use of ibuprofen through its inhibitive action on cyclooxygenase (COX) enzymes. In addition, NSAIDs are known to exacerbate asthma and therefore caution must be exercised if they are to be used with an asthmatic patient. The treatment plan will be very different for inflammatory conditions, such as rheumatoid arthritis or gout. An acute episode of gout—whereby there is a deposition of uric acid crystals in joints—causes a painful and inflammatory response facilitated by white blood cells. While this inflammation cannot be relieved by paracetamol because it has no known anti-inflammatory properties[2]

[2] This claim has not been completely confirmed as the pharmacological action of paracetamol is not entirely understood. Some literature suggests that the centrally mediated analgesic properties of paracetamol are due to its action on cannabinoid receptors (Mallet et al., 2008) while other literature suggests it does exhibit some anti-inflammatory activity via the cyclooxygenase 3 (COX-3) enzyme (Munoz et al., 2010).

ibuprofen (and other NSAIDs) are extremely effective. In such cases, while some patients may have a history of gastrointestinal disorders or asthma, the benefits obtained from a *short* course (up to a few days as appropriate for current treatment recommendations) of an NSAID such as ibuprofen to relieve the acute inflammation may outweigh the long-term risks associated with their use.

2.4.2 Patient medical history

As discussed above, the patient is at the center of the prescribing process. Questions need to be raised about past illness or disease and special physiological states that require caution in prescribing, such as when a patient is pregnant. For example, the anti-acne agent isotretinoin must not be used by pregnant women and any therapy involving the use of this drug will generally be supervised in some countries by a specialist dermatologist (Lagan et al., 2014). It is also vital that prescribers check for any patient sensitivities or allergies to medications. A single dose of amoxicillin to a patient with a β-lactam antibiotic sensitivity or allergy can result in a potentially life-threatening anaphylactic shock event. In cases of β-lactam antibiotic sensitivity or allergy an alternative antibiotic with similar bacterial coverage should be prescribed such as the broad-spectrum antibiotic erythromycin (Robinson et al., 2002). Although the general first-line therapy for mild hypertension in many countries is bendrofluazide (bendroflumethiazide) and other thiazide diuretics (e.g., hydrochlorothiazide, chlorthalidone), these drugs prevent the excretion of uric acid from the body (Handler, 2010). For someone with gout or elevated uric acid levels, these drugs are inappropriate given that gout can be further exacerbated upon initiation of thiazide therapy (Bruderer et al., 2014; McAdams-DeMarco et al., 2014).

2.4.3 Concomitant medications

Along with the patient history, the possibility of a prescribed medication affecting the efficacy of another medication taken by a patient needs to be checked to prevent underprescribing or overprescribing. For example, caution needs to be exercised when the antidepressant fluoxetine is prescribed at the same time as the synthetic weak opioid tramadol. Fluoxetine is a selective serotonin reuptake inhibitor (SSRI) and while tramadol is a pain-killer affecting the μ- and other opioid-receptors, it also exhibits SSRI properties. The concomitant use of both these drugs may lead to a condition known as *serotonin syndrome* (a type of seizure and loss of coordination) (Rizzoli et al., 2013). A patient's medical history may not alert the prescriber to inappropriate concomitant medications. A drug-drug interaction can occur between prescription and OTC medications, and so it is important for prescribers to ask patients to advise them of any other OTC medications, vitamins, or even dietary supplements that they may be taking before prescribing any new medication. For example, St. John's Wort (*Hypericum perforatum*) is a herbal supplement commonly taken for mild depression and is generally available in non-pharmacy outlets in many countries. However, it increases the activity of the enzyme CYP3A4 (see discussion in the following section) which is also responsible for the metabolism of many other drugs, including warfarin, an anticoagulant that requires regular monitoring (Di et al., 2008; Izzo and Ernst, 2009;

Kimura et al., 2010). Therefore, without knowing that a particular patient is taking St. John's Wort, a prescriber may have difficulty establishing the reason why they cannot stabilize a patient on a particular dosage of warfarin. Because St. John's Wort has a similar pharmacological action to SSRIs, caution must also be exercised by prescribers who prescribe tramadol for St. John's Wort users as the development of serotonin syndrome is also possible from this combination.

2.4.4 Dose titrations

As we have discussed above, there are various patient and drug factors that need to be considered before a drug is considered acceptable and effective in a patient. A particular medicine that is efficacious for one particular patient at a certain dose may not achieve the same therapeutic effect in another patient as a result of variation in: (1) drug–receptor responses and (2) the length of time the drug remains in the body. These variations are based on the *pharmacodynamics* and *pharmacokinetics* of the drug in question. Pharmacodynamics is the study of what a drug does to the patient—typically this refers to the therapeutic response obtained from a particular dose of medicine through its drug–receptor binding at the target tissue(s). Pharmacokinetics is the study of what the body does to the drug, that is the rate and the extent to which the drug is absorbed, distributed, metabolized, and eliminated or excreted by a patient.

During the initiation of a new pharmacotherapy, the prescriber needs to determine the optimal dose for the patient. Often, patients will experience side effects or adverse reactions to medications that will require a change in dose or medicine within the first two weeks (Langley et al., 2005). Initiation may require trialing a drug at various doses until target symptoms are treated adequately with tolerable and minimal side effects for the patient and this process is known as *dose titration*. If a dose titration process for a particular drug is not successful, that is the prescriber cannot find an appropriate therapeutic dose with tolerable side effects, another drug must be chosen for the pharmacotherapy. The original prescribed drugs are no longer used and in this way, the dose titration process can contribute to wastage of pharmaceuticals (Braund et al., 2007, 2008, 2009) as dispensed medication from previous dose titrations cannot be ethically reissued and reused by another patient.

Dose titrations are necessary due to the variability of drug action in different individuals. The clinical response of patients to a particular medication is determined by: (1) how well the drug is absorbed into the systemic circulation or bloodstream; (2) the therapeutic response arising from drug–receptor binding at the site of drug action; and (3) the efficacy of drug elimination and excretion processes in individuals. Any interpatient variation in these pharmacodynamic and pharmacokinetic processes will contribute to differences between the patients in their therapeutic response to a particular drug. As noted above, a common pharmacodynamic problem in clinical therapy is the interaction between two or more drugs when they bind to the same receptor in target and nontarget tissues, making the palpable effects of both drugs stronger than expected. An example of this type of pharmacodynamic interaction is tramadol and fluoxetine (see above). When both these drugs are prescribed to be taken together, it is necessary for the prescriber to exercise caution and clinical judgment in determining appropriate dosages for both medications so that they are still effective but any risks from their combination are minimized.

Pharmacokinetics also play a significant role in dose titrations as a major site of drug–drug interaction occurs with liver enzymes that are physiologically responsible for the elimination and degradation of drugs in the body. The predominant group of liver enzymes involved in drug degradation is Cytochrome P450 (CYP) (Francis and Delgoda, 2014). There are many variants (isoforms) of CYP and common ones include CYP1A2, CYP2E1, CYP2D6, and CYP3A4. Certain aspects of CYP elimination mechanisms are important for dose titrations. CYP enzymes can be *induced* or *inhibited* (Badal and Delgoda, 2014). Because these are drug degradation enzymes, medications which induce a CYP enzyme allow the enzyme to remove a drug more efficiently. This process results in lowered concentrations of the drug in the bloodstream and also affects drug–drug binding at receptor sites in target tissues. Drug–drug binding is established with a chemical equilibrium between bound drug and free drug in the systemic circulation. If the concentration of free drug in the systemic circulation is lowered, the chemical system compensates by creating more free drug from bound drug. This loss of bound drug also leads to a loss of therapeutic response (Pan et al., 2013). On the other hand, elevated drug concentrations from the inhibition of CYP enzymes will strengthen a therapeutic response. CYP3A4 is a common CYP enzyme that can be induced or inhibited and a classic example of a pharmacokinetic interaction involving CYP3A4 is St. John's Wort and warfarin (see above). Such an interaction can lead to problematic stabilization of a patient on warfarin anticoagulant therapy if the prescriber is ignorant of a patient's supplement- or OTC medication-taking practices.

There are many variants of CYP isoforms known as *polymorphisms*, which are genetically determined between different individuals and influence how well an individual eliminates a drug (Kumar et al., 2014). CYP2D6 is a particular CYP isoform with genetically predetermined polymorphism and individuals with inefficient CYP2D6 are known as *poor metabolizers* (Samer et al., 2012). Drugs which undergo metabolism by CYP2D6 will be eliminated slowly in poor metabolizers and can result in elevated systemic drug concentrations which can reach toxic levels in severe cases (Andreassen et al., 2012). A lack of efficacy will be experienced by individuals possessing efficient CYP2D6 (extensive metabolizers). These individuals risk a reduced therapeutic efficacy from CYP2D6-metabolized drugs because the clearance of these drugs from the systemic circulation may be much faster than the normal response assumed by the prescriber. Lower heart rate and blood pressures were observed in poor and intermediate metabolizers compared to extensive metabolizers during early initiation of antihypertensive therapy with metoprolol which is a drug metabolized by CYP2D6 (Batty et al., 2014). Blood metoprolol concentrations in poor and intermediate metabolizers were 2.1–4.6 times higher than that of extensive metabolizers.

Initially primary care prescribers usually do not have immediate access to the diagnostic means for determining pharmacodynamic and pharmacokinetic parameters in a patient and therefore prescribers should "start low" and "go slow" when initiating a chronic or long-term medication therapy (Basler et al., 2005). By using the lowest effective therapeutic dose initially and then increasing the dose when necessary for further symptomatic relief, prescribers will not only reduce the possibility of adverse effects in patients but also minimize any future wastage of medications later considered to be in excess of the needs of the patient. Such a "start low, go slow" approach is particularly recommended for patients on polypharmacy due to the risk of drug–drug interactions (Falzone et al., 2013) and is especially important when the clinical response achieved is not proportional to the dose taken by the patient. For example, the usual starting dose of bendrofluazide for mild hypertension is 2.5 mg daily with 1.25 mg daily being adequate for some patients (Vanmolkot et al., 1999). Because bendrofluazide is not a 'high-ceiling' diuretic (Bruderer et al., 2014), a dose of bendrofluazide greater than 2.5 mg daily would not enhance blood pressure control but rather increase the risk of the onset of undesirable side effects such as gout. Because the side effects of gout are extremely painful and intolerable (see above), patients on an excessive dose of bendrofluazide will most likely stop using the medication, and then go back to their prescriber and ask for alternative medication to treat their hypertension condition. To avoid the creation of unused medication, not only is the type of medication important for optimal therapy, but also careful consideration of the initial dose during titration may prevent adverse effects that discourage the patients from taking their medication.

2.5 "How much are those antibiotics? I want 'em!"

Aside from the involvement of clinical issues and medication nonadherence, strategies to control the marketing and supply of medications can also contribute to unused pharmaceuticals. For the benefit of the patient, medications may sometimes be supplied in bulk quantities irrespective of whether the condition to be treated is short-term (acute) or long-term (chronic). The quantity of medication dispensed may also depend on subsidy issues (see PHARMAC above). In New Zealand, PHARMAC only subsidizes certain medications if they are dispensed from pharmacies in 90-day bulk lots. This allows the patient to collect a three-month supply of medication in one dispensing act and ultimately reduces the total cost of dispensing fees payable by the New Zealand Government to pharmacies. While this scheme, known as "stat" or "all-at-once" dispensing, has increased the availability and accessibility of medicines used in New Zealand, because patients can save time and collect medications on a less frequent basis, there is little, if any, distinction as to whether the "stat-dispensed" medication is used for an acute or chronic condition. Paracetamol and ibuprofen are examples of "stat" medications on the New Zealand *Pharmaceutical Schedule* (see above) and significant wastage of both drugs could occur if either are used on a short-term or "as required" basis for the management of acute pain (Braund et al., 2009).

Oftentimes, prescribers also prefer to prescribe quantities in excess of the likely requirements rather than have the patient run out of medication. While the patient may consider such practices "value for money" by reducing the number of visits to their prescriber, this practice can again lead to accumulation of unused pharmaceuticals (Langley et al., 2005). In addition, due to resource constraints in many parts of the world, there is now more pressure for prescribers in primary care to see patients less frequently and to have standing orders or repeat prescriptions for patients on chronic medications. While this may be appropriate for patients on long-term stable therapy, it may not be suitable for patients initiating treatment (see above). Unused medications may also result from a patient's expectations of a prescriber to issue them with a prescription irrespective of whether this is warranted or not (Britten et al., 2002). This type of irrational prescribing can occur because the prescriber and or/patient do not feel the outcome of a consultation is satisfactory unless they leave with a prescription.

There have been many reports of irrational prescribing of antibiotics for the treatment of viral infections, despite the fact that antibiotics have been clearly demonstrated to do nothing to shorten or prevent viral illness (Glass et al., 2010). Wang et al. (1999) analyzed computerized records of Canadian preschool children's visits to a doctor for instances of antibiotic prescriptions and diagnoses of respiratory disease. Antibiotics were prescribed for 49% of these instances, where only 24% of them were for a diagnosis of influenza (Wang et al., 1999). Furthermore, it was estimated that prescribing costs of antibiotics when they were of questionable benefit totalled CAD $423,693. Linder et al. (2005) reported the inappropriate prescribing of antibiotics in 26% of clinic visits undertaken by study participants and commented that such practices were more likely to occur during the influenza season.

Bucher et al. (2003) conducted a placebo-controlled, double-blind randomized control trial on 252 adults treated at primary health care centers in Switzerland for a severe common cold (acute rhinosinusitis). Each participant exhibited symptoms for 48 h or more with mucus discharge and frontal pain as a result of nasal congestion. The treatment group was given the β-lactam, broad-spectrum antibiotic amoxicillin in combination with clavulanic acid for six days. The results suggested that amoxicillin–clavulanic acid had no significant effect in reducing the length and severity of the cold compared to the control (placebo) group. In addition, participants who had taken the antibiotics were more likely to experience diarrhea, an adverse and not uncommon effect arising from antibiotic therapy. These results are consistent with those of a French study which suggested that antibiotic treatment of influenza does not significantly reduce the length of illness, the development of secondary bacterial complications associated with influenza, or the number of lost workdays (Carrat et al., 2004).

The use of empirical ("better safe than sorry") antibiotic prescribing practices may also contribute to irrational antibiotic prescribing. Often, antibiotics are prescribed even if there is no evidence of a secondary bacterial infection (Chidiac and Maulin, 2006) and they are considered to be the ultimate solution to a respiratory infection. Over 99% of participants in a US study did not receive follow-up from the prescriber after a course of antibiotics (Misurski et al., 2011). In addition, 79% of treated participants did not show any symptoms of secondary bacterial infection. These results are supported by those of a UK study where no difference in recovery time was observed between participants exhibiting symptoms of viral infections and being prescribed an antibiotic to treat that infection (Harnden et al., 2007). Inappropriate antibiotic prescribing for influenza in the United States alone has been estimated to cost US$211 million (Misurski et al., 2011).

Irrational use of antibiotics can also occur for the intended treatment of diagnosed bacterial infections. The use of fluoroquinolones, which are potent broad-spectrum antibacterials previously reserved for the treatment of severe pneumonia or complicated urinary tract infections (UTI), are now commonly used for less complicated UTI (Honderlick et al., 2006). This observation suggests that many antibiotics previously perceived to be useful against a complicated bacterial infection are now being used to treat more general infections and prescribers should be more open minded about using a variety of antibiotics in their treatment regimens to prevent antibiotic resistance (Honderlick et al., 2006). The irrational use of antiviral drugs such as oseltamivir and zanamivir is not uncommon and has also been reported when a negative influenza test is observed (Linder et al., 2009).

It is evident from the above studies that primary care prescribers need to minimize the irrational use of many drugs, particularly antibiotics. The use of diagnostic tests where possible to determine the presence of a bacterial infection may minimize inappropriate antibiotic prescribing (Mazzaglia et al., 2003). Also, the prescriber should ensure that the patient receives optimal pharmacotherapy and put aside the perception that a simpler but more appropriate pharmacotherapy regimen (or no pharmacotherapy regimen at all) may concern the patient. A US survey of 244 patients from a primary care practice in Georgia suggested that most of the participants were willing to accept reassurance or simple over-the-counter treatment if such therapies were deemed most appropriate by their doctor, even if the doctor's visit was paid for out-of-pocket by the patient (Schwartz and Wagner, 2000). However, the advent of *direct to consumer advertising* or *marketing* (DTCA) of prescription-only medications has placed additional pressure on prescribers to first choose such medications rather than nonpharmaceutical options, including, but not limited to, diet and lifestyle modifications (Donohue et al., 2007).

DTCA enables medication information to be more widely available to the everyday consumer but at the same time has the ability to influence what patients want from their treatment rather than what is clinically necessary (Ruhoy and Daughton, 2008). The perceptions and feelings about DTCA and its use in the provision of health information for patients and prescribers have changed over time. While DTCA was generally seen as a positive influence on healthcare until around 2000, by 2012 more people had become a lot more pessimistic about medical information accompanying DTCA (Wood and Cronley, 2014). Currently DTCA has its supporters and detractors (Arney et al., 2013). A group of US pharmacy students believe that overall DTCA is positive because patients can be more proactive about their own healthcare (Harrington et al., 2013) in contrast to the negative sentiment earlier expressed by a group of 15 seniors aged 63–82 years (Grenard et al., 2011). Many of these senior participants believed that advertisements often omitted information on medication adverse effects and used too many semi-accurate metaphors (Delbaere, 2013). In addition, they believed the effectiveness of the drug was often inflated above and beyond what clinical data had claimed (Grenard et al., 2011). An Australian study suggested that patients from lower social-economic backgrounds were more likely to believe DTCA information (Chaar and Lee, 2012).

While DTCA has the potential to change patient behavior and promote good health habits including medication adherence (Bradford et al., 2006), it is nonetheless a potential source of coercion for patients to obtain a particular pharmacotherapy or a particular brand of medication (Kamenica et al., 2013). It has never been more important for prescribers to act responsibly for their patients by only prescribing what is necessary to a patient's pharmacotherapy and help dispel myths and information that are not based on sound clinical evidence. Given the wide variety of media including the Internet, patients are constantly being exposed to a wide range of medication information, some of which can be highly contradictory (Kravitz and Bell, 2013). The problem is not a lack of information but rather knowing what information to trust (Kravitz and Bell, 2013). Therefore, prescribers and pharmacists should aim to be that authoritative source of accurate medication information that patients can indeed look to and trust.

However, not all patient medication use can be managed by health care professionals. Medications are commonly seen as commodity items for which increasingly large "bulk buy" quantities are often available with substantial discounts (Jackson et al., 2012). The rise of online pharmacies have made medications more accessible than before the advent of the Internet (Letkiewicz and Gorski, 2010). The wide availability of these "cloud" premises has meant that counterfeit medications of variable quality or strength can be sold without control to consumers (Orizio et al., 2011; Cicero and Ellis, 2012; Fittler et al., 2013) and many prescription and controlled drugs are also available for purchase without prescription (Bachhuber and Cunningham, 2013). Antibiotics are one group of medications commonly sold over the Internet and available to customers after an average of seven days' shipping time (Mainous et al., 2009). Not only does this situation pose a therapeutic problem, as the drug treatment of any infections cannot be initiated until the drugs are received, but also quantities in excess of that required for a typical 7- or 14-day course can be purchased (Mainous et al., 2009). Such practices will undoubtedly increase medication wastage as well as increase the possibility of antibiotic resistance in the future. Online pharmacies and the Internet are a new frontier for medication supply that has still not been adequately regulated on an international level (Brown and Li, 2014), given the variation in classifications of medications in different countries (see above). Until stricter international laws are imposed on the supply of medications between different countries, the consumer satiety for "dirt cheap" and "without prescription" medications will probably continue to rise, potentially leading to increased medication wastage (Ivanitskaya et al., 2010).

2.6 Accumulation of unused medications

The interplay of patient nonadherence to medication, rational and irrational prescribing, issues pertaining to medication supply, expenditure and subsidy, and the widely promoted availability of medications ultimately can result in the significant accumulation of unused medications (Ruhoy and Daughton, 2008).

Accumulation of medications inevitably occurs wherever they are used in society. Other than the pharmacy, hoarding of medications is usually only required for emergency purposes in first-aid kits (Ruhoy and Daughton, 2008). First-aid kits are often used in homes and workplaces as a readily accessible source of medications and emergency supplies valuable in severe trauma or life-threatening situations. The range of products required in a first-aid kit must be compliant with relevant occupational health and safety legislation (Gaydos et al., 2011). Although medications may not be necessary for a basic first-aid kit, their inclusion may be beneficial for first-aid kits serving high-risk situations (Pye and Greenhalgh, 2010). For example, Pye and Greenhalgh (2010) list several antibiotics, antihistamines, corticosteroids, and analgesics in a recommended first-aid kit for recreational dive boats involving participants with variable deep-water diving skills (Table 2.4).

Table 2.4 **Medications recommended for the first-aid kit of a deep-sea recreational dive boat**

Medication	Formulation	Therapeutic purpose
Paracetamol (acetaminophen)	500 mg tablets	Mild analgesic/pain-killer
Diclofenac	25 mg tablets	Anti-inflammatory (NSAID)
Pethidine (or other opioid)	100 mg injection	Stronger pain-killer
Hydrocortisone	1% cream	Topical anti-inflammatory
Loperamide	2 mg capsules	Prevents and treats diarrhea
Sodium alginate/bicarbonate	tablets (e.g., Gaviscon®)	Treats reflux and heartburn
Prednisone/prednisolone	5 mg or other strength tablets	Anti-inflammatory
Amoxicillin	250 mg tablets or capsules	Oral broad-spectrum antibiotic
Erythromycin	250 mg tablets or capsules	Oral antibiotic (penicillin-intolerant patients)
Penicillin G (or other antibiotic)	Intramuscular injection	Antibiotic for severe bacterial infection
Lignocaine	5% injection	Local anesthetic
Chlorpheniramine	10 mg ampoules	For allergies and anaphylactic shock
Adrenaline	1/1,000 ampoules	For anaphylactic shock

Source: Pye and Greenhalgh (2010).

It is inevitable that most of these medications will be left unused on board the boat to ultimately expire and necessitate disposal. However, the health benefits of having them in a first-aid kit are generally recognized to be more important than their risk of expiry and wastage. For example, the use of adrenaline (epinephrine) is lifesaving for people prone to anaphylactic shocks. Subsequently, it is important for boat owners to regularly maintain accurate inventories of items to avoid excessive accumulation of medication in first-aid kits. Preventing the chemical or microbial degradation of medications in first-aid kits by storing them in cool places free from the effects of sunlight and humidity will also ensure medication quality before the stipulated expiry date (Ruhoy and Daughton, 2008).

Primary care practices and retail/community pharmacies are a reservoir for the accumulation of unused medications. Many medications are given free-of-charge to primary care practices and pharmacies as a promotional item by pharmaceutical sales representatives. A pharmaceutical company may also evaluate the efficacy of a representative's sales pitch by the number of free samples given away during a particular timeframe. Commonly, primary care practices and pharmacies do not have the incentive to resist the temptation of receiving gifts or other gratis services. Given these apparent benefits, general practices and pharmacies are usually only too happy to receive free promotional material from pharmaceutical sales representatives. The receipt and use of promotional medication samples can be both positive and negative. It may enhance the ability of prescribers to perform less wasteful dose titrations by offering free samples as short-term trial courses of pharmacotherapy. However, free samples can also accumulate and expire if unused (Ruhoy and Daughton, 2008). Additionally, the free samples offered to patients may also end up accumulating in households (see below) with their use ultimately influenced by medication adherence behavior and practice or clinical issues (as discussed above). Retail/community pharmacies in many countries also have the incentive to give away free samples even when these samples are possibly not necessary for the health or cosmetic care of a customer. However, the provision of free samples (as in a product promotion or advertising special) to customers may potentially increase the number of likely customers and/or revenue for a pharmacy.

Other facilities that can lead to the accumulation of unused medications include rest homes and prisons as a result of the death of residents or from confiscations, respectively (Ruhoy and Daughton, 2008). Veterinary offices regularly accumulate medications for the treatment of animals, and closed systems, such as cruise ships and the armed forces, usually manage and maintain a stockpile of medications for the use of their members (Ruhoy and Daughton, 2008). However, the most widely reported place on a global scale for medication storage is in the domestic household. While the amount of unused medication that accumulates in households can never be accurately estimated, there have been studies which have attempted to quantify unused pharmaceuticals that have been returned to pharmacies by the consumer, caregiver or patient. We will address these types of unused medications in the following section.

2.6.1 Return of unused medications to pharmacies and general medical practices

Unwanted, unused, and expired medications are either destined for disposal (Chapter 3) or collected by a take-back programme (popular in the United States) (Chapter 3), a general practice, or pharmacies (Ruhoy and Daughton, 2008). Unused pharmaceuticals returned to pharmacies or general practices are in some ways unique, as they can be readily classified or quantified and the resulting data can form a ready source of information to predict drug usage patterns in a community. In addition, the reason for their return can usually be recorded during cataloging and such data can be used to determine the circumstances which have lead to a pharmaceutical not being used.

A study of medication returns in the United Kingdom conducted with eight community pharmacies and five general practice surgeries reported the return of 340 medication items over a four-week period (Langley et al., 2005). Pharmacies were the more popular and collected 87.6% of the overall quantity of returned medications. The most common reason recorded for medication returns was prescribers changing therapy, while the most common type of returned medication were cardiovascular drugs (28.5% of the total returned). These results were confirmed in another, more extensive UK study undertaken over two months (Mackridge and Marriott, 2007) in which 4,000 returned medication items were recordered and of which the majority had not expired. The most common type of returned medication was again cardiovascular medications (27%) followed by psychiatric and central nervous systems agents

(24%). A similar study undertaken in an Omani hospital pharmacy over a five-month study period resulted in the return of a total of 1,071 drug items with the most common group (24%) also being cardiovascular medications (Al-Siyabi and Al-Riyami, 2007). Cardiovascular medications were also commonly returned in a New Zealand study (James et al., 2009). All these studies suggest that cardiovascular medications may make up a large fraction (~25–30%) of all unused medications. Given that many of these medications are returned to pharmacies or general practices often as a result of changes in the prescribed therapeutic regime, it is likely that these wastes resulted from patient issues such as: (1) problems with dose titration, (2) a lack of perceived or actual efficacy, and (3) adverse effects. Medications used for cardiovascular conditions generally manage the condition on a chronic basis (i.e., hypertension); therefore, they are generally given in large quantities in a single dispensing and increase the possibility for bulk wastage when the therapy changes or is prematurely terminated.

In contrast to the above studies, the identification and quantitation of 159 kg of medication returned to several Otago region community pharmacies in New Zealand over a nine-month period indicated that the most commonly returned medication was paracetamol (92%) (Braund et al., 2007). In all, 6,059 units of 500 mg paracetamol tablets were collected and included a selection of generic brands (usually only available as dispensary items) as well as the Panadol® brand. However, only 18% (1,114 units) of these paracetamol tablets returned were of the Panadol® brand, the rest being generic-brand tablets that were fully subsidized at the time of the study (for discussion on pharmaceutical expenditure issues in New Zealand, see above). From this observation, most of the paracetamol tablets returned probably came from excessive prescribing, although a minor proportion may have originated from an OTC purchase. This study suggests that the use of data for pharmacy returns of drugs may enable the prediction of the types of medications that are being prescribed excessively. The excessive prescribing of paracetamol in New Zealand may be due to the bulk prescribing commonly practiced for subsidy purposes, as paracetamol is currently (2014) a "stat" medication (see discussion on PHARMAC above).

It is important to realize that the above examples of identification and quantitation studies only represent a snapshot of the types and quantities of unused medications that exist in the community and may not be entirely representative of the extent of medication wastage on a global scale. However, this type of data collected at the local level can still help predict the types of pharmaceutical hazards that need to be managed.

2.7 Strategies to prevent accumulation of unused pharmaceuticals

While some level of pharmaceutical waste in the form of unused medication is unavoidable, the quantity of such material can be minimized by healthcare providers and patients. From the above discussion, it is clear that patients, prescribers, and other healthcare professionals can all have a role in reducing the quantity of unused pharmaceutical material necessitating disposal.

The ideal "environmentally friendly" prescriber:

- Always practices evidence-based medicine. They stay up-to-date with current prescribing guidelines so that the best choice, dose, and duration of medication can be prescribed for the patient.
- Adopts a "start low, go slow" therapeutic approach to the treatment of chronic medical conditions. Given physiological, biochemical, and genetic differences (e.g., age, body weight, and ethnicity) between individuals, it is impossible to completely determine how a drug will behave upon initiating a new course of pharmacotherapy for a patient. A conscientious prescriber will start a new patient on the lowest effective dose and titrate when necessary.
- Prescribes short courses or trial prescriptions where possible to determine efficacy or adverse effects upon initiation of pharmacotherapy.
- Considers medical history and interacting medications in prescribing.
- Fights against unnecessary polypharmacy but uses multiple pharmacotherapies when best evidence suggests significant health benefits from such an approach.
- Understands the reasons for the behavior of patients and establishes concordance with them in a manner that encourages the patient to be comfortable with the pharmacotherapy prescribed and minimizes medication nonadherence.
- Informs patients of the most cost-effective treatment choices but also provides information on the costs involved if a patient chooses an alternative form of a prescribed medication (e.g., a non-subsidized brand of medication).
- Provides patients with accurate, objective information about therapy options independent of DTCA. Is not coerced by such promotional materials when determining the prescription needs but will prescribe when there is justified merit for the medication.
- Practices "green" principles and considers the environmental impact of pharmaceuticals (*pharmacovigilance* and *stewardship*, to be discussed later in Chapter 8).

An "environmentally friendly" patient/consumer:

- Adheres to the pharmacotherapy as prescribed and agreed upon with the prescriber; follows dosage instructions accurately.
- Minimizes the number of missed doses (and hence potentially unused medication).
- Establishes an open and honest relationship with the prescriber and will ask questions about their pharmacotherapy.
- Utilizes the prescriber as a source of health information. Critically discusses information presented in DTCA promotionals relevant to their condition or pharmacotherapy.
- Informs the prescriber when their medical condition changes as the prescribed medication may then be no longer required and hence reduce the incidence of unused medication that must ultimately be disposed of.
- Obtains medications from legitimate sources (i.e., a physical pharmacy premises supervised by a registered pharmacist).
- Does not purchase more medication than necessary over the counter from both pharmacies and non-pharmacy outlets. While the purchase of some medications for emergency storage (e.g., in first-aid kits) is unavoidable, careful attention to quantities, storage conditions, and expiry dates will ensure that medications of a single type are not unnecessarily hoarded.
- When medication is no longer required, returns any unused material to a take-back programme, the dispensing pharmacy, or other state/national facility for appropriate disposal (Chapter 3).

In addition to the above ideal behavior for prescribers and patients, it should be noted that pharmacists also have a role in being responsible for ensuring that patients understand the reasons for the use of their prescribed medications, maintaining health literacy and promoting medication adherence in the community.

While not all of these behavioral guidelines can be followed exactly, the implementation of some of them will at least initiate a reduction of pharmaceutical waste arising from unused medication that necessitates disposal. Even if data on medication usage patterns cannot be accurately collected, reducing the quantity of unused medication decreases the possibility of significant quantities of pharmaceuticals being discharged into the environment.

References

Abu Shaqra, Q.M., Shawaqfah, M.T., et al., 2014. Microbiological quality of blister pack tablets in community pharmacies in Jordan. Trop. J. Pharm. Res. 13 (2), 261–266.

Ahmed, R., Aslani, P., 2014. What is patient adherence? A terminology overview. Int. J. Clin. Pharm. 36 (1), 4–7.

Alkadi, H.O., 2007. Antimalarial drug toxicity: a review. Chemotherapy 53 (6), 385–391.

Al-Siyabi, K., Al-Riyami, K., 2007. Value and types of medicines returned by patients to Sultan Qaboos University hospital pharmacy, Oman. Sultan Qaboos Univ. Med. J. 7 (2), 109–115.

Alvarez-Jimenez, M., Alcazar-Corcoles, M.A., et al., 2014. Online, social media and mobile technologies for psychosis treatment: a systematic review on novel user-led interventions. Schizophr. Res. 156 (1), 96–106.

Andreassen, T.N., Eftedal, I., et al., 2012. Do CYP2D6 genotypes reflect oxycodone requirements for cancer patients treated for cancer pain? A cross-sectional multicentre study. Eur. J. Clin. Pharmacol. 68 (1), 55–64.

Andres, E., Dali-Youcef, N., et al., 2009. Recognition and management of drug-induced cytopenias: the example of idiosyncratic drug-induced thrombocytopenia. Expert Opin. Drug Saf. 8 (2), 183–190.

Arney, J., Street, R.L., et al., 2013. Consumers' various and surprising responses to direct-to-consumer advertisements in magazine print. Patient Prefer. Adherence 7, 95–102.

Bachhuber, M.A., Cunningham, C.O., 2013. Availability of buprenorphine on the Internet for purchase without a prescription. Drug Alcohol Depend. 130 (1–3), 238–240.

Badal, S., Delgoda, R., 2014. Role of the modulation of CYP1A1 expression and activity in chemoprevention. J. Appl. Toxicol. 34 (7), 743–753.

Bailey, S.C., Oramasionwu, C.U., et al., 2014. Rethinking adherence: a health literacy-informed model of medication self-management. J. Health Commun. 18, 20–30.

Bains, S.S., Egede, L.E., 2011. Associations between health literacy, diabetes knowledge, self-care behaviors, and glycemic control in a low income population with type 2 diabetes. Diabetes Technol. Ther. 13 (3), 335–341.

Basler, H.D., Griessinger, N., et al., 2005. Pain assessment and pain treatment in the geriatric patient. Part II: pain treatment. Schmerz 19 (1), 65–73.

Bassols, A., Bosch, F., et al., 2002. How does the general population treat their pain? A survey in Catalonia, Spain. J. Pain Symptom Manag. 23 (4), 318–328.

Batty, J.A., Hall, A.S., et al., 2014. An investigation of CYP2D6 genotype and response to metoprolol CR/XL during dose titration in patients with heart failure: a MERIT-HF substudy. Clin. Pharmacol. Ther. 95 (3), 321–330.

Bennett, H., Laird, K., et al., 2009. The effectiveness of health coaching, home blood pressure monitoring, and home-titration in controlling hypertension among low-income patients: protocol for a randomized controlled trial. BMC Public Health 9, 6.

Bobo, W.V., Miller, S.C., et al., 2005. The abuse liability of dextromethorphan among adolescents: a review. J. Child Adoles. Subst. Abuse 14 (4), 55–75.
Bon, A., Jongejan, R.C., 2001. Treatment of non-malignant pain in the Dutch GP population: an analysis of treatment standards, prescription habits and basic attitudes. Clin. Rheumatol. 20, 2–5.
Boudreau, D.M., Wirtz, H., et al., 2013. A survey of adult awareness and use of medicine containing acetaminophen. Pharmacoepidemiol. Drug Saf. 22 (3), 229–240.
Braae, R., McNee, W.et al., 1999. Managing pharmaceutical expenditure while increasing access. The pharmaceutical management agency (PHARMAC) experience. Pharmacoeconomics 16 (6), 649–660.
Bradford, W.D., Kleit, A.N., et al., 2006. Effects of direct-to-consumer advertising of hydroxymethylglutaryl coenzyme a reductase inhibitors on attainment of LDL-C goals. Clin. Ther. 28 (12), 2105–2118.
Braich, P.S., Almeida, D.R., et al., 2011. Effects of pictograms in educating 3 distinct low-literacy populations on the use of postoperative cataract medication. Can. J. Ophthalmol. 46 (3), 276–281.
Braund, R., Abbott, J.H., 2007. Analgesic choice when treating musculoskeletal sprains and strains. NZ J. Physiother. 35 (2), 54–60.
Braund, R., Yuen, Y.-C., et al., 2007. Identification and quantification of medication returned to Otago pharmacies. NZ J. For. Sci. 31 (4), 258–262.
Braund, R., Chuah, F., et al., 2008. Identification of the reasons for medication returns. NZ J. For. Sci. 35, 248–252.
Braund, R., Gn, G., et al., 2009. Investigating unused medications in New Zealand. Pharm. World Sci. 31 (6), 664–669.
Britten, N., Ukoumunne, O.C., et al., 2002. Patients' attitudes to medicines and expectations for prescriptions. Health Expect. 5 (3), 256–269.
Brosnan, S., Barron, E., et al., 2012. Health literacy and the Clozapine patient. Perspect. Public Health 132 (1), 39–42.
Brown, J., Li, C., 2014. Characteristics of online pharmacy users in a nationally representative sample. J. Am. Pharm. Assoc. 54 (3), 289–294.
Bruderer, S., Bodmer, M., et al., 2014. Use of diuretics and risk of incident gout a population-based case-control study. Arthritis Rheum. 66 (1), 185–196.
Bucher, H.C., Tschudi, P., et al., 2003. Effect of amoxicillin-clavulanate in clinically diagnosed acute rhinosinusitis: a placebo-controlled, double-blind, randomized trial in general practice. Arch. Intern. Med. 163 (15), 1793–1798.
Bundgaard, H., Larsen, C., 1978. Kinetics and mechanism of sucrose-accelerated degradation of penicillins in aqueous-solution. Int. J. Pharm. 1 (2), 95–104.
Burillo-Putze, G., Aldea-Perona, A., et al., 2013. Emergent drugs (II): the Pharming phenomenon. An. Sist. Sanit. Navar. 36 (1), 99–114.
Carrat, F., Schwarzinger, M., et al., 2004. Antibiotic treatment for influenza does not affect resolution of illness, secondary visits or lost workdays. Eur. J. Epidemiol. 19 (7), 703–705.
Carter, S.R., Moles, R., et al., 2013. Medication information seeking behavior of patients who use multiple medicines: how does it affect adherence? Patient Educ. Couns. 92 (1), 74–80.
Chaar, B.B., Lee, J., 2012. Role of socioeconomic status on consumers' attitudes towards DTCA of prescription medicines in Australia. J. Bus. Ethics 105 (4), 447–460.
Chen, A.M.H., Yehle, K.S., et al., 2014. Relationships between health literacy and heart failure knowledge, self-efficacy, and self-care adherence. Res. Social Adm. Pharm. 10 (2), 378–386.
Chidiac, C., Maulin, L., 2006. Using antibiotics in case of influenza. Med. Maladies Infect. 36 (4), 181–189.
Chui, J., Tordoff, J., et al., 2004. Trends in accessibility to medicines for children in New Zealand: 1998–2002. Br. J. Clin. Pharmacol. 57 (3), 322–327.

Cicero, T.J., Ellis, M.S., 2012. Health outcomes in patients using no-prescription online pharmacies to purchase prescription drugs. J. Med. Internet Res. 14 (6), 347–359.
Correll, C.U., Shaikh, L., et al., 2011. Antipsychotic polypharmacy: a survey study of prescriber attitudes, knowledge and behavior. Schizophr. Res. 131 (1–3), 58–62.
Cramer, J.A., 1998. Enhancing patient compliance in the elderly—role of packaging aids and monitoring. Drugs Aging 12 (1), 7–15.
da Silva, M.G.C., Soares, M.C.F., et al., 2012. Self-medication in university students from the city of Rio Grande, Brazil. BMC Public Health 12, 7.
de Oliveira, M.A., Francisco, P., et al., 2012. Self-medication in the elderly population of Campinas, Sao Paulo State, Brazil: prevalence and associated factors. Cad. Saude Publica 28 (2), 335–345.
de Vries, S.T., Keers, J.C., et al., 2014. Medication beliefs, treatment complexity, and non-adherence to different drug classes in patients with type 2 diabetes. J. Psychosom. Res. 76 (2), 134–138.
Delbaere, M., 2013. Metaphors and myths in pharmaceutical advertising. Soc. Sci. Med. 82, 21–29.
Di, Y.M., Li, C.G., et al., 2008. Clinical drugs that interact with St. John's wort and implication in drug development. Curr. Pharm. Des. 14 (17), 1723–1742.
Dominguez, M.D.G., Gonzalez, M.I.L., et al., 1992. Drug-use pattern in an outpatient geriatric population. Rev. Clin. Esp. 191 (8), 412–415.
Donohue, J.M., Cevasco, M., et al., 2007. A decade of direct-to-consumer advertising of prescription drugs. N. Engl. J. Med. 357 (7), 673–681.
Duong, M., Salvo, F., et al., 2014. Usage patterns of ‘over-the-counter’ vs. prescription-strength nonsteroidal anti-inflammatory drugs in France. Br. J. Clin. Pharmacol. 77 (5), 887–895.
Falzone, E., Hoffmann, C., et al., 2013. Postoperative analgesia in elderly patients. Drugs Aging 30 (2), 81–90.
Feas, X., Brasic, J.R., et al., 2009. Quinine potential toxicological implications: analysis of tonic waters in Spain. Rev. Esp. Nutr. Comunit. 15 (2), 97–102.
Fittler, A., Lanko, E., et al., 2013. Behaviour analysis of patients who purchase medicines on the internet: can hospital pharmacists facilitate online medication safety? Eur. J. Hosp. Pharm. 20 (1), 8–12.
Fletcher, D.C., 1976. Can quinine in tonic water be hazardous. J. Am. Med. Assoc. 236 (3), 305.
Francis, S., Delgoda, R., 2014. A patent review on the development of human cytochrome P450 inhibitors. Expert Opin. Ther. Pat. 24 (6), 699–717.
Gaydos, M., Bhatia, R., et al., 2011. Promoting Health and Safety in San Francisco's chinatown restaurants: findings and lessons learned from a pilot observational checklist. Public Health Rep. 126, 62–69.
Gazmararian, J.A., Ziemer, D.C., et al., 2009. Perception of barriers to self-care management among diabetic patients. Diabetes Educ. 35 (5), 778–788.
Gerber, B.S., Cano, A.I., et al., 2010. A pharmacist and health promoter team to improve medication adherence among Latinos with diabetes. Ann. Pharmacother. 44 (1), 70–79.
Ghosh, A., Charlton, K.E., et al., 2014. Addressing the deficiencies in the evidence-base for primary practice in regional Australia—sentinel practices data sourcing (SPDS) project: a pilot study. BMC Fam. Pract. 14, 11.
Glass, S.K., Pearl, D.L., et al., 2010. A province-level risk factor analysis of fluoroquinolone consumption patterns in Canada (2000–06). J. Antimicrob. Chemother. 65 (9), 2019–2027.
Gould, I.M., 2009. Antibiotic resistance: the perfect storm. Int. J. Antimicrob. Agents 34, 4.
Grenard, J.L., Uy, V., et al., 2011. Seniors' perceptions of prescription drug advertisements: a pilot study of the potential impact on informed decision making. Patient Educ. Couns. 85 (1), 79–84.

Hailey, J.H.N., Arscott, J., 2014. Using Technology to effectively engage adolescents and young adults into care: STAR TRACK adherence program. J. Assoc. Nurse. Aids Care 24 (6), 582–586.

Handler, J., 2010. Managing hypertensive patients with gout who take thiazide. J. Clin. Hypertens. 12 (9), 731–735.

Hansen, R.A., Kim, M.M., et al., 2009. Comparison of methods to assess medication adherence and classify nonadherence. Ann. Pharmacother. 43 (3), 413–422.

Harnden, A., Perera, R., et al., 2007. Respiratory infections for which general practitioners consider prescribing an antibiotic: a prospective study. Arch. Dis. Child. 92 (7), 594–597.

Harrington, A.R., Desselle, S.P., et al., 2013. Pharmacy students' opinions of direct-to-consumer advertising: a pilot study at one university. Res. Social Adm. Pharm. 9 (4), 458–466.

Hatah, E., Braund, R., et al., 2012. General practitioners' perceptions of pharmacists' new services in New Zealand. Int. J. Clin. Pharm. 34 (2), 364–373.

Hatah, E., Tordoff, J., et al., 2014. Pharmacists' performance of clinical interventions during adherence support medication reviews. Res. Social Adm. Pharm. 10 (1), 185–194.

Hawthorne, A.B., Rubin, G., et al., 2008. Review article: medication non-adherence in ulcerative colitis—strategies to improve adherence with mesalazine and other maintenance therapies. Aliment. Pharmacol. Ther. 27 (12), 1157–1166.

Heaton, P.C., Tundia, N.L., et al., 2013. U.S. emergency departments visits resulting from poor medication adherence: 2005–07. J. Am. Pharm. Assoc. 53 (5), 513–519.

Heneghan, C.J., Glasziou, P., et al., 2006. Reminder packaging for improving adherence to self-administered long-term medications. Cochrane Database Syst. Rev. 1, 23.

Honderlick, P., Callen, P., et al., 2006. Uncomplicated urinary tract infections, what about fosfomycin and nitrofurantoin in 2006? Pathol. Biol. 54 (8–9), 462–466.

Hopkinson, J.H., 1978. Vicodin, a new analgesic—clinical evaluation of efficacy and safety of repeated doses. Curr. Ther. Res. Clin. Exp. 24 (6), 633–645.

Hovstadius, B., Petersson, G., 2011. Non-adherence to drug therapy and drug acquisition costs in a national population—a patient-based register study. BMC Health Serv. Res. 11, 326. http://dx.doi.org/10.1186/1472-6963-11-326.

Hu, H., Wu, F.L.L., et al., 2012. Effectiveness of education programs about oral antibiotic suspensions in pediatric outpatient services. Pediatr. Neonatol. 54 (1), 34–42.

Huang, H.Y., Maguire, M.G., et al., 2000. Impact of pill organizers and blister packs on adherence to pill taking in two vitamin supplementation trials. Am. J. Epidemiol. 152 (8), 780–787.

Hugtenburg, J.G., Borgsteede, S.D., et al., 2009. Medication review and patient counselling at discharge from the hospital by community pharmacists. Pharm. World Sci. 31 (6), 630–637.

Ivanitskaya, L., Brookins-Fisher, J., et al., 2010. Dirt cheap and without prescription: how susceptible are young US consumers to purchasing drugs from rogue internet pharmacies? J. Med. Internet Res. 12 (2), 15.

Izzo, A.A., Ernst, E., 2009. Interactions between herbal medicines and prescribed drugs an updated systematic review. Drugs 69 (13), 1777–1798.

Jackevicius, C.A., Li, P., et al., 2008. Prevalence, Predictors, and outcomes of primary nonadherence after acute myocardial infarction. Circulation 117, 1028–1036.

Jackson, G., Patel, S., et al., 2012. Assessing the problem of counterfeit medications in the United Kingdom. Int. J. Clin. Pract. 66 (3), 241–250.

James, T.H., Helms, M.L., et al., 2009. Analysis of medications returned to community pharmacies. Ann. Pharmacother. 43 (10), 1631–1635.

Jana, A.H., Joshi, N.S.S., et al., 1994. Sweeteners for frozen success—a review. Aust. J. Dairy Technol. 49 (2), 98–109.

Jatrana, S., Crampton, P., 2009. Primary health care in New Zealand: who has access? Health Policy 93 (1), 1–10.

Jatrana, S., Crampton, P., et al., 2011. Ethnic differences in access to prescription medication because of cost in New Zealand. J. Epidemiol. Community Health 65 (5), 454–460.

Kamenica, E., Naclerio, R., et al., 2013. Advertisements impact the physiological efficacy of a branded drug. Proc. Natl. Acad. Sci. U. S. A. 110 (32), 12931–12935.

Kaufman, D.W., Kelly, J.P., et al., 2002. Recent patterns of medication use in the ambulatory adult population of the United States: the Slone survey. JAMA 287 (3), 337–344.

Kemp, A., Roughead, E., et al., 2010. Determinants of self-reported medicine underuse due to cost: a comparison of seven countries. J. Health Serv. Res. Policy 15 (2), 106–114.

Kimura, Y., Ito, H., et al., 2010. Inhibitory effects of polyphenols on human cytochrome P450 3A4 and 2C9 activity. Food Chem. Toxicol. 48 (1), 429–435.

Kravitz, R.L., Bell, R.A., 2013. Media, messages, and medication: strategies to reconcile what patients hear, what they want, and what they need from medications. BMC Med. Inform. Decis. Mak. 13, 8.

Kumar, V., Banerjee, B.D., et al., 2014. Association of CYP1A1, CYP1B1 and CYP17 gene polymorphisms and organochlorine pesticides with benign prostatic hyperplasia. Chemosphere 108, 40–45.

Laffer, M.S., Feldman, S.R., 2014. Improving medication adherence through technology: analyzing the managing meds video challenge. Skin Res. Technol. 20 (1), 62–66.

Lagan, B.M., Dolk, H., et al., 2014. Assessing the availability of the teratogenic drug isotretinoin outside the pregnancy prevention programme: a survey of e-pharmacies. Pharmacoepidemiol. Drug Saf. 23 (4), 411–418.

Langley, C., Marriott, J., et al., 2005. An analysis of returned medicines in primary care. Pharm. World Sci. 27 (4), 296–299.

Latif, A., Pollock, K., et al., 2011. The contribution of the medicines use review (MUR) consultation to counseling practice in community pharmacies. Patient Educ. Couns. 83 (3), 336–344.

Laufs, U., 2011. Medication adherence in patients with chronic diseases. Significance and new approaches. Nervenarzt 82 (2), 153–154, 156-8.

Lee, E., Braund, R., et al., 2009. Examining the first year of medicines use review services provided by pharmacists in New Zealand: 2008. NZ Med. J. 122 (1293), 3566.

Lehmann, A., Aslani, P., et al., 2014. Assessing medication adherence: options to consider. Int. J. Clin. Pharm. 36 (1), 55–69.

Letkiewicz, S., Gorski, A., 2010. The potential dual use of online pharmacies. Sci. Eng. Ethics 16 (1), 59–75.

Levy, G., 2001. Medication non-compliance—when hard science meets soft science. Optimal Dose Identification: Excerta Medica. Amsterdam, Elsevier Science Bv. 1220: 125–133.

Li, C., Martin, B.C., 2011. Trends in emergency department visits attributable to acetaminophen overdoses in the United States: 1993–2007. Pharmacoepidemiol. Drug Saf. 20 (8), 810–818.

Linder, J.A., Bates, D.W., et al., 2005. Antivirals and antibiotics for influenza in the United States, 1995–2002. Pharmacoepidemiol. Drug Saf. 14 (8), 531–536.

Linder, J.A., Nieva, H.R., et al., 2009. Antiviral and antibiotic prescribing for influenza in primary care. J. Gen. Intern. Med. 24 (4), 504–510.

Linjakumpu, T., Hartikainen, S., et al., 2002. Use of medications and polypharmacy are increasing among the elderly. J. Clin. Epidemiol. 55 (8), 809–817.

Lipchock, S.V., Reed, D.R., et al., 2012. Relationship between bitter-taste receptor genotype and solid medication formulation usage among young children: a retrospective analysis. Clin. Ther. 34 (3), 728–733.

Logan, B.K., Yeakel, J.K., et al., 2012. Dextromethorphan abuse leading to assault, suicide, or homicide. J. Forensic Sci. 57 (5), 1388–1394.

Lounsbery, J.L., Green, C.G., et al., 2009. Evaluation of pharmacists' barriers to the implementation of medication therapy management services. J. Am. Pharm. Assoc. 49 (1), 51–58.

Mackridge, A.J., Marriott, J.F., 2007. Returned medicines: waste or a wasted opportunity? J. Public Health 29 (3), 258–262.

Madorell, B.S., Ortiz, C.J., et al., 2006. Tonic water as cause of severe hypoglycemia in non-diabetic subject. Med. Clin. 127 (16), 639.

Mainous, A.G., Everett, C.J., et al., 2009. Availability of antibiotics for purchase without a prescription on the Internet. Ann. Fam. Med. 7 (5), 431–435.

Mallet, C., Daulhac, L., et al., 2008. Endocannabinoid and serotonergic systems are needed for acetaminophen-induced analgesia. Pain 139 (1), 190–200.

Marques, J.G., Calado, G., et al., 2013. Tonic water: a rare cause of exanthema. Allergol. Immunopathol. 40 (1), 60–61.

Marshall, I.J., Wolfe, C.D.A., et al., 2012. Lay perspectives on hypertension and drug adherence: systematic review of qualitative research. Br. Med. J. 345, 16.

Matthes, J., Albus, C., 2014. Improving adherence with medication a selective literature review based on the example of hypertension treatment. Dtsch. Arztebl. Int. 111 (4), 41–47.

Mazzaglia, G., Caputi, A.P., et al., 2003. Exploring patient- and doctor-related variables associated with antibiotic prescribing for respiratory infections in primary care. Eur. J. Clin. Pharmacol. 59 (8–9), 651–657.

McAdams-DeMarco, M.A., Maynard, J.W., et al., 2014. A urate gene-by-diuretic interaction and gout risk in participants with hypertension: results from the ARIC study. Ann. Rheum. Dis. 72 (5), 701–706.

Mennella, J.A., Beauchamp, G.K., 2008. Optimizing oral medications for children. Clin. Ther. 30 (11), 2120–2132.

Mennella, J.A., Spector, A.C., et al., 2013. The bad taste of medicines: overview of basic research on bitter taste. Clin. Ther. 35 (8), 1225–1246.

Misurski, D.A., Lipson, D.A., et al., 2011. Inappropriate antibiotic prescribing in managed care subjects with influenza. Am. J. Manage Care 17 (9), 601–608, U54.

Modi, A.C., Ingerski, L.M., et al., 2012. White coat adherence over the first year of therapy in pediatric epilepsy. J. Pediatr. 161 (4), 695–699.

Mohan, A., Riley, M.B., et al., 2012. PictureRx: illustrated medication instructions for patients with limited health literacy. J. Am. Pharm. Assoc. 52 (5), E122–E129.

Mohan, A.V., Riley, M.B., et al., 2013. Illustrated medication instructions as a strategy to improve medication management among Latinos: a qualitative analysis. J. Health Psychol. 18 (2), 187–197.

Molloy, G.J., O'Carroll, R.E., et al., 2014. Conscientiousness and medication adherence: a meta-analysis. Ann. Behav. Med. 47 (1), 92–101.

Moore, N., Diris, H., et al., 2004. NSAID use profiles derived from reimbursement data in France. Therapie 59 (5), 541–546.

Morrow, D.G., Conner-Garcia, T., et al., 2012. An EMR-based tool to support collaborative planning for medication use among adults with diabetes: design of a multi-site randomized control trial. Contemp. Clin. Trials 33 (5), 1023–1032.

Mosher, H.J., Lund, B.C., et al., 2012. Association of health literacy with medication knowledge, adherence, and adverse drug events among elderly veterans. J. Health Commun. 17, 241–251.

Mukhtar, O., Weinman, J., et al., 2014. Intentional non-adherence to medications by older adults. Drugs Aging 31 (3), 149–157.

Munoz, J., Navarro, C., et al., 2010. Synergism between COX-3 inhibitors in two animal models of pain. Inflammopharmacology 18 (2), 65–71.
Muso, Y., Kentarou, O., et al., 2007. Fixed eruption due to quinine: report of two cases. J. Dermatol. 34 (6), 385–386.
Nachega, J.B., Parienti, J.J., et al., 2014. Lower pill burden and once-daily antiretroviral treatment regimens for HIV infection: a meta-analysis of randomized controlled trials. Clin. Infect. Dis. 58 (9), 1297–1307.
Neubert, A., Verhamme, K., et al., 2011. The prescribing of analgesics and non-steroidal anti-inflammatory drugs in paediatric primary care in the UK, Italy and the Netherlands. Pharmacol. Res. 62 (3), 243–248.
Nishtala, P.S., Bagge, M.L., et al., 2014. Potentially inappropriate medicines in a cohort of community-dwelling older people in New Zealand. Geriatr. Gerontol. Int. 14 (1), 89–93.
Orizio, G., Merla, A., et al., 2011. Quality of online pharmacies and websites selling prescription drugs: a systematic review. J. Med. Internet Res. 13 (3), 25.
Orton, L., Barnish, G., 2005. Unit-dose packaged drugs for treating malaria. Cochrane Database Syst. Rev. 2, 25.
Osterberg, L., Blaschke, T., 2005. Adherence to medication. N. Engl. J. Med. 353 (5), 487–497.
Pan, A.C., Borhani, D.W., et al., 2013. Molecular determinants of drug-receptor binding kinetics. Drug Discov. Today 18 (13–14), 667–673.
Pasina, L., Brucato, A.L., et al., 2014. Medication non-adherence among elderly patients newly discharged and receiving polypharmacy. Drugs Aging 31 (4), 283–289.
Peuskens, J., Rubio, G., et al., 2014. Dosing and switching of paliperidone ER in patients with schizophrenia: recommendations for clinical practice. Ann. Gen. Psychiatry 13, 10.
Peyrot, M., Rubin, R.R., 2011. Factors associated with persistence and resumption of insulin pen use for patients with type 2 diabetes. Diabetes Technol. Ther. 13 (1), 43–48.
Phatak, H.M., Thomas, J., 2006. Relationships between beliefs about medications and nonadherence to prescribed chronic medications. Ann. Pharmacother. 40 (10), 1737–1742.
Platt, I., Green, H.J., et al., 2014. Understanding adherence in patients with coronary heart disease: illness representations and readiness to engage in healthy behaviours. Aust. Psychol. 49 (2), 127–137.
Postiglione, K., Ferris, R., et al., 1995. Immune-mediated agranulocytosis and anemia associated with Thymoma. Am. J. Hematol. 49 (4), 336–340.
Pye, J., Greenhalgh, T., 2010. First aid kits for recreational dive boats, what should they contain? Travel Med. Infect. Dis. 8 (5), 311–317.
Reeve, E., Wiese, M.D., 2014. Benefits of deprescribing on patients' adherence to medications. Int. J. Clin. Pharm. 36 (1), 26–29.
Reeve, E., To, J., et al., 2014. Patient barriers to and enablers of deprescribing: a systematic review. Drugs Aging 30 (10), 793–807.
Remondi, F.A., Cabrera, M.A.S., et al., 2014. Non-adherence to continuous treatment and associated factors: prevalence and determinants in adults 40 years and older. Cad. Saude Publica 30 (1), 126–136.
Rizzoli, P., Burch, R., et al., 2013. The risk of serotonin syndrome with concomitant use of triptan antimigraine drugs and SSRI/SNRI antidepressants: a population-based surveillance study. Cephalalgia 33 (S8), 22–23.
Robinson, J.L., Hameed, T., et al., 2002. Practical aspects of choosing an antibiotic for patients with a reported allergy to an antibiotic. Clin. Infect. Dis. 35 (1), 26–31.
Ruhoy, I.S., Daughton, C.G., 2008. Beyond the medicine cabinet: an analysis of where and why medications accumulate. Environ. Int. 34 (8), 1157–1169.

Samer, C.F., Lorenzini, K.I., et al., 2012. Applications of CYP450 testing in the clinical setting. Mol. Diagn. Ther. 17 (3), 165–184.
Schaecher, K.L., 2013. The importance of treatment adherence in HIV. Am. J. Manage Care 19 (12), S231–S237.
Schillinger, D., Wang, F., et al., 2006. Language, literacy, and communication regarding medication in an anticoagulation clinic: a comparison of verbal vs. visual assessment. J. Health Commun. 11 (7), 651–664.
Schneider, M.P., Gertsch, A., et al., 2014. Cyberhealth serving to support individual intake of medication. Swiss Med. Wkly. 143:w13827.
Schwartz, G.R., Quigley, H.A., 2008. Adherence and persistence with glaucoma therapy. Surv. Ophthalmol. 53, S57–S68.
Schwartz, M.P., Wagner, P.J., 2000. Which medicines do our patients want from us? J. Fam. Pract. 49 (4), 339–341.
Simoens, S., Sinnaeve, P.R., 2014. Patient co-payment and adherence to statins: a review and case studies. Cardiovasc. Drugs Ther. 28 (1), 99–109.
Sleath, B., Blalock, S.J., et al., 2014. Provider education about glaucoma and glaucoma medications during videotaped medical visits. J. Ophthalmol. 7, 238939.
Solomon, M.D., Majumdar, S.R., 2010. Primary non-adherence of medications: lifting the veil on prescription-filling behaviors. J. Gen. Intern. Med. 25 (4), 280–281.
Stellefson, M., Chaney, B., et al., 2014. Web 2.0 chronic disease self-management for older adults: a systematic review. J. Med. Internet Res. 15 (2), e35.
Storm, A., Andersen, S.E., et al., 2008. One in 3 prescriptions are never redeemed: primary nonadherence in an outpatient clinic. J. Am. Acad. Dermatol. 59 (1), 27–33.
Stosic, R., Dunagan, F., et al., 2011. Responsible self-medication: perceived risks and benefits of over-the-counter analgesic use. Int. J. Pharm. Pract. 19 (4), 236–245.
Sutherland, R., Vincent, P.C., et al., 1977. Quinine-induced agranulocytosis—toxic effect of quinine bisulfate on bone-marrow cultures in vitro. Br. Med. J. 1 (6061), 605–607.
Tesfa, D., Keisu, M., et al., 2009. Idiosyncratic drug-induced agranulocytosis: possible mechanisms and management. Am. J. Hematol. 84 (7), 428–434.
Thurman, E.M., Ferrer, I., 2012. Liquid chromatography/quadrupole-time-of-flight mass spectrometry with metabolic profiling of human urine as a tool for environmental analysis of dextromethorphan. J. Chromatogr. A 1259, 158–166.
Tordoff, J.M., Bagge, M.L., et al., 2010. Medicine-taking practices in community-dwelling people aged>= 75 years in New Zealand. Age Ageing 39 (5), 574–580.
Touchette, D.R., Burns, A.L., et al., 2006. Survey of medication therapy management programs under medicare part D. J. Am. Pharm. Assoc. 46 (6), 683–691.
Tsiantou, V., Pantzou, P., et al., 2010. Factors affecting adherence to antihypertensive medication in Greece: results from a qualitative study. Patient Prefer. Adherence 4, 335–343.
Urquhart, J., 1998. Pharmacodynamics of variable patient compliance: implications for pharmaceutical value. Adv. Drug Deliv. Rev. 33 (3), 207–219.
Urquhart, J., 2002. The odds of the three nons when an aptly prescribed medicine isn't working: non-compliance, non-absorption, non-response. Br. J. Clin. Pharmacol. 54 (2), 212–220.
van Onzenoort, H.A., Neef, C., et al., 2012. Determining the feasibility of objective adherence measurement with blister packaging smart technology. Am. J. Health Syst. Pharm. 69 (10), 872–879.
Vanmolkot, F.H.M., de Hoon, J., et al., 1999. Impact of antihypertensive treatment on quality of life: comparison between bisoprolol and bendrofluazide. J. Hum. Hypertens. 13 (8), 559–563.

Vermeire, E., Hearnshaw, H., et al., 2001. Patient adherence to treatment: three decades of research. A comprehensive review. J. Clin. Pharm. Ther. 26, 331–342.

Vervloet, M., Spreeuwenberg, P., et al., 2013. Lazy Sunday afternoons: the negative impact of interruptions in patients' daily routine on adherence to oral antidiabetic medication. A multilevel analysis of electronic monitoring data. Eur. J. Clin. Pharmacol. 69 (8), 1599–1606.

Vrijens, B., De Geest, S., et al., 2012. A new taxonomy for describing and defining adherence to medications. Br. J. Clin. Pharmacol. 73 (5), 691–705.

Vrijens, B., Claeys, M.J., et al., 2014. platelet aggregation with ticagrelor twice daily vs. clopidogrel once daily based on patient adherence data (the TWICE project). Br. J. Clin. Pharmacol. 77 (5), 746–755.

Wamala, S., Merlo, J., et al., 2007. Socioeconomic disadvantage and primary non-adherence with medication in Sweden. Int. J. Qual. Health Care 19, 134–140.

Wang, E.E.L., Einarson, T.R., et al., 1999. Antibiotic prescribing for Canadian preschool children: evidence of overprescribing for viral respiratory infections. Clin. Infect. Dis. 29 (1), 155–160.

Weissenfeld, J., Stock, S., et al., 2010. The nocebo effect: a reason for patients' non-adherence to generic substitution? Pharmazie 65 (7), 451–456.

Wilcox, C.M., Cryer, B., et al., 2005. Patterns of use and public perception of over-the-counter pain relievers: focus on nonsteroidal antiinflammatory drugs. J. Rheumatol. 32 (11), 2218–2224.

Williams, M.V., Baker, D.W., et al., 1998. Inadequate literacy is a barrier to asthma knowledge and self-care. Chest 114 (4), 1008–1015.

Wood, K.S., Cronley, M.L., 2014. Then and now: examining how consumer communication and attitudes of direct-to-consumer pharmaceutical advertising have changed in the last decade. Health Commun. 29 (8), 814–825.

Yamanouchi, T., Mizoguchi, M., et al., 2010. Idea for unit-dose packaging in Japan. In: Proceedings of the 17th Iapri World Conference on Packaging. Irvin, Sci. Res. Publ., Inc-Srp, pp. 165–167.

Yamazaki, N., Taya, K., et al., 2010. The most appropriate storage method in unit-dose package and correlation between color change and decomposition rate of aspirin tablets. Int. J. Pharm. 396 (1–2), 105–110.

Zigman, D., Blier, P., 2012. A framework to avoid irrational polypharmacy in psychiatry. J. Psychopharmacol. 26 (12), 1507–1511.

Zullig, L.L., McCant, F., et al., 2014. A health literacy pilot intervention to improve medication adherence using Meducation (R) technology. Patient Educ. Couns. 95 (2), 288–291.

Disposal of unused medications

3

3.1 Overview

In the previous chapter, we discussed some of the clinical, social, economic, and legislative aspects concerning the prescription and use of medication and the circumstances that can lead to unused medications. The next part of the environmental life-cycle of pharmaceuticals explores how patients discard their unused medications, how this can be influenced by their beliefs and attitudes toward unused medication, and how these disposal practices contribute to environmental impacts.

3.2 The origins of unused medication

As explained in the previous chapter, when a patient is prescribed a medication, they may either comply with the therapy and take all the medication or choose not to take at least part of it (Ruhoy and Daughton, 2008). When patients fail to take all of their medications, they usually resort to storing unused medications in a home cabinet for use at a later date should their medical condition return (De Bolle et al., 2008). A ready supply of medications usually empowers the patient to self-medicate when they feel it is necessary (De Bolle et al., 2008). However, this behavior can lead to an irrational accumulation of unused medications (Ruhoy and Daughton, 2008) because many patients subsequently forget about the types and quantities of medications they have stored over time.

We have already discussed instances where storage of medication may be necessary, such as the stockpiling of medication in first-aid kits for emergency use (Chapter 2). Medications stored in the home as a result of bulk dispensing can also be beneficial to the patient particularly if the patient has a chronic medical condition. However, *irrational* storage of medications that are no longer needed for the treatment of a particular medical condition is a potential environmental concern. As discussed previously, health literacy and medication adherence practices of patients may be poor. Many patients view medications as a "health currency" that can be exchanged between family members and friends. In Greece, antibiotics and analgesics are seen as potential treatments for future ailments and are commonly shared between individuals. As such, there is incentive for individuals to stockpile these types of medications for later use (Tsiligianni et al., 2012). The act of saving medication for future use is also reportedly a popular practice in Sweden (Persson et al., 2009), New Zealand (Braund et al., 2009b), Ireland (Vellinga et al., 2014), and Malaysia (Fatokun, 2014). Elderly patients who were prescribed strong opioid pain killers have also reported that they share medication with their peers when they feel that their contemporaries are not receiving an adequate level of analgesia (Ellis et al., 2011). Poor medicines management coupled

The Life-Cycle of Pharmaceuticals in the Environment. http://dx.doi.org/10.1016/B978-1-907568-25-1.00003-7

with medication nonadherence and polypharmacy (Chapter 2) also contribute to medications that are simply abandoned in the cupboard (Sorensen et al., 2005) and unintentional accumulation from such forgetfulness often occurs (Stewart and Pearson, 1999). The collection of goods and objects in a haphazard manner as a result of *hoarding disorder* (HD)—a disordered accumulation of belongings in the household—may also give rise to an accumulation of unused medications (Nordsletten et al., 2013).

Irrespective of the factors that contribute to such accumulation, the ultimate result is that the medications risk becoming unused over time. When such medications are not needed any longer, they are usually discarded by the patient, caregiver, or householder. An estimated 2.8 million pounds (1,270 metric tons) of prescribed medication has been estimated to be unused annually in the United States alone (Taras et al., 2014). How do patients and caregivers dispose of unused medications, and what impact do these practices have on the environment? The following section will explore some common disposal practices for unused medications around the world.

3.3 Medication disposal by households

The disposal of unused medications in the rubbish bin has been the predominant disposal method for a long time (Bound and Voulvoulis, 2005; Abahussain et al., 2006; BAPPG, 2006; Abahussain and Ball, 2007) and remains the most common practice, given the accessibility of this route of disposal to the ordinary householder (Table 3.1). For example, out of 208 households in the South Baka District of Serbia, only 8.7% of urban and 6.4% of rural households rinsed unused medications down the toilet compared to 85.6% of urban and 74.5% of rural households who preferred the rubbish bin (Kusturica et al., 2012). These results are consistent with those reported by Abahussain et al. (2006), where 76.5% of the study participants commonly rubbished their unwanted medications compared to only 11.2% who disposed of them down the sink. However, the use of the toilet or sink as the predominant route for disposal of unused medication is not unheard of in selected communities, such as the practices reported in a US survey of outpatient hospital pharmacy participants (Seehusen and Edwards, 2006).

These recent findings are in contrast to results obtained from earlier studies from 1980 to 2000 when unused medications were more frequently rinsed down the sink or toilet. Ridout et al. (1986) reported that 46% of the 443 participants surveyed rinsed unused medications through a portal to their municipal sewage system. This finding was confirmed in later surveys by Kuspis and Krenzelok (1996), Sullivan and George (1996), and Boivin (1997) who reported that 35.4%, 27%, and 46% respectively of the participants in each of these studies disposed of unused medications in the toilet or sink.

Some studies have reported the use of less common disposal routes other than the rubbish bin, toilet, or sink. Kusturica et al. (2012) reported the backyard burning of unused medications by rural families in Serbia (13.8%), although this practice was not reported for urban families. Such observations confirm those of an earlier study involving urban and rural participants in Lithuania (Krupiene and Dvarioniene, 2007)

Table 3.1 Methods used to dispose of unused medications as reported in recent 2012–2014 studies. Multiple selections (i.e., toilet, sink and/or rubbish bin) were generally allowed in a single response to a questionnaire.

Year of study	Country	Sample size	Dosage form/ study setting[b]	Method of disposal				Author
				Rubbish bin	Sink	Toilet	Other	
2012	Nigeria	188	–	96	6	17	13	Auta et al. (2012)
	Malaysia	885	Liquid	25	31[a]	19	14	Auta et al. (2012)
	Malaysia	885	Solid	65	42	17.7	11.5	Azad et al. (2012)
	Malaysia	885	Semi-solid	78	5.8	0.79	20.5	Azad et al. (2012)
	Serbia	104	Urban	85.6	0.68	8.7	5.8	Kusturica et al. (2012)
	Serbia	94	Rural	74.5	–	6.4	19.2	Kusturica et al. (2012)
2013	Ireland/Malta	1130	Liquid	57	28[a]	28[a]	–	Fenech et al. (2013)
	Ireland/Malta	1130	Solid	68	40[a]	40[a]	–	Fenech et al. (2013)
	India	236	–	94	32	12	31	Aditya (2013)
	Tanzania	409	–	66.6	–	19.7	9.1	Makatta (2013)
	USA	444	–	59	–	31[a]	52	Wieczorkiewicz et al. (2013)
2014	Malaysia	250	–	78.8[b]	–	5[b]	48[b]	Fatokun (2014)
	Ireland	202[c]	–	51	29	14	6	Vellinga et al. (2014)

[a] Combined disposal in toilet and sink (sewage system).
[b] Antibiotic disposal.
[c] Number of participants utilizing "environmentally inappropriate" disposal practices (72% of total participants).

where accessibility and the creation of open fires is a lot more commonplace as a means of destroying unwanted domestic waste.

The variability of these disposal practices is influenced by the convenience, availability, and ease of access to disposal or waste facilities. In other instances, the "esthetics" or ergonomics of the dosage form design may persuade a patient to dispose of it in a particular manner.

3.4 Attitudes toward medication disposal

3.4.1 Influence of dosage form design

The dosage form design, that is whether a medication is a solid tablet or capsule, liquid solution or suspension, or a cream or ointment, can affect disposal practice. In Germany (Gotz and Keil, 2007), only 16% of study participants reported the disposal of solid dosage forms in the toilet; however, 43% disposed of liquid formulations in the sink or toilet. In New Zealand, Braund et al. (2009b) reported that tablets and capsules were predominantly returned to the pharmacy by study participants, whereas liquid dosage forms were more commonly rinsed down the toilet or sink. Azad et al. (2012) reported the rinsing of liquid medications into the sewage system by 42% of participants compared to only 25% of participants who reported their disposal in the rubbish bin. Again, such disposal practices can potentially be associated with convenience.

Tablets, capsules, and other solid dosage forms are less likely to be placed in the sink or toilet as patients may want to avoid any perceived plumbing issues resulting from waste pipe blockages that might potentially arise from the disposing of such bulk solid material in this manner. On the other hand, bottles of liquid formulations may be too bulky to fit into a rubbish bin; hence, it may be generally easier for the householder to empty the contents of the bottle into the sink or toilet and dispose of the empty container in a rubbish bin. However, there can be exceptions to this generalized observation. Fenech et al. (2013) reported that 57% of their Maltese study participants discarded liquid medications in their containers into the rubbish with only 28% of study participants discarding the liquid contents into either the sink or the toilet. This more recent observation may be skewed by the fact that the sewage system is increasingly perceived by the general public as the disposal route most likely to lead to an adverse environmental impact and that participants are becoming increasingly aware of such potential environmental issues (Fenech et al., 2013). However, some tablets and capsules are deliberately rinsed down the sink by pharmacies for legal reasons and such medications include controlled and illicit drugs of abuse (*see* discussion below).

3.4.2 Medication disposal methods and environmental awareness

The awareness of an individual of the potential environmental harm associated with environmentally inappropriate medication disposal practices may influence their preferred route for disposal of unused medication (Persson et al., 2009). An increase in environmental awareness through mass media advertising campaigns has been reported to reduce the likelihood of direct household disposal of unused medications via the sink, toilet, or rubbish bin (BAPPG, 2006; Abahussain and Ball, 2007). In a US study, the major reason why consumers would use take-back programmes (*see* below) to dispose of their unwanted medications was reported to be to protect the environment (Thach et al., 2013). However, other studies such as that of Kusturica et al. (2012) reported that although half of the households surveyed thought that the disposal of medications in the toilet, sink, or rubbish bin may have a detrimental effect on the environment, this was not reflected in their *actual* disposal practices with many participants still using the rubbish bin.

A more recent US study (Wieczorkiewicz et al., 2013) has reported that 75% of the participants did not consider the flushing of medications down the toilet or sink to be an appropriate disposal practice. Eighty-six percent (86%) of those participants who did not consider the flushing of medications environmentally acceptable, cited that sewage disposal of unwanted medications was not appropriate because of environmental pollution. Nevertheless, 31% of the participants still reported the use of this route for unwanted medication disposal. In Tanzania, nearly 67% and 20% of the participants in a questionnaire study reported the disposal of medications in the rubbish and toilet, respectively (Makatta, 2013). Nonetheless, 62% of the participants were reportedly aware that the improper disposal of unused medications could result in environmental harm.

It is clear from these studies that environmental awareness plays a role in the attitudes and medication disposal practices of study participants; however, awareness alone does not seem to correlate with *actual* medication disposal practices of participants.

3.4.3 Medication disposal and advice received

Patient education and advice may be more useful in influencing appropriate disposal practices than awareness of environmental issues alone. Seventy-five percent of participants in a recent study in Ireland who appropriately returned their unused medications to the doctor, pharmacist, or hospital had received advice from health professionals on acceptable disposal practices (Vellinga et al., 2014). Even so, 17% of the study participants were reported to still accumulate unused medications simply because they did not know how to properly discard them.

In India, 94% of participants reported the disposal of unused medications in rubbish while 32% used the sink (Aditya, 2013).[1] Ninety-two percent of them never received any advice from authorities or other official information sources on how to dispose of unwanted medications. A similar observation was reported for the United States (Wieczorkiewicz et al., 2013) where 80% of participants never received any information from the media or healthcare provider on environmentally acceptable disposal practices for unused medications.

[1] Note multiple responses were allowed in this study. A particular study participant can claim to dispose of medications in the sink and in the rubbish bin. As such, the percentages do not tally to 100%.

A lack of advice from health professionals usually means patients are unaware of best disposal practices. The practice of disposing of medications was commonly reported in a combined sample from Malta and Ireland while Fenech et al. (2013) reported that only 5% and 22% of the study participants aged 45 years or younger and over 45 years, respectively, had received any previous advice from a doctor or pharmacist on how to appropriately dispose of their unused medications. Overall, only 33% and 14% of Irish and Maltese participants, respectively, received any advice from a pharmacist concerning appropriate medication disposal practices and 78% of all the participants reported a lack of awareness of how to "properly" dispose of unused medications. The lack of such formal advice in a particular country is usually associated with a lack of official protocols established to deal with pharmaceutical waste (*see* below). Many school nurses in the United States have recently been reported to be unsure of how medications could be discarded in an environmentally safe manner and wanted further education on best disposal practices (Taras et al., 2014).

3.4.4 Medication disposal practices and demographics

The demographics of a study sample can influence medication disposal practices with older study participants more likely to report the disposal of unused medications down the toilet or sink (WCRC, 2006). However, there is also considerable variability between the results of different studies. For example, Kotchen et al. (2009) reported no significant difference between younger (under 55 years) and older (over 55 years) participants in choosing the rubbish bin as their preferred disposal route. No significant demographic trend was also observed in a study of unused medication disposal practices involving military outpatients utilizing a US hospital pharmacy (Seehusen and Edwards, 2006). However, these results may not be particularly representative of the general medication disposal practices observed in a more general population.

3.4.5 Availability of formalized protocols for unused medication disposal

The lack of appropriate advice may stem from a lack of formalized disposal and destruction protocols for the disposal and destruction of unused and unwanted medications in many countries. In fact, there is much contradictory advice and confusing recommendations from various authorities to pharmacists, healthcare workers, patients, and caregivers regarding the disposal of unused medications and it can sometimes be difficult for healthcare workers to determine what is the most environmentally sound advice (Vellinga et al., 2014).

Confusion arising from a lack of standard operating procedures for medication disposal can occur in many countries at a state or local level (Glassmeyer et al., 2009a, 2009b; Koshy, 2013). The lack of standard operating procedures in schools has resulted in US school nurses utilizing a wide range of disposal practices to eliminate stockpiles of abandoned medication left over from school children (Taras et al., 2014). Such practices included the flushing of medications down the toilet, the discarding of medications in the rubbish, or the addition of the waste into containers labeled for the disposal of "sharps" (Taras et al., 2014).

In the past, the US Food and Drug Administration (FDA) recommended the flushing of prescription medications down the drain to prevent the risk of household poisonings; however, this disposal practice has since been discouraged by the US EPA (Taras et al., 2014). While flushing medications down a sink or a toilet also has merits in preventing diversion of controlled medications (*see* discussion below), it must be clarified by state and local authorities that such a practice is only to be encouraged for certain medications.

No formalized protocols for the disposal of unused medications exist in many countries (Kusturica et al., 2012). However, in Kuwait a Ministry of Health guideline exists for the disposal of medical wastes from hospitals and healthcare centers (Abahussain et al., 2012). This guideline stipulates that pharmacies return unwanted medications back to a central state-run depot where medication waste is destroyed by high-temperature incineration according to specifications issued by the Kuwait Environmental Protection Agency. However, this collection and destruction system for unwanted medications is inaccessible to the general public.

Irrespective of whether formalized protocols for medication disposal are established or not within a country, many participants in such studies report their desire for a centralized system for the collection and destruction of unwanted medications so that a single official protocol for medication disposal can be implemented (Taras et al., 2014). To date, there is no internationally recognized single protocol for collecting and destroying medication waste from end users, but many different short-term and more permanent systems for unused medication disposal have been implemented in various parts of the world (Fass, 2011).

3.5 Feasibility of take-back programmes for unused medication

A take-back programme is a coordinated scheme organized at the national, state, or local level designed to reclaim unused and unwanted medications from the public (Perry et al., 2014). It can be run and funded by various organizations and bodies, including medical practitioners, pharmacists, the government and law enforcement agencies (i.e., federal and provincial police departments), and pharmaceutical manufacturers. However, use of these systems may not necessarily be enforced through legislation such as that governing medical waste management (Chapter 7).

In the United States, the state of Maine has traditionally been active in funding state and local drug take-back programmes through its Department of Environmental Protection (DEP) (Glassmeyer et al., 2009a, 2009b). The DEP also provides free mailing envelopes for the public to send back unwanted medications to the national drug enforcement agency. Despite the success of this programme for collecting more unwanted medications per resident than any other state in the United States, running such a programme long-term has been challenging for the DEP. This is because the countless medicines collected have to be sent outside of the state for destruction which incurs significant freight costs (Tucker, 2011). Strategies are currently being trialed to deal with the waste collected, including the notion that medication waste can be composted and landfilled rather than incinerated out-of-state at high cost (*Maine might compost unused drugs*—available http://www.onlinesentinel.com/news/maine-might-compost-unused-drugs_2011-10-01.html, accessed 17-10-13).

Take-back programmes have also been trialed for limited periods of time in Colorado and other US states, but despite these initiatives at the state level, there is no official nationwide programme that offers regular collection of unwanted medications in the United States. However, federally funded "national days" for the collection of unwanted medications do occur on an infrequent basis. One such notable campaign was initiated in San Francisco, California, under the direction of the Bay Area Pollution Prevention Group (BAPPG) and involved a two-day collection of unwanted medications in 39 pharmacies at a cost of around US$100,000 (BAPPG, 2006). The first nationwide collection day for unwanted medications occurred in September 2010 with the participation of law enforcement officers from 3,000 California state and local law enforcement agencies (Tucker, 2011). The national take-back initiative usually occurs twice every year in spring and autumn.

Similar collection programmes have been established in other western countries including Canada (Kotchen et al., 2009). In Australia, the Return of Unused Medicine (RUM) campaign was established as early as 1998 and was funded by the Australian government at a cost of AUD$3 million initially for three years and AUD$6 million for an additional four years from 2005 (*see* www.returnmed.com.au, accessed 9-8-14). The programme allows community pharmacies to take back unwanted medications from patients at no cost to the patient or pharmacy.

The popularity and efficacy of any take-back scheme for the return of unused medication to a pharmacy is very dependent on how it is administered. A successful take-back scheme must be as convenient and resilient as any other waste management programme. Currently, a pilot take-back programme in the US state of Texas suggests that there is more unused medication requiring disposal than current take-back programmes can deal with (Thach et al., 2013). A recent study undertaken in the United States (Wieczorkiewicz et al., 2013) reported that 65% of the participants would be agreeable to return unused medications to their prescribing doctor's office if the medications were deposited into a locked safe. Only 13% of study participants supported the idea of mail-back programmes simply because they did not consider the postal system to be a secure repository for holding unwanted medications during transit. In fact, 93.8% of the study participants preferred to instead drop off their unused medications at a retail/community pharmacy.

Similar sentiments concerning take-back programmes have been reported in Croatia (Jonjic and Vitale, 2014). Here, community pharmacies are the depots for the take-back of unused medications and each pharmacy is financially responsible for the disposal of this returned pharmaceutical material. This usually means that the state-owned pharmacies subsidized by the government are more willing to accept unused medications than are privately owned pharmacies which must bear the full costs of disposal. However, because the state government requires private citizens and those that are responsible for take-back programmes to manage the waste (i.e., a "polluter pays" system), advertising of pharmacy take-back programmes is still poor in Croatia and many citizens are unaware of their existence. Even though some Croatian citizens may be aware of the take-back programmes available in their local area, they may still refuse to return medications to pharmacies. This is because take-back containers are only accessible to pharmacy staff and the lack of anonymity dissuades patients from handing over unwanted medications to these staff. The provision of secure and locked drop boxes which preserve the anonymity of personal and health information of patients would significantly improve the likely success of any such take-back programme.

A take-back system is only effective if it is economically viable. In Kuwait, it has been suggested that medications are commonly deposited in the household rubbish because an official scheme for the collection of unused medications is unlikely to be cost-effective (Abahussain and Ball, 2007). Current evidence suggests that cost is a major barrier to the successful implementation of take-back programmes. Forty-four percent of participants in a US survey (Wieczorkiewicz et al., 2013) were not keen to pay any fees when using a take-back system and only 37.5% were prepared to pay a maximum cost of US$2 per use. In addition, mail-back envelopes sold from Walgreens pharmacies at a cost of US$2.99 per bag have not had as much success as free federally funded take-back initiatives for consumers (Tucker, 2011). Clearly, a free take-back programme would be the most desirable scenario for the consumer; however, such programmes will need public funding in some manner. One study suggests that the average consumer would be willing to pay an additional US$1.53 per prescription dispensed (Kotchen et al., 2009) to fund a take-back scheme; however, participant demographics affected how much they were willing to pay. Participants with more socialist political views were reportedly more likely than those with conservative political views to pay a higher premium per returned prescription for funding a take-back programme (Kotchen et al., 2009).

Finally, a permanent take-back programme may be more effective than occasional events organized to collect unused medication. A study by Kotchen et al. (2009) determined that 68% of the participants would be willing to participate in a take-back programme if it was run permanently at their local pharmacy. Thach et al. (2013) also reported that participants were more likely to use a pharmacy offering an additional take-back programme for unused medications than one that did not offer the service. Furthermore, participants who actively used a take-back scheme considered it to be a valuable service to meet their medication disposal needs.

In conclusion, if take-back programmes are to be effective in the community in reducing inappropriate disposal practices for unused medication, it is clear that they need to:

- be available on a permanent, long-term basis;
- be well advertised at the state and/or local level;
- be free to a user if at all possible but if a charge is to be made, it should be low cost rather than a completely user-pays "out-of-pocket" system (i.e., through taxes or small co-payments in prescriptions);
- be streamlined and able to handle the return of variable quantities of pharmaceutical waste from the public;
- provide easy access to drop-off points at pharmacies or other depots in the community and allow for anonymous collection;
- have secure depots to avoid tampering, theft, or diversion of collected medications.

3.6 Management of pharmaceutical waste at take-back depots

Once collected, medication waste still has to be disposed of at a take-back depot. However, destruction of pharmaceutical waste is often not a straightforward process. Some depots may not necessarily have the required facilities to transform pharmaceutical wastes into environmentally acceptable by-products.

Take-back depots act as the intermediary between the public and facilities which destroy unwanted medications. Generally, medications collected by take-back programmes are shipped off to a licensed medical waste contractor for destruction by high-temperature incineration (Tong et al., 2011). Pharmacies are the major take-back depots for the return of unwanted medications in many countries and these pharmacies are relied on to store and manage the returned medication. Usually, hundreds and thousands of kilograms of unwanted medications are brought to take-back depots which then necessitate destruction (Glassmeyer et al., 2009a, 2009b). Over a one-month period, 505 kg of unwanted medications were collected from 91 pharmacies in Croatia (Jonjic and Vitale, 2014). In the Otago region of New Zealand, 1,294 kg of this material was returned to pharmacies over nine months (Braund et al., 2007). A disposal of unused medication (DUMP) campaign carried out in conjunction with community pharmacies in New Zealand resulted in the collection of over 1,600 bags of unwanted medication during a one-month period (Braund et al., 2009a) and another study reported the collection of over 3,700 individual prescription medication items (James et al., 2009).

Many patients and other end-users assume that returned medications are "out of sight and mind" and never consider what will ultimately happen to these medications after their return. A laudable but erroneous idea often perceived by patients is that their returned medications will be reused by the pharmacy (Mackridge and Marriott, 2007) but the ethical legislation in most countries does not allow this to occur. While some countries, such as Sweden, have an extensive take-back scheme run by the national pharmaceutical society (Apoteket AB) which collects unwanted medications returned to pharmacies and arranges for their destruction by high-temperature incineration (Persson et al., 2009), the collection and destruction system in many other countries is often much less sophisticated. A 1996 US study found that only 3% of pharmacies surveyed had any specific policies for the disposal of expired medication once collected by the pharmacy (Kuspis and Krenzelok, 1996). In addition, 68% of the pharmacies surveyed deposited unwanted medication in the rubbish or rinsed them down the toilet or sink. Furthermore, it was not uncommon for pharmacists to recommend to their customers that they should dispose of any unwanted medications themselves down the toilet or sink, or in the rubbish. However, the environmental awareness among pharmacists has seemingly improved since that time with a more recent US study (Jarvis et al., 2009) reporting that 80% of the pharmacists surveyed had been informed about environmentally-acceptable disposal practices before any educational intervention was carried out in the study. In addition, 47% of the participants considered inappropriate disposal of pharmaceuticals to be a major environmental problem.

Pharmacists face the same disposal challenges as the general public in countries that lack any legally binding and formalized protocols to deal with pharmaceutical wastes. In Kuwait, over 73% of pharmacists deposited unwanted medications in the rubbish despite the availability of guidelines (*see* above) prescribed by the Kuwait Ministry of Health for medical waste disposal (Abahussain et al., 2012). While there is a contractor system which collects unused medications from New Zealand pharmacies, some medications such as illicit or controlled drugs (e.g., morphine, methylphenidate, methadone, and oxycodone; Chapter 2) are not collected by these contractors (Tong et al., 2011). Also many contractors who collect unused medications from pharmacies provide relatively compact collection bins that can only be used for the disposal of small tablets and capsules. Subsequently, large bottles of liquid suspensions are more commonly rinsed down the sink, with the empty container being disposed of in the rubbish bin (Tong et al., 2011).

Many pharmacies do not want to risk any expired controlled drugs from being diverted into the black market during storage in the pharmacy or during transit to appropriate disposal contractors (Fischer et al., 2010). A common destruction practice for this type of returned medication is to crush and soak the drugs in hot water and then rinse them down the pharmacy sink (Tong et al., 2011), as most community pharmacies do not have convenient access to contractors able to perform the high-temperature incineration necessary to completely degrade this type of drug. In New Zealand, pharmacists are held legally responsible for the destruction of any expired or returned controlled drugs. The quantity of drug destroyed is annotated in a controlled drugs register and signed for by the pharmacist in the presence of a witness (Tong et al., 2011). However, controlled drugs are only destroyed occasionally in most New Zealand pharmacies; therefore, it is not known whether high-temperature incineration procedures are cost-effective or not for the destruction of controlled drugs at least in New Zealand.

Legal restrictions concerning the use and supply of controlled drugs in other countries can make the destruction of unwanted quantities of them rather problematic. Many take-back programmes established for the return of unused ordinary prescription medications cannot be used in the same manner for controlled drugs without the involvement of law enforcement (Vivian, 2009). In the United States, the Food and Drug Administration (FDA) lists a host of controlled medications that can be "safely" flushed down the sink (Table 3.2). While such a practice is legally safe, as it avoids any risk of illegal abuse and diversion, whether it is also environmentally safe is another matter and clearly depends on other factors such as the efficiency of removal in any subsequent sewage treatment facilities (Christenson, 2008; Mitka, 2009).

While some take-back depots and pharmacies in the United States have legitimate rights to accept controlled medications, as a team of law enforcement officers are present to collect these drugs, not all pharmacies employ law enforcement to accept medications from the public (Herring et al., 2008). However, in 2010, US President Barack Obama signed off on the *Safe and Responsible Drug Disposal Act* which allows certain authorized non-law enforcement parties, including take-back programme personnel and specified pharmacies, to accept and dispose of unwanted controlled drugs from end users without incurring the cost involved in oversight by law enforcement personnel (Tucker, 2011).

Table 3.2 Controlled drugs considered appropriate by the US FDA to be disposed of by rinsing into the sewage system (Mitka, 2009)

Brand Name and Dosage Form (US)	Active Pharmaceutical Ingredient
Abstral, tablets	Fentanyl
Actiq, lozenge	Fentanyl citrate
Avinza, capsules	Morphine sulfate
Daytrana, patch	Methylphenidate
Demerol, tablets	Meperidine(pethidine) hydrochloride
Demerol, oral solution	Meperidine hydrochloride
Diastat/diastat acudial, rectal gel	Diazepam
Dilaudid, oral liquid	Hydromorphone hydrochloride
Dolophine hydrochloride, tablets	Methadone hydrochloride
Duragesic, patch	Fentanyl
Embeda, capsules	Morphine sulfate; Naltrexone hydrochloride
Exalgo, tablets	Hydromorphone hydrochloride
Fentora, tablets	Fentanyl citrate
Kadian, capsules	Morphine sulfate
Methadone hydrochloride, oral solution	Methadone hydrochloride
Methadose, tablets	Methadone hydrochloride
Morphine sulfate, tablets	Morphine sulfate
Morphine sulfate, oral solution	Morphine sulfate
Nucynta ER, tablets	Tapentadol
Onsolis, patch	Fentanyl citrate
Opana, tablets	Oxymorphone hydrochloride
Opana ER, tablets	Oxymorphone hydrochloride
Oramorph SR, tablets	Morphine sulfate
Oxecta, tablets	Oxycodone hydrochloride
Oxycodone hydrochloride, capsules	Oxycodone hydrochloride
Oxycodone hydrochloride, oral solution	Oxycodone hydrochloride
Oxycontin, tablets	Oxycodone hydrochloride
Percocet, tablets	Acetaminophen; oxycodone hydrochloride
Percodan, tablets	Aspirin; oxycodone hydrochloride
Xyrem, oral solution	Sodium oxybate

Data adapted from the US FDA Web site, http://www.fda.gov/Drugs/ResourcesForYou/Consumers/BuyingUsingMedicineSafely/EnsuringSafeUseofMedicine/SafeDisposalofMedicines/ucm186187.htm, accessed 9-7-14).

In summary, if the operating costs of take-back programmes were reduced globally, they could be the definitive answer to minimizing the environmental discharge of pharmaceutical contaminants from unused medications. In addition, if contractor costs to incinerate unwanted waste at high temperatures were to be reduced, this would allow pharmacies to provide end users and patients with a more efficient service that could deal with any quantity of medication waste. As destruction costs are regularly borne by private pharmacies and take-back programmes (Jonjic and Vitale, 2014), governmental bodies around the world should reexamine how these costs can be reduced and whether an appropriate level of funding can be provided from a national source to meet the demands of these programmes.

3.7 The "best" method for the disposal of unused medications?

The current recommendation for the disposal of unused medications is for end users to take advantage of take-back or mail-back programmes wherever they are available. As pharmaceutical waste from such take-back programmes is generally incinerated at high temperatures, the environmental risks posed by the unwanted medications are minimized when these programmes are used. In addition, the ever-increasing environmental awareness by the public is an incentive for authorities to formulate and enact legislation to provide more legally and environmentally acceptable disposal opportunities to the end user, such as those provided by the *Safe and Responsible Drug Disposal Act* in the United States. However, in areas without such take-back programmes, one recommendation offered in the United States as part of the *SmartRx Disposal*™ programme is to mix unwanted pharmaceuticals with cat litter or coffee grounds, seal the contents in a waterproof and leakproof container, and then dispose of the container in the rubbish (Daughton and Ternes 1999; Daughton and Ruhoy 2007; Becker et al., 2010). However, this disposal approach raises its own issues, such as how leakproof the container is and how long it will remain sealed during both transit to the landfill and once it is stored there.

Although the direct disposal of unused pharmaceuticals in the rubbish bin is generally not considered to be more environmentally acceptable than take-back programmes (Taras et al., 2014), a life-cycle assessment of many pharmaceuticals (Chapter 8) has renewed interest in the idea that the most environmentally acceptable way to manage unwanted pharmaceuticals may indeed be for the user to throw them directly into the rubbish (Cook et al., 2012b). In this study, a theoretical assessment of the contaminant output resulting from three common disposal routes (via sewers, via the rubbish bin, and via take-back leading to high-temperature incineration) was undertaken under the International Organization for the Standardization of Environmental Management (ISO 14040–Geneva, Switzerland) specifications. It was found that if pharmaceuticals were the *only* cause of environmental pollution, then the best option would be to use a take-back programme followed by incineration. However, taking into account the broader picture including the associated energy consumption and pollution arising from a typical incineration procedure, it may not necessarily be environmentally acceptable. While pharmaceuticals are completely destroyed in high-temperature incineration procedures, the emissions of carcinogens and other airborne contaminants increase by a factor of 700% relative to baseline levels defined in the study. In addition, the model predicted a substantial increase in the amount of global-warming, ozone-depleting, and smog-forming compounds from such high-temperature incineration. Furthermore, many users need to travel to the nearest depot involved in take-back programmes and the resulting consumption of fossil fuels can also have environmental consequences. Therefore, the results from this environmental life-cycle assessment could negate the feasibility of high-temperature incineration based on the arguments that: (1) the emissions from incineration arising from *non-pharmaceutical* airborne pollutants could be more detrimental to the environment than untreated pharmaceutical residues; and (2) the costs of establishing mainstream take-back programmes in many countries may not be economically feasible.

While people report that they are prepared to use a take-back programme provided it involves only minimal, if any, cost to them (Kotchen et al., 2009), this may also not necessarily be true. A report of a pilot take-back programme in Clark County, Washington (United States) observed that only 0.1% of the people from that particular county participated in the programme (Cook et al., 2012b). In addition, while the take-back programme established for over 40 years by Apoteket AB in Sweden is usually considered to be the global "gold standard" of take-back programmes, it is used regularly by only less than half (43%) of the nation's population (Persson et al., 2009). While this could be considered to be a lot more successful than the pilot programme established in Clark County, Washington (United States), it is not necessarily the most optimal solution that governmental and healthcare authorities are looking for if all the contamination from the incineration of the unwanted medications and the cost of running such a programme were to be taken into account (Cook et al., 2012b). The authors contend that a 50% take-back programme participation rate would lead to the emission of the same quantity of pharmaceuticals into the environment as if 100% of all end users were to dispose of their unused medications in the rubbish bin. Whatever the costs involved in running take-back programmes are, the fact remains that the disposal of medications in the rubbish is a well-established and convenient process that can be undertaken in a home at no extra costs. As such, any widespread acceptance of a take-back programmes would require it to be extremely popular and environmentally and economically feasible (Cook et al., 2012b).

In conclusion, while take-back programmes leading to the high-temperature incineration of pharmaceutical waste do reduce the emission of pharmaceuticals into the natural environment, their overall environmental benefits are still largely unknown and yet to be fully evaluated. However, the best advice to patients, caregivers, and consumers consistent with US FDA and EPA guidelines (Cook et al., 2012a; Daughton, 2012) would be:

(1) Use drug take-back programmes and subsequent incineration of unused medications when these facilities are available.

(2) Rinse controlled drugs as specified by the FDA or other authorities to be illicit or drugs of abuse down the toilet or sink only when no take-back depot capable of receiving controlled drugs is available in your area.

(3) Do not rinse any other types of unused medications down the sink or toilet but if no take-back programme is available, it is better to dispose of them in the rubbish bin.

3.8 Conclusions

The ultimate solution to minimizing or preventing pharmaceuticals from polluting the environment apart, from responsible prescribing practices (Chapter 2), is for patients and health professionals to be proactive in their use of environmentally safe methods for the disposal of unwanted/unused medications. As take-back programmes rely on participation to be cost-effective, it has never been more important than now for health professionals to educate and inform the public about environmental issues arising from the disposal of medical materials. These types of educational initiatives can be easily incorporated into take-back events. For example, Gray-Winnett et al. (2010) describe a year-long take-back initiative where student pharmacists and officials co-ordinated, planned, and executed a take-back programme for unwanted medications in Knoxville, Tennessee. The student community initiative consisted of tents and de-pots for the collection of unused medications and unwanted mercury thermometers in high-traffic areas for ease of public access and drop off. More than 1,100 pounds (~500 kg) of unused medications were collected from the public during a year.

Another ongoing take-back campaign involving pharmacists and pharmacy students has been operating in Broward County, Florida, United States since 2009 (Fass, 2011). Operation Medicine Cabinet involves the participation of 30 police stations and community centers in the area and allows pharmacists and pharmacy students the opportunity to provide medicines information to the public. In turn, pharmacy students are educated by law en-forcement personnel on various legal, administrative, clinical, and pharmacological aspects regarding both licit and illicit prescription medications before participation in the event. It has been reported that over 150,000 doses of medications have been taken back and sub-sequently incinerated since the initiation of this take-back programme over five years ago.

A further educational initiative as described by Eissen and Backhaus (2011) is to inte-grate the topic of "Pharmaceuticals in the environment" into the school or undergraduate curriculum so that young people of the next generation will have the knowledge, skills, and attitudes to minimize environmental pollution associated with the human use of pharmaceuticals. The study describes the creation of various worksheets which elabo-rate on various aspects of the environmental presence of pharmaceuticals, their sources and pathways to the environment, how wastewater treatment plants work, hazards for human health associated with pharmaceuticals, and the measures taken to minimize their environmental discharge. While some of the participants were reported to be somewhat bewildered and confused by the amount of scientific terminology associated with some of the topics covered, many participants appreciated the informative nature of the work-sheets. In addition, many participants felt that changes to their individual medication dis-posal practices for the common good can lead to a positive impact on the environment.

The use of returned medications arising from take-back programmes has been ex-plored by Mackridge and Marriott (2007). As discussed above, returned medications that have already been circulated among the public cannot be ethically redispensed. However, many packets of medications collected in take-back programmes are returned before the end of their shelf-life and in their original unopened packaging. Considering that many medications reported in a UK study by Mackridge and Marriott (2007) were classified as World Health Organization (WHO) essential medications, it can be argued that it is a waste of resources to dispose of these perfectly usable returned medications (Mackridge and Marriott, 2007). However, the reuse of returned medications can lead to a loss of public confidence in the safety and efficacy of medications dispensed from a pharmacy (Mackridge and Marriott, 2008). In addition, while an estimated cost of UK£70 million can be potentially recovered by the reuse of medications, this saving could potentially dwin-dle to UK£18 million given that only approximately 25% of the medications returned to UK pharmacies are in a usable and unopened state. In addition, the use of better tamper-proof packaging technologies may lead to additional manufacturing costs for medicines and pose problems for medication compliance (Mackridge and Marriott, 2008).

As health professionals with expertise in the use of medicines, pharmacists are well placed to offer sound advice on the most environmentally safe options for the disposal of unused medications. It is therefore important for pharmacists to encourage patients to return unused medications to pharmacies or make use of take-back programmes wherever available, as well as encourage conscientious use of prescribed medication to minimize the need to dispose of any unused pharmaceutical material. Hopefully, the individual efforts of the patient and health professional with the encouragement of regional, state, and national bodies will eventually lead to a reduction in the quantity of pharmaceutical waste potentially released into the environment.

References

Abahussain, E.A., Ball, D.E., 2007. Disposal of unwanted medicines from households in Kuwait. Pharm. World Sci. 29, 368–373.

Abahussain, E.A., Ball, D.E., et al., 2006. Practice and opinion towards disposal of unused medication in Kuwait. Med. Princ. Pract. 15 (5), 352–357.

Abahussain, E., Waheedi, M., et al., 2012. Practice, awareness and opinion of pharmacists toward disposal of unwanted medications in Kuwait. Saudi Pharm. J. 20 (3), 195–201.

Aditya, S., 2013. Safe medication disposal: Need to sensitize undergraduate students. Int. J. Pharm. Life Sci. 4 (3), 2475–2480.

Auta, A., Banwat, S.B., et al., 2012. Medicines in pharmacy students' residence and self-medication practices. J. Young Pharmacists. 4 (2), 119–123.

Azad, M.A.K., Ansary, M.R.H., et al., 2012. Disposal practice for unused medications among the students of the International Islamic University Malaysia. J. Appl. Pharm. Sci. 2 (7), 101–106.

BAPPG, 2006. Report on the San Francisco Bay Area's Safe Medicine Disposal days. Bay Area Pollution Prevention Group, San Francisco.

Becker, J.A., Ortner, P.M., et al., 2010. Don't rush to flush: Safer pharmaceutical practices for hospice home care and home health nurses. Home Health Care Manage. Practice 22 (3), 202–206.

Boivin, M, 1997. The cost of medication waste. Can. Pharm. J, May 1997.

Bound, J.P., Voulvoulis, N., 2005. Household disposal of pharmaceuticals as a pathway for aquatic contamination in the United Kingdom. Environ. Health Perspect. 113 (12), 1705–1711.

Braund, R., Yuen, Y.-C., et al., 2007. Identification and quantification of medication returned to Otago pharmacies. NZFP 31 (4), 258–262.

Braund, R., Gn, G., et al., 2009a. Investigating unused medications in New Zealand. Pharm. World Sci. 31 (6), 664–669.

Braund, R., Peake, B.M., et al., 2009b. Disposal practices for unused medications in New Zealand. Environ. Int. 35 (6), 952–955.

Christenson, T., 2008. Fish on morphine: Protecting Wisconsin's natural resources through a comprehensive plan for proper disposal of pharmaceuticals. Wisconsin Law Rev. 2008 (1), 141–179.

Cook, S.M., Love, N.G., et al., 2012a. Response to comment on life cycle comparison of environmental emissions from three disposal options for unused pharmaceuticals. Environ. Sci. Technol. 46 (15), 8521–8522.

Cook, S.M., VanDuinen, B.J., et al., 2012b. Life cycle comparison of environmental emissions from three disposal options for unused pharmaceuticals. Environ. Sci. Technol. 46 (10), 5535–5541.

Daughton, C.G., 2012. Comment on life cycle comparison of environmental emissions from three disposal options for unused pharmaceuticals. Environ. Sci. Technol. 46 (15), 8519–8520.

Daughton, C.G., Ternes, T.A., 1999. Pharmaceuticals and personal care products in the environment: agents for subtle change? Environ. Health Perspect. 107, 907–938.

Daughton, C.G., Ruhoy, I.S., 2007. Types and quantities of leftover drugs entering the environment via disposal to sewage—revealed by Coroner Records. Sci. Total. Environ. 388 (1–3), 137–148.

De Bolle, L., Mehuys, E., et al., 2008. Home medication cabinets and self-medication: A source of potential health threats? Ann. Pharmacother. 42 (4), 572–579.

Eissen, M., Backhaus, D., 2011. Pharmaceuticals in the environment: An educational perspective. Environ. Sci. Pollut. Res. 18 (9), 1555–1566.

Ellis, J.C., Mullan, J., et al., 2011. Prescription medication hoarding and borrowing or sharing behaviours in older residents in the Illawarra, New South Wales, Australia. Australas. J. Ageing. 30 (3), 119–123.

Fass, J.A., 2011. Prescription drug take-back programmes. Am. J. Health Syst. Pharm. 68 (7), 567–570.

Fatokun, O., 2014. Exploring antibiotic use and practices in a Malaysian community. Int. J. Clin. Pharm. 36 (3), 564–569.

Fenech, C., Rock, L., et al., 2013. Attitudes towards the use and disposal of unused medications in two European Countries. Waste Manag. 33 (2), 259–261.

Fischer, B., Bibby, M., et al., 2010. The global diversion of pharmaceutical drugs: non-medical use and diversion of psychotropic prescription drugs in North America: a review of sourcing routes and, control measures. Addiction 105 (12), 2062–2070.

Glassmeyer, S.T., Hinchey, E.K., et al., 2009. Disposal practices for unwanted residential medications in the United States. Environ. Int. 35, 566–572.

Gotz, K., Keil, F., 2007. Drug disposal in private households: Does the disposal of pharmaceuticals via domestic sanitary devices contribute to water contamination? Z Umweltchem Okotox 19 (3), 180–188.

Gray-Winnett, M.D., Davis, C.S., et al., 2010. From dispensing to disposal: The role of student pharmacists in medication disposal and the implementation of a take-back programme. J. Am. Pharm. Assoc. 50 (5), 613–618.

Herring, M.E., Shah, S.K., et al., 2008. Current regulations and modest proposals regarding disposal of unused opioids and other controlled substances. J. Am. Osteopath. Assoc. 108 (7), 338–343.

James, T.H., Helms, M.L., et al., 2009. Analysis of medications returned to community pharmacies. Ann. Pharmacother. 43 (10), 1631–1635.

Jarvis, C.L., Seed, S.M., et al., 2009. Educational campaign for proper medication disposal. J. Am. Pharm. Assoc. 49 (1), 65–68.

Jonjic, D., Vitale, K., 2014. Issues around household pharmaceutical waste disposal through community pharmacies in Croatia. Int. J. Clin. Pharm. 36 (3), 556–563.

Koshy, S., 2013. Disposal of unwanted medications: Throw, bury, burn or just ignore? Int. J. Pharm. Pract. 21 (2), 131–134.

Kotchen, M., Kallaos, J., et al., 2009. Pharmaceuticals in wastewater: Behavior, preferences, and willingness to pay for a disposal programme. J. Environ. Manag. 90, 1476–1482.

Krupiene, J., Dvarioniene, J., 2007. Pharmaceutical Pathways to the Environment in Lithuania. Environ. Res. Eng. Manag. 41 (3), 33–39.

Kuspis, D.A., Krenzelok, E.P., 1996. What happens to expired medications? A survey of community medication disposal. Vet. Hum. Toxicol. 38 (1), 48–49.
Kusturica, M.P., Sabo, A., et al., 2012. Storage and disposal of unused medications: knowledge, behavior, and attitudes among Serbian people. Int. J. Clin. Pharm. 34 (4), 604–610.
Mackridge, A.J., Marriott, J.F., 2007. Returned medicines: Waste or a wasted opportunity? J. Public Health (Oxf) 29 (3), 258–262.
Mackridge, A., Marriott, J., 2008. Reply to: Returned medicines: Waste or wasted opportunity? J. Public Health 30 (2), 209–210.
Makatta, G.B., 2013. Assessment of knowledge and practices for disposal of unfinished, unwanted and expired medications from households in Iringa Municipal Council, Tanzania. OJS 4 (1), 587–594.
Mitka, M., 2009. FDA: Flush certain unused medications. J. Am. Med. Assoc. 302 (19), 2082.
Nordsletten, A.E., de la Cruz, L.F., et al., 2013. Finders keepers: The features differentiating hoarding disorder from normative collecting. Compr. Psychiatry 54 (3), 229–237.
Perry, L.A., Shinn, B.W., et al., 2014. Quantification of an ongoing community-based medication take-back programme. J. Am. Pharm. Assoc. 54 (3), 275–279.
Persson, M., Sabelstrom, E., et al., 2009. Handling of unused prescription drugs—knowledge, behaviour and attitude among Swedish people. Environ. Int. 35, 771–774.
Ridout, S., Waters, W.E., et al., 1986. Knowledge of and attitudes to medicines in the Southampton community. Brit. J. Clin. Pharm. 21, 701–712.
Ruhoy, I.S., Daughton, C.G., 2008. Beyond the medicine cabinet: an analysis of where and why medications accumulate. Environ. Int. 34 (8), 1157–1169.
Seehusen, D.A., Edwards, J., 2006. Patient practices and beliefs concerning disposal of medications. J. Am. Board Fam. Med. 19 (6), 542–547.
Sorensen, L., Stokes, J.A., et al., 2005. Medication management at home: Medication-related risk factors associated with poor health outcomes. Age Ageing 34 (6), 626–632.
Stewart, S., Pearson, S., 1999. Uncovering a multitude of sins: medication management in the home post acute hospitalisation among the chronically ill. Aust. NZ J. Med. 29 (2), 220–227.
Sullivan, M.J., George, C.F., 1996. Medicine taking in Southampton: A second look. Br. J. Clin. Pharmacol. 42, 567–571.
Taras, H., Haste, N.M., et al., 2014. Medications at school: disposing of pharmaceutical waste. J. Sch. Health 84 (3), 160–167.
Thach, A.V., Brown, C.M., et al., 2013. Consumer perceptions about a community pharmacy-based medication take back programme. J. Environ. Manag. 127, 23–27.
Tong, A.Y., Peake, B.M., et al., 2011. Disposal practices for unused medications in New Zealand community pharmacies. J. Primary, Health Care 3 (3), 197–203.
Tsiligianni, I.G., Delgatty, C., et al., 2012. A household survey on the extent of home medication storage. A cross-sectional study from rural Crete, Greece. Eur. J. Gen. Prac. 18 (1), 3–8.
Tucker, C., 2011. Drug takebacks aim to prevent abuse, protect environment: Disposing of medications the right way. The Nation's Health 41 (2), 1–15.
Vellinga, A., Cormican, S., et al., 2014. Public practice regarding disposal of unused medicines in Ireland. Sci. Total Environ. 478, 98–102.
Vivian, J.C., 2009. Disposal of controlled substances. U.S. Pharmacist 34 (3), 38–41.
WCRC, 2006. SoundStatsTM research programme.
Wieczorkiewicz, S.M., Kassamali, Z., et al., 2013. Behind closed doors: Medication storage and disposal in the home. Ann. Pharmacother. 47 (4), 482–489.

Detection and presence of pharmaceuticals in the environment

4

4.1 Overview

Trace levels of human pharmaceuticals arising from both their prescribed use (Chapter 2) and the disposal of unused medication (Chapter 3) can be detected in many waste streams such as sewage and landfill leachate from where they can enter the environment. There is an increasing public awareness of the real or potential environmental issues arising from such micropollutants. This has come about in part from the advent of sophisticated analytical techniques in the last few decades that have enabled their detection at increasingly lower levels. In particular, the development of efficient chromatography and sensitive mass spectrometry procedures has facilitated the routine detection of environmental contaminants in the ng/L (approximately 1 part in 10^{12}) range. The data obtained from such routine analyses have given environmental scientists the opportunity for strategic risk assessment and informed management of such contaminants. Certain pharmaceuticals may pose more risks to human and animal health than others and often management decisions are made based on their individual and collective environmental presence, fate, persistence, bioaccumulation, and toxicity. The use of sensitive analytical instrumentation is therefore critical in obtaining experimental data that help environmental scientists and governmental authorities make such informed risk management decisions. However, even with such sophisticated and sensitive instrumentation, the process is rarely straightforward as accurate and precise collection, sample preparation, and analysis procedures that avoid any artifacts must be strictly adhered to before accurate data can be obtained. In this chapter, practical aspects of the sample preparation and analytical techniques necessary for detecting pharmaceuticals at trace concentrations and the challenges faced by analysts to ensure the quality of the analytical data determined for environmental samples are explored. These are illustrated with some typical concentrations that have been reported for selected pharmaceuticals in the environment.

4.2 Environmental detection of pharmaceuticals

The environmental detection of pharmaceuticals (also referred to as the *target analytes* in this chapter) involves the progressive steps of sample collection, sample storage in those cases when samples are not immediately analyzed, sample preparation and extraction, and sample analysis (Figure 4.1). Because environmental concentrations

The Life-Cycle of Pharmaceuticals in the Environment. http://dx.doi.org/10.1016/B978-1-907568-25-1.00004-9

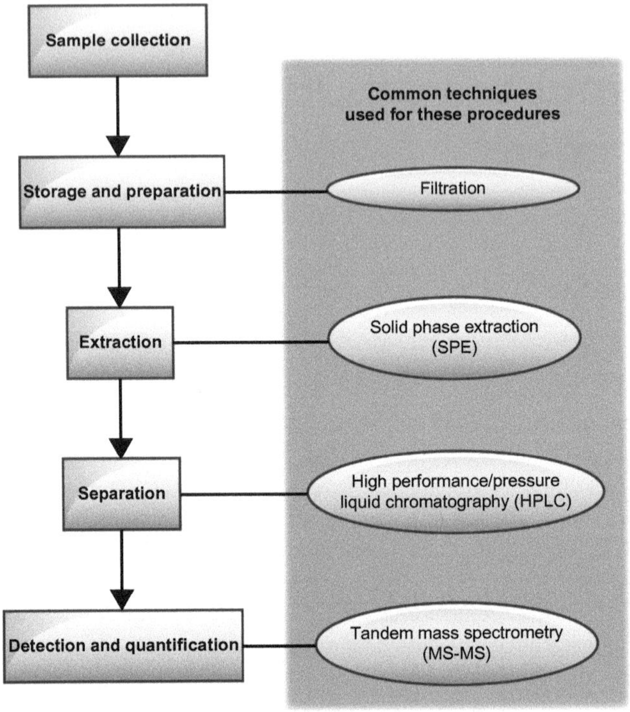

Figure 4.1 Procedures and the associated techniques commonly involved in the preparation and analysis of environmental samples.

of pharmaceutical contaminants are typically extremely low, precautions must be taken during each of these steps to minimize any accidental loss of analyte or the accidental addition of analyte from an external source of sample contamination (Baker and Kasprzyk-Hordern, 2011a, 2011b). For example, caffeine can easily contaminate sample containers due to its ubiquitous presence arising in large part from its widespread use in coffee and some soft drinks. Therefore, an analyst who determines caffeine concentrations in an environmental sample must be methodical in all aspects of cleanliness to prevent false positives and/or artificially elevated levels of this drug arising from sample mishandling (Vanderford et al., 2011).

The initial step in any determination of the concentrations of micropollutants such as pharmaceuticals is to carefully consider all aspects of sample collection. These include the choice of the most appropriate containers for the target analytes and/or the nature of the environmental sample type (e.g., biosolid, sediments, wastewater, surface water). Other decisions often have to be made on methodological approaches in the anticipated field investigations, such as the amount of water, sediment, or soil necessary for an acceptable analysis given the sensitivity of the instrumentation available to the analyst and/or the method and frequency of collecting samples.

4.3 Sample collection

Once decisions have been made on sampling approaches, an analyst should visit the proposed sampling site to determine any limitations or other aspects such as ease of access (*see* discussion in Section 4.3.2). The initial methodological approach may now need to be revised to allow for the realities of the sampling site. For example, there may be limitations on where an analyst can obtain samples in a sewage treatment plant due to health and safety concerns or standard operating procedures of the facility. The proposed sampling containers and other associated equipment such as flow meters (*see* below) may also need to be reconsidered after such a visit.

The sampling container type used by an analyst should be the most appropriate one for the type of sample. Typically, water sampling is undertaken using containers such as polyethylene flagons or glass bottles. Solid sludges can be collected with metal buckets, placed in glass bottles, lyophilized (freeze-dried), and ground to consistent-sized particles typically of 250 μm before storage, extraction, and analyses (Chen et al., 2012; Peysson and Vulliet, 2013). Any container must also be clean enough not to introduce any detectable analyte in the subsequent analysis.

4.3.1 To silanize or not to silanize–that is the sample container

The most common sampling container for environmental water samples is made from amber glass in order to minimize any light-induced (photochemical) reactions that might occur during storage of the sample between collection and analysis (Table 4.1). Glass can be presilanized before use to minimize loss of basic pharmaceuticals, such as fluoxetine, which adsorb to the free hydroxyl silicate surface of the glass (Vanderford et al., 2003). Silanization involves rinsing the glassware with a 10% (v/v) solution of dimethyldichlorosilane in toluene followed by flushing with pure toluene and methanol and then drying the glassware in an oven for 3 h at 160 °C (Ahrer et al., 2001).

Other commonly used types of containers for water samples include those constructed from high-density polyethylene (HDPE) or polyethylene terephthalate (PET) (Conley et al., 2008; Wu et al., 2008; Liu et al., 2009). Steel containers have also been used for sampling wastewater (Lajeunesse et al., 2008).

The choice of sampling container is also dependent on the target analyte of interest, given that pharmaceuticals vary immensely in physicochemical properties and may therefore not be possible to specify a single "best" container material for sampling all of the different possible target analytes. Vanderford et al. (2011) tested the recovery efficiencies of 16 pharmaceuticals and personal care products and five steroidal compounds after spiking these analytes into 950 mL deionized water samples contained in different types of storage vessels: amber glass, silanized amber glass, clear glass, and high-density polyethylene (HDPE) bottles. After storage of the water samples for two weeks at 4 °C and storage of the resulting extracts at –20 °C, amber glass bottles were found to be the most acceptable. Interestingly, more trimethoprim was reported in water stored in HDPE bottles than in glass containers of any type, suggesting that a contaminant from the HDPE bottles had potentially leached out into the sample and interfered with the instrumental analysis of trimethoprim. In addition, the signal intensity arising

Table 4.1 **Sample bottle types used for the collection of water samples in some field investigations of pharmaceutical analytes**

Target analyte(s)	Water type	Sample bottle type	References
Acidic/neutral drugs	Wastewater, surface (river) water	Solvent-washed glass	Metcalfe et al. (2003)
Antibiotics Macrolide	Wastewater, river water	Grab samplers	Yang and Carlson (2004)
Antidepressants	Wastewater	Stainless steel container	Lajeunesse et al. (2008)
Chloramphenicol, tetracyclines	Wastewater, river water	Polyethylene	Liu et al. (2009)
Fluoroquinolones, macrolides, sulphonamides, trimethoprim	River water, sea water	Amber glass	Zhang et al. (2012)
Gemfibrozil	Groundwater, wastewater	Amber glass	Fang et al. (2012)
Glucocorticoid activity[a]	Surface water	Amber glass	Schriks et al. (2013)
Ibuprofen, Iopromide	Wastewater, river water	Amber glass	Wang et al. (2013)
Multi-residue	Lake water, stream water	HDPE bottles	Wu et al. (2008)
	Surface water	Amber HDPE bottles	Conley et al. (2008)
	Wastewater	Glass bottles	Okuda et al. (2008)
	Wastewater, seawater	Amber glass	Togola and Budzinski (2008)
	River water	Amber glass	Madureira et al. (2009)
	Lake water	Amber glass	Morasch et al. (2010)
	Wastewater	Amber glass	Sim et al. (2011)
	Reservoir water, drinking water	Amber glass	Kleywegt et al. (2011)
	River water	Amber glass	Pojana et al. (2011)
	Wastewater	Autosampler (refrigerated)	Rua-Gomez and Puttmann (2012)
	Wastewater, surfacewater, groundwater	Amber glass	Tran et al. (2013)
	Groundwater	Amber glass	Van Stempvoort et al. (2013)

[a] Did not detect specific glucocorticoids.

from the detection of sulfamethoxazole was substantially lower in water samples stored in HDPE containers compared to glass bottles. This result suggested the possibility of *ion suppression* (*see* discussion later in this chapter) by a contaminant originating from the container. The authors also detected diclofenac in a blank sample stored in one of their HDPE bottles and this may have arisen from the cross-transfer of diclofenac from another sample into the blank via adsorption onto HDPE. From all these results, Vanderford et al. (2011) recommended that HDPE bottles should be avoided for sample storage purposes when the target analytes include trimethoprim, sulfamethoxazole, or diclofenac but may be perfectly acceptable for many other target analytes.

Analyte losses were also observed by Suri et al. (2012) during storage of estrogen solutions in low-density polyethylene containers. However, some compounds, such as 17α-estradiol, 17β-estradiol, 17α-dihydroequilin, and levonorgestrel, exhibited higher losses in a silanized glass bottle than in a HDPE bottle. Such adsorption might have been anticipated from a physicochemical point of view because most estrogens are highly lipophilic (i.e., oil-soluble and have a log $K_{ow} > 4$, where K_{ow} is the octanol-water partition coefficient parameter) (Mompelat et al., 2013). Analyte loss resulting from sample storage in a scratched glass bottle after repetitive use has also been reported, but was lessened by pre-silanization (Suri et al., 2012).

The effect of glassware silanization has been reported in various other studies. Baker and Kasprzyk-Hordern (2011a, 2011b) investigated the effect of glassware silanization on the recovery of drugs of abuse from water samples. Silanization was observed to be particularly important for efficient recovery of amine-group drugs as they adsorb to the free silanol groups present on unsilanized glass surfaces. For example, the recovery of amphetamine from methanol was 16% after sample storage and extraction in nonsilanized glassware but improved to 88% after silanization. However, silanization did not significantly improve the recovery of heroin, 6-acetylmorphine, buprenorphine, and fentanyl (Baker and Kasprzyk-Hordern, 2011a, 2011b). Furthermore, silanization generally did not have any effect on the recovery of benzodiazepines (Baker and Kasprzyk-Hordern, 2011a, 2011b). All these results suggest that silanization is generally beneficial to analyte recovery (Ahrer and Buchberger, 2001) even though there are no significant improvements in recovery reported for at least some analytes.

Despite the apparent benefits of silanization of glass container surfaces, Vanderford et al. (2011) found that progesterone was more efficiently recovered from nonsilanized amber glass than silanized amber glass containers by 10–20%. However, a more recent study supports container silanization in some sex hormone analyses. Suri et al. (2012) studied the reproducibility of data arising from storage of 50 µg L^{-1} aqueous Milli-Q solutions of a range of estrogens in plain glass, HDPE, low-density polyethylene, polypropylene, light polypropylene, and silanized glass containers all stored in the dark for one week at 4 °C. For the range of estrogens analyzed, the best reproducibility was observed for silanized glass containers followed by polypropylene and plain glass containers while low-density polyethylene and HDPE containers exhibited the poorest analytical performance. The huge variability observed in the results observed when using containers made of plastic materials may be due to possible adsorption and/or leaching of estrogens from the insides of the containers. However, whether to silanize glass or not for estrogen analysis remains equivocal.

Because silanization does not seem to *necessarily* improve recovery of analytes in stored solutions, the US EPA does not recommend silanization of glassware in all situations

(*Stability of Pharmaceuticals, Personal Care Products, Steroids, and Hormones in Aqueous Samples, POTW Effluents, and Biosolids*, available http://water.epa.gov/scitech/methods/cwa/upload/methodsppcp.pdf, accessed 20-11-13). However, it does endorse the use of silanized glass containers in sample storage for the analysis of some pharmaceuticals, such as sulfonamides and quinolone antibiotics. In addition, water samples intended for the analysis of macrolide antibiotics should not be placed in HDPE containers, as the macrolide antibiotic azithromycin was observed to be completely lost after sample storage in this type of container. Tetracyclines can also be stored in plain (unsilanized) glass and HDPE containers for up to seven days; thus for many pharmaceuticals classes, the use of unsilanized glass in containers would seem to suffice (Mompelat et al., 2013).

HDPE bottles may be acceptable for storage of sewage water, but not drinking water. Analyte concentrations in sewage water are expected to be typically higher than those present in drinking water. Even minor changes in concentration due to adsorption to HDPE bottles containing drinking water samples may mean the difference between detectable and nondetectable levels of some analytes (*Stability of Pharmaceuticals, Personal Care Products, Steroids, and Hormones in Aqueous Samples, POTW Effluents, and Biosolids*, available http://water.epa.gov/scitech/methods/cwa/upload/methodsppcp.pdf, accessed 20-11-13).

In summary, before a type of sample container is chosen, it is important to first determine the analytes of interest and then consider the type of environmental sample in which they are to be collected and analysed.

4.3.2 Modes and frequency of sampling

The manner and frequency with which samples are collected can have a profound effect on the quality of the subsequent analytical results and the development of valid conclusions (Ort et al., 2010). For water samples, the simplest type of sampling involves obtaining *grab samples* in which single, discrete samples of water are taken from a study site irrespective of the ambient environmental conditions at that time. However, the level of the target analyte in a water body and the ambient environmental conditions may vary over time. To better allow for this variation and collect data that is more representative of real concentrations, a number of grab samples can be pooled over a 24-h period to give a *composite sample* for which the observed concentration is an average over this time interval. Such a technique is often used in sampling at sewage treatment facilities to normalize for fluctuating instantaneous levels of the target analytes arising from diurnal human excretory patterns (Castiglioni et al., 2006).

A *methodical* sample collection procedure takes into account the time of day when larger amounts of an analyte are likely to be present in the environmental sample. In the case of sewage, pharmaceuticals excreted in urine and feces will generally appear in pulses in the sewage over a 24-h period given that humans are more likely to excrete at certain times, such as late morning rather than in the small hours of the night. In addition, weather conditions may drastically affect the results obtained. For example, rain can indirectly increase sewage flow volumes through shared sewage and stormwater pipes and hence dilute the concentrations of any pharmaceuticals in the sewage. Without being aware of this situation, an analyst may underestimate the concentration of a target analyte in a sample of sewage collected under these conditions unless appropriate corrections are made (Ort et al., 2010). A sampling campaign that is responsive to such ambient changes and collects samples at a higher frequency is likely to yield a more accurate result and minimize the probability that a "pulse" of target analyte in the sewage will be missed (Ort et al., 2010). Knowledge of the *hydraulic retention time*, which is the time that a target analyte remains in a particular sewage treatment facility, is also necessary for successful sampling to maximize the possibility of collecting relevant pulses of sewage (Ort et al., 2010).

The analyst or investigator also has to keep in mind the overall goals of the study when considering the mode and frequency of sampling (Ort et al., 2010). Grab samples are considered adequate in many instances for method development or qualitatively demonstrating the presence or absence of a target analyte, and to assess the applicability of a given instrumental method (*see* below) to detect specific analytes in a particular type of environmental sample. However, the absence of an analyte in a single grab sample of wastewater does not necessarily mean it is not present at other times of the day or during another season. Many medications used to treat specific seasonal ailments, such as oseltamivir phosphate for influenza in winter and cetirizine hydrochloride for hay fever in springtime, are more likely to be detected in sewage at these times of the year. This seasonal dependence is clearly illustrated by the reported increases in caffeine concentration detected in Lake Simcoe (Ontario, Canada) water in summer due to tourism in the area at this time of the year (Kurissery et al., 2012).

Seasonal variations to environmental conditions and weather patterns are also important considerations in surface water sampling (Allan et al., 2006). The amount of rainfall and the location of sewage discharge outflows relative to the sampling site can affect the measured target analyte concentrations (Baker and Kasprzyk-Hordern, 2013). These are likely to be much higher downstream than upstream from the point of discharge into a river of the effluent from a sewage treatment facility. However, the cumulative concentration increase downstream of a river is counterbalanced by dilution from the extra water introduced into the river by the sewage treatment facility. Increases in river volume arising from the discharge of tributaries located between the discharge point and the sampling site can also result in artificially low estimates of analyte concentrations at the discharge point.

In summary, there are many factors which can influence the level of a given analyte that is detected in a water sample collected under specific spatial and temporal conditions. These factors should be taken into account as much as possible in designing the mode and frequency of sampling in order that the observed analyte concentration is as representative as possible of the true concentration.

4.3.3 Sample storage conditions and preservation techniques

Often, samples collected in the field cannot be immediately analyzed. To minimize analyte decomposition during storage, a number of preservation techniques can be employed including (Vanderford et al., 2011):

(1) The addition of chemical reagents, such as preservatives or quenchers, to prevent microbial growth or reduce analyte degradation through reaction with other chemical species present in the sample such as chlorine in treated drinking water and sewage. Antioxidants commonly used to reduce free chlorine levels in water samples include ascorbic acid, sodium thiosulfate, and sodium sulfite.

(2) The use of appropriate light-proof storage containers such as amber glass bottles to minimize any light-induced (photochemical) reaction of the analyte.
(3) The optimum sample storage temperature to minimise biologically- or chemically-induced degradation of the target analyte(s).

Liquid samples can be stored at 4 °C for up to 72 h before extraction, without the addition of any preservatives (Vanderford et al., 2011). However, if sodium azide (1 g/L) and ascorbic acid (50 mg/L) are used as antimicrobial and quenching agents, respectively, the samples can be satisfactorily stored at 4 °C for up to 28 days before extraction with less than 15% loss in analyte in these circumstances (Vanderford et al., 2011).

If samples are to be extracted for analysis within a short time after collection (i.e., less than three days), it may not be necessary to freeze samples. Chiaia et al. (2008) reported no changes to the physicochemical stability of several illicit drugs and their metabolites in wastewater samples refrigerated for 24 h at 4 °C and then frozen at −20 °C for three weeks compared to those that were immediately frozen to −20 °C after collection. In addition, the freezing of samples can also have drawbacks. For example, the defrosting of water samples has been observed to produce insoluble aggregates which have the potential to adsorb target analytes (Vanderford et al., 2011).

Microbial activity leading to analyte degradation during sample storage should also be avoided. Acidification of a sample to $pH \leq 2$ by the addition of hydrochloric acid has been reported to be effective in reducing microbial degradation of many target analytes (Vanderford et al., 2003).

It is clear from the above discussion that there is considerable variability in the optimal sample storage conditions to minimize analyte degradation during storage. A general guide would be that when analysis is undertaken within three days of collection, water samples can be left refrigerated at 4 °C but for any longer periods they must be frozen to avoid any analyte degradation before analysis. For a recent and more detailed review of optimal storage conditions for analysis of a wide range of individual pharmaceuticals measured in natural water samples, see Mompelat et al. (2013).

4.3.4 *Sample filtration*

A natural water sample such as sewerage invariably contains a significant quantity of particulate matter that can interfere with the extraction and subsequent analysis of pharmaceuticals. This particulate matter can be removed by filtration of the water samples after collection and before any storage, using glass fiber or nylon membrane filter papers of ≤1 μm pore size (Camacho-Munoz et al., 2009; Sturini et al., 2009; Morasch et al., 2010a, 2010b; Yu and Wu, 2011). However, the type of filter paper can be critical for some types of pharmaceuticals. For example, hydrogen bonding between the hydroxyl groups in estrone and cellulose has been observed to lower the recovery of this analyte and instead, glass fiber filters are recommended (Liz et al., 2012). For a detailed discussion of filtration materials, the reader should consult Heath et al. (2010).

4.4 Sample extraction

Even after utilizing the optimal collection, filtration, and storage techniques, the levels of most pharmaceuticals in a natural water sample are generally still below the minimum detection levels of most readily available analytical instrumentation. To be able to detect a signal, it is therefore necessary to increase the concentrations of the analytes after collection of the samples and before separation and detection. This is typically achieved by extracting the analytes from the original volume of water sample using either solid-phase extraction (SPE) or liquid–liquid extraction (LLE) techniques and then redissolving the extracted material in a much smaller volume of a solvent for subsequent separation and instrumental analysis. At the same time as increasing the concentrations of the target analytes, other chemical species present in the original sample that could interfere with the separation and/or analysis can often be removed in this step.

4.4.1 Solid-phase extraction

In SPE, the natural water sample (the *mobile* phase) is passed through a cartridge containing a specialized solid material (the *solid* phase) to which the different analytes dissolved or suspended in the water sample bond in a selective manner depending on their physical and chemical properties. The target analytes are then removed from the solid phase by eluting with small volumes of solvents and the resulting solutions can be further concentrated by progressive evaporation before subsequent instrumental analysis.

A wide variety of SPE cartridges with different types of solid phases are commercially available and allow the analyst to be selective in what analytes are separated from the original water sample. This selectivity arises from varying the functional groups that are built onto the solid support of the cartridge. *Normal-phase* cartridges have exposed silica or silanol groups that attract polar compounds which can be eluted using a more polar solvent. *Reversed-phase* cartridges have hydrocarbon endcapped silanol groups that offer enhanced attraction for nonpolar or hydrophobic compounds which are eluted with nonpolar solvents. *Anion* and *cation* exchange cartridges have positively- and negatively-charged functional groups to attract analytes present in the sample as anions and cations, respectively, depending on the pH.

Mixed-mode polymeric cartridges are also commonly used such as the Oasis® Hydrophilic-Lipophilic Balance (HLB™), the Oasis® Mixed-mode Anionic (MAX™), and Cationic (MCX™) exchange cartridges. These mixed-mode cartridges improve extraction selectivity for certain target analytes compared to normal- and reversed-phase cartridges. For example, the Oasis® MAX™ cartridge is commonly used for the extraction of NSAIDs as many of the representative compounds in this class have a carboxylic acid functional group and are negatively charged at pH=7 (Gibson et al., 2007).

In a typical SPE procedure, the extraction cartridge is mounted in a vacuum manifold and initially conditioned using a small volume of solvent (typically water) before the water sample to be analyzed is passed through the cartridge. Subsequently, different solvents are used to wash the cartridge and selectively remove any contaminants from the solid support. Finally, one or more solvent(s) are used to elute the remaining target analytes from the cartridge. The US EPA's *Method 1694* (available http://water.epa.gov/scitech/methods/cwa/bioindicators/upload/2008_01_03_methods_method_1694.pdf, accessed 2-1-14) describes some standardized extraction procedures for the choice of extraction solvents and the analysis of particular pharmaceuticals in environmental samples.

Before establishing an extraction protocol, an analyst will usually evaluate the quantity of target analyte that is successfully obtained from different types of SPE cartridges, given a standardized extraction method and a sample spiked with a known amount of the analyte. This enables the percentage of *recovery efficiency* to be calculated and usually the preferred cartridge is the one exhibiting the highest value for this parameter. However the best cartridge for the extraction of two similar target analytes may be different. For example, atorvastatin was best recovered when extracted with Oasis HLB cartridges relative to Bond Elute C8 and DSC-18 cartridges (≤ 50%); however, lovastatin and pravastatin were recovered better when extracted with DSC-18 cartridges (90%) (Miao and Metcalfe, 2003). In contrast, a Bond Elute C8 cartridge was considered the better choice for the extraction of simvastatin. Overall, HLB cartridges were preferred over the other two because the extracts achieved using this type of cartridge did not suffer from as much ion suppression during subsequent LC-MS/MS analysis as extracts from the other two cartridges (*see* below). Kolpin et al. (2002) and Camacho-Munoz et al. (2009) reported a poor recovery of acidic pharmaceuticals, sex steroids, sulfamethoxazole, trimethoprim, caffeine, and carbamazepine using Oasis MCX cartridges even in acidified water samples, but recoveries were generally improved using Oasis HLB cartridges.

The variable recoveries of different analytes achieved using different types of SPE cartridges has led to the use of more than one type of cartridge in series with each other, in an attempt to maximize the recovery of more than one analyte. For example, the recovery of several sulfonamides and the recovery of basic analytes was improved by using two Oasis HLB cartridges in series (Na et al., 2011). However, the use of more than one cartridge per sample also increases the cost of analysis.

The large number of different types of SPE cartridges available commercially can make it difficult to select just a single one which gives the highest recoveries across a range of target analytes. This selection process can be helped to some extent by considering the physicochemical interactions between the analytes and the nature of the solid-phase material in a cartridge (Bäuerlein et al., 2011). The adsorption of a target analyte onto a cartridge can involve van der Waals forces, coulombic (ionic) bonding, π-π interactions, and hydrogen bonding. In selecting an appropriate cartridge, it is therefore helpful to predict the likely relative importance of these interactions between a given analyte and the cartridge. For example, HLB cartridges are based on an unmodified backbone of nonpolar (aliphatic chains and benzyl groups) and polar groups (pyrrolidone), while the other mixed-mode Oasis cartridges consist of the same polymers but with additional functional groups. MAX cartridges have additional quaternary ammonium groups, while MCX cartridges have sulfonate groups. These two cartridges are positively- and negatively-charged irrespective of the pH of the loaded sample. On the other hand, Oasis weak anionic exchange (WAX) and weak cationic exchange (WCX) cartridges have additional piperazine and carboxylic acid groups, respectively. The ionization of these groups is pH-dependent and thus the pH of the sample determines whether ionic interactions, hydrogen bonding, or polar interactions predominate.

However, this type of 'theoretical' exercise is no substitute for actual measurements of recovery efficiencies for target analytes in a given sample using a particular type of cartridge, as analyte losses inevitably occur even with the most appropriate extraction cartridges. Furthermore, it may be virtually impossible for the optimal recovery of each target analyte in a high throughput (i.e., the ability to analyze many samples in a short timeframe, *see* chromatography below), multiresidue screening of pharmaceutical contaminants. Given that the physicochemical properties of pharmaceuticals can vary significantly even within one class of pharmaceuticals (e.g., among NSAIDs such as diclofenac, ibuprofen), any extraction methodology using a single type of SEP cartridge will often be a compromise so that all target analytes have satisfactory if not optimal recoveries. In addition, target analyte recovery efficiencies may differ between various environmental samples (e.g. sewage water vs. surface water) and may be dependent on the initial sample cleanliness or the amount of pre-extraction clean-up performed on the sample. For accurate and meaningful results, an analyst must follow an extraction protocol consistently in order to maintain consistent recovery efficiencies.

4.4.2 Liquid–liquid extraction

Liquid–liquid extraction (LLE) is a more traditional method of separating analytes in a sample based on their variable solubility in two different liquid solvents. Although this separation technique is commonly used for the isolation of specific compounds after a chemical synthesis, it is less commonly used for environmental analyses as large volumes of solvent are needed and the recoveries of target analytes are often low (Primel et al., 2012).

4.5 Chromatographic separation of environmental sample extracts

Once an appropriate extraction procedure has been performed on an environmental sample, the next step is to analyze the extract to identify and quantify the target analytes. However, before these identification and quantitation steps, the analyst has to separate the various chemical components in the concentrated sample extract. This is usually carried out using some form of chromatography with the most common being high performance (pressure) liquid chromatography (HPLC). The basis of this separation method is similar to that of SPE in that a small volume of the concentrated sample extract is injected into a liquid *mobile phase* which is then flowed under high pressure (50–350 bar) conditions through a solid support column of silica or modified silica (the *stationary phase*) made of a similar material as the solid supports in SPE cartridges. The target analytes are selectively adsorbed onto the column depending on their physicochemical properties. Individual types of target analytes are retained on the stationary phase for a characteristic period of time (the *retention time*) dependent on their physicochemical properties, the nature of the stationary phase, and the flow conditions before they are selectively eluted back into the mobile phase. In this way, the analytes are separated out in a time-dependent manner in the mobile phase exiting the column. Further information about HPLC separation can be found in texts such as McMaster (2005) and Jandera (2012).

Various types of stationary phase (column) materials are selective for different analytes. For example, ion exchange columns separate analytes based on their charge, whereas silica columns and reversed-phase silica columns separate analytes primarily on their polarities. The columns most commonly used for environmental analyses are reversed-phase C18 and C12, which have modified silanol groups that have been endcapped with nonpolar hydrocarbon groups. As a result, nonpolar induced dipole-induced dipole forces are primarily responsible for adsorption of target analytes onto these columns. Typically, a polar mobile phase (such as water) is used as the mobile phase with nonpolar C18 and C12 stationary phases.

An organic (less polar than water) solvent such as methanol and acetonitrile which is miscible with water and known as a *modifier* can also be added to the water mobile phase in various proportions to decrease the retention time of the target analyte. In addition, the eluent (mobile phase) can be used in either *isocratic* or *gradient* modes of flow. Under isocratic flow, the mobile phase retains a constant chemical composition throughout the chromatographic run, whereas under gradient flow conditions, two (or more) different solvents often with different polarities, such as water and acetonitrile, are mixed in various proportions in a programmed, time-dependent manner over the period of chromatography. In environmental analyses, gradient mobile phases are often employed to increase the difference between the retention times of two (or more) target analytes with similar physicochemical properties.

A common problem in confidently identifying individual components occurs when two analytes have similar physicochemical properties and the mobile phase, even using a gradient flow, fails to achieve different retention times for each component (McMaster, 2005). Given the complex nature of environmental samples, the failure to achieve different retention times may make subsequent identification and quantification of the target analytes difficult. To help ameliorate these problems, the US EPA's *Method 1694* for the analysis of pharmaceuticals and personal care products (available http://water.epa.gov/scitech/methods/cwa/bioindicators/upload/2008_01_03_methods_method_1694.pdf, accessed 2-1-14) provides some guidance on appropriate mobile-phase compositions that can be used for the separation of selected target analytes and details of their expected retention times.

A modified version of HPLC, known as UPLC or UHPLC (ultra performance liquid chromatography or ultra-high performance liquid chromatography), has recently gained popularity with environmental analysts (Batt et al., 2008). This is because UPLC/UHPLC columns are shorter (5 cm) in physical length than the conventional HPLC columns (typically 20–30 cm) and the higher pressures applied in UHPLC can separate target analytes faster leading to shorter analyte retention times (Baranowska and Kowalski, 2010; Shelver et al., 2010; Yu et al., 2012). In addition, although the capital cost of UHPLC equipment is currently higher than that used for conventional HPLC, it is more cost-effective during its operation as the shorter chromatographic run times lead to less solvent use for the mobile phase and faster sample throughput (Anumol et al., 2013).

4.6 Mass spectrometric detection and quantitation

Once target analytes in environmental samples have been separated by HPLC (or UHPLC) into individual components as a function of their retention times, it is necessary to identify and quantify them. This is typically undertaken using the

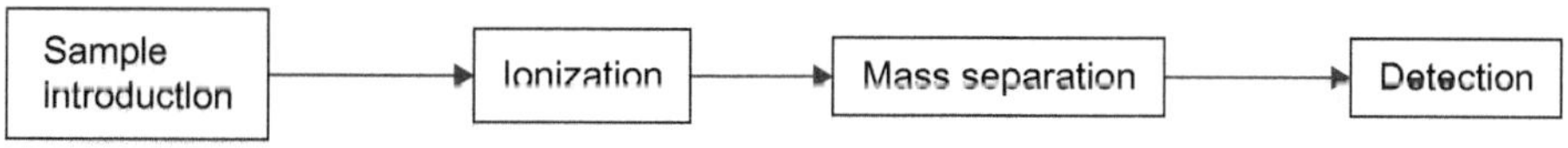

Figure 4.2 Basic sequence of events in a mass spectrometer.

technique of mass spectrometry (MS) (*see* Figure 4.2) in which the analyte species are fragmented and ionized and then passed through a mass filter before their detection as a function of their mass-to-charge (*m/z*) ratios.[1]

The main components of a mass spectrometer (Figure 4.3) perform a sequence of events under computer control:

(1) vaporization of the sample species for entry to the ion source where ionization and fragmentation occur;
(2) acceleration of the mixture of ions into the mass analyzer where they are separated either "in time" or "in space" using a variable electromagnetic field;
(3) detection and quantification of the abundance of resolved ions based on their *m/z* ratios.

A number of these components are under vacuum in order to minimize the loss of ions through collisions with neutral gas molecules and with the walls of the ion source, mass analyzer, and detector.

The first step involving sample ionization is required because almost all pharmaceutical analytes in the HPLC output stream are present as neutral molecules but to then separate them into different molar weights using a variable electric field in the mass analyser, requires the analyte molecules to be charged, i.e., to exist as positive or negative ions. There is a wide variety of sample ionization methods that have been used (see Section 2.2 of Greaves and Roboz (2014)) but the most common for pharmaceutical

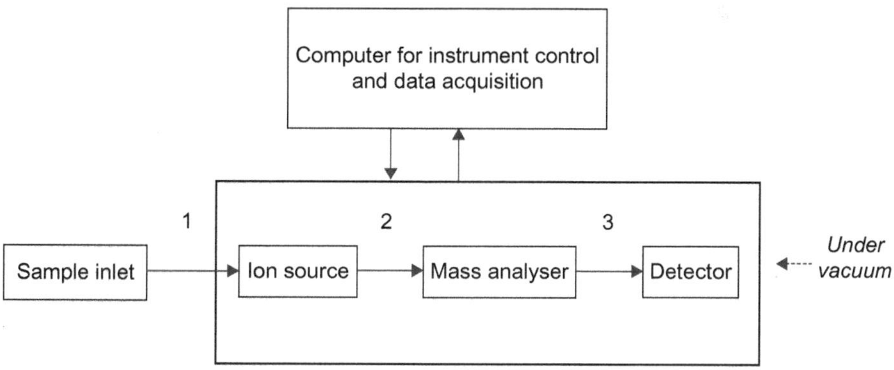

Figure 4.3 Block diagram of a mass spectrometer.

[1] The mass-to-charge (*m/z*) ratio is determined by the molecular mass of a fragment divided by its charge. For example, a fragment with a mass of 200 atomic mass units (amu) and a charge of +2 will have an *m/z* value of 100.

analysis is *electrospray ionization* (ESI) in which a liquid sample is subjected to a high voltage (2–5 kV) of either +ve or –ve polarity leading to the respective formation of positive ions of the form $[M+H]^+$, i.e., M+ proton, or negative ions of the form $[M–M]^-$, i.e., loss of a proton. ESI is considered a "soft" ionization procedure in comparison with other ionization in that it leads to the least decomposition of the parent neutral molecules and it is also very efficient. The choice of which ESI mode is used (i.e., ESI+ or ESI–) depends in part on the chemical structure of the target analyte and their relative ionization efficiencies. Many pharmaceuticals have polar groups such as carboxyl or amine groups which lend themselves more to the formation of anions by the loss of a proton and hence the use of the ESI-mode but this is by no means universal. This point is illustrated in the examples of different chemical structures for the common pharmaceutical target analytes listed in Table 4.2.

A common type of mass analyzer involves a *quadrupole* in which four rods arranged precisely in a square configuration are subjected to combinations of rf and dc voltages to create specific electric fields that allow transmission only of ions with a specific *m*/*z* to the detector. The results are recorded as a *mass spectrum* which is typically a plot of ion abundance as a function of *m*/*z* by changing the rod voltages sequentially over time. Alternatively, the abundance of an ion with just a single *m*/*z* value can be measured using a specific set of rod voltages.

For environmental analysis, a mass spectrometer is typically combined with an HPLC system so that the input sample to the MS consists of the target analytes eluted from the HPLC column as a function of time. This analytical process is known as LC–MS (*liquid chromatography–mass spectrometry*), and the essential components of a typical LC-MS instrument are illustrated in Figure 4.4.

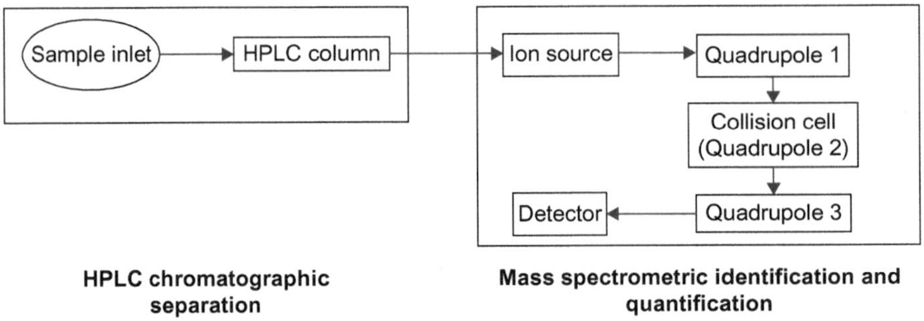

Figure 4.4 Basic components of a triple quadrupole (QqQ) LC–MS/MS system used for environmental analysis.
Copyright (2014) From Mass Spectrometry for the Novice by J. Greaves and J. Roboz. Reproduced by permission of Taylor and Francis Group, LLC, a division of Informa plc.

This particular configuration is called a *triple quadrupole* LC-MS/MS instrument in which the initial mixture of primary fragment ions formed in the ion source is separated in Q1 and then each *one* of these ion types is further fragmented in a collision cell (q2) to form smaller secondary product ions by collision of the precursor ion with an inert gas typically in a quadrupole electric field (*collision-induced dissociation* (CID)). These secondary product ions arising from just a single type of primary fragment ion are then separated in Q3 before their detection in the final step.

Table 4.2 **LC-MS/MS instrumental and method parameters used for the quantitation of some common pharmaceuticals**

Pharmaceutical target analyte	Paracetamol	Diclofenac	Metoprolol	17α-Ethinylestradiol	Simvastatin
Therapeutic class	Analgesic	NSAI	β-blocker	Oral contraceptive	Lipid control
Chemical structure	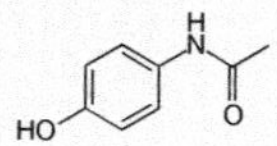	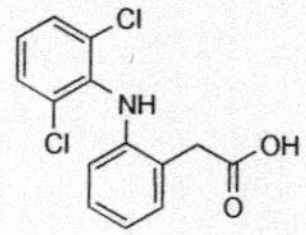	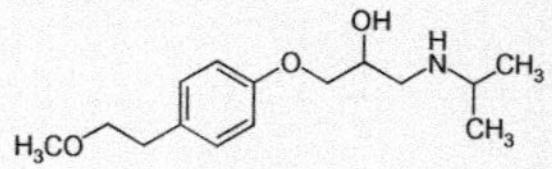	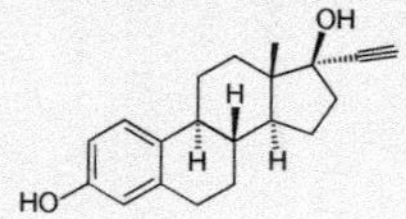	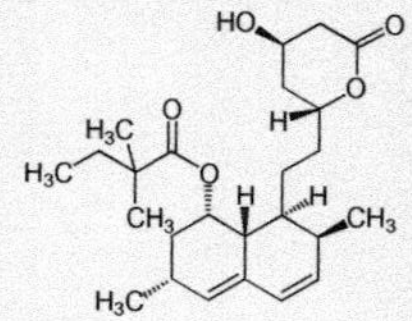
Molar mass	151.1	296.1	267.4	296.4	418.5
ESI mode	+	–	+	–	+
Primary MRM transition	152.0>110.0	294.0>250.1	268.2>116.3	295.1>145.1	419.5>199.3
Instrument detection limit (IDL) (ng/L)[a]	0.5	0.1	0.05	0.1	1
Instrument quantification limit (IQL) (ng/L)[a]	1	1	0.5	5	10
Method detection limit (MDL) (ng/L)[a]					
River water	9	10	67	32	140
STP effluent	31	33	106	22	131
Recovery % (RSD%)[a]					
River water	18 (15)	68 (5)	71 (4)	59 (9)	48 (13)
STP effluent	39 (10)	119 (12)	87 (5)	61 (4)	56 (12)
Matrix effect (ME %)[a] (absolute/relative)					
River water	11.1/–4.6	–235.6/–26.7	–31.3/–16.2	–37.4/–28.1	–67.6/35.2
STP effluent	56.5/23.4	–158.2/34.5	38.5/–19.4	2.6/–19.2	46.5/–26.6

[a] Data from Ali-Odaini et al. (2010).

A particular advantage of using a triple quadrupole (QqQ) MS for environmental analyses such as the detection and quantification of a range of different pharmaceuticals in a natural water sample, arises from an improvement in the signal-to-noise (S/N) ratios and hence lower detection limits compared to using a more conventional single quad LC-MS instrument operating in *single ion monitoring* (SIM) mode. This situation arises from the fact that ions emerging from Q1 can include not only those from ionization of the target analyte but also ions with the same *m/z* but arising from coeluting species present in the HPLC output stream. In spite of using selective extraction processes such as SPE (*see* Section 4.4.1), it is also possible to coextract other compounds such as solvent contaminants. However, by subjecting the ions with the same *m/z* to further fragmentation in a collision cell (q2), it is very unlikely that these coeluting interfering species will give the same *secondary* fragment ions as those *parent* ions derived from the target analyte. Hence, after further separation of these secondary ions in Q3, the resulting mass spectrum will exhibit less noise and hence a higher S/N ratio leading to a lower limit of detection. This *selected reaction monitoring* (SRM) mode of mass spectrometric measurement is illustrated schematically in Figure 4.5 for a primary fragment ion (m_1/z) being further fragmented into a secondary fragment ion (m_2/z).

As well as improving the selectivity and sensitivity, MS/MS minimizes the probability of observing false positives because the instrument can be set to detect only

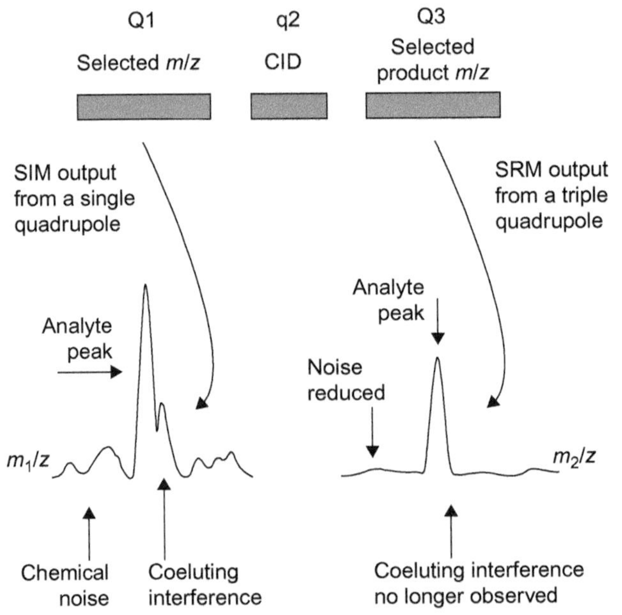

Figure 4.5 Comparison of mass spectra measured using single ion monitoring (SIM) and selected reaction monitoring (SRM) modes with (QqQ) LC-MS/MS instrumentation. Copyright (2014) From Mass Spectrometry for the Novice by J. Greaves and J. Roboz. Reproduced by permission of Taylor and Francis Group, LLC, a division of Informa plc.

a specific combination of m_1/z and m_2/z known as a *transition* (e.g., the transition 294.0 > 250.1 has been identified by Al-Odaini et al. (2010) as suitable to monitor the levels of diclofenac in sewage samples). Although the observation of a single transition, where the identity of a unique precursor and daughter ion is known, is usually adequate to prevent the observation of false positives, they can still sometimes occur (Segura et al., 2007). For example, a false positive result from the use of only one transition for the detection of 17α-ethinylestradiol in sewage influent has been reported by Schlusener and Bester (2005). To avoid this possibility, analysts often employ *multiple reaction monitoring* (MRM) whereby the MS/MS instrument is set to observe additional transitions (e.g., 294 > 178 as well as 294 > 250 for diclofenac (Al-Odaini et al., 2010)) to confirm the presence of a target analyte in a sample. If it is not possible to monitor such additional transitions, other identifying features of the target analyte signal, such as the HPLC retention time, can also be used to confirm identification of the analyte (Nebot et al., 2007). Some of the major transitions used for MRM in the LC-MS/MS analysis of four commonly analyzed pharmaceuticals are listed in Table 4.2.

4.6.1 Forms of output from LC–MS/MS

The output from a mass spectrometric detector is usually in the form of a *mass spectrum* consisting of an ion peak intensity versus a wide range of *m/z* values. However, when combined with the initial chromatographic separation step in LC–MS/MS, mass spectra can be measured for each of the peaks in the chromatographic trace and this gives rise to a time-dependent output in the form of different types of *mass chromtogram*. The *x*-axis is now the chromatographic (retention) time rather than the *m/z* value of the detected ion(s) while the *y*-axis can vary (see Murray et al., 2013):

(1) A *total ion current* (TIC) chromatogram where the value plotted on the *y*-axis is the sum of all the separate ion currents measured by the detector for all the different ions detected over a selected range of *m/z* values.
(2) An *extracted-ion* (EI or XI) chromatogram where the *y*-axis is now the peak intensity of the signal observed for a *selected m/z* value as extracted from the mass spectra data recorded for the entire mass range.
(3) A *selected-ion monitoring* (SIM) chromatogram in which the mass spectrometer is operated in the SIM mode and the intensity plotted on the *y*-axis now refers to just the signal detected for a specific *m/z* value corresponding to the precursor or daughter ions of a particular analyte (within an adjustable mass tolerance typically of one mass unit). An SIM chromatogram differs from an XI chromatogram in that in the SIM mode only data for specific ions are collected compared to the collection of data for the entire mass range and subsequent analysis for specific ions (EI).
(4) A *selected-reaction monitoring* (SRM) chromatogram in which the mass spectrometer is operated in the SRM mode and the intensity plotted on the *y*-axis refers to the signal detected for a specific combination (transition) of precursor and product ion for a particular analyte (again within an adjustable mass certain tolerance).
(5) A *multiple-reaction monitoring* (MRM) chromatogram is also derived from operation in the SRM mode but now contains signals from *multiple* transitions involving *product ions* arising from one or more *precursor ions*.

TICs are of limited used in the analysis of a mixture of pharmaceuticals that typically occurs in an environmental sample (i.e., *multi-residue* analysis), because the single peak intensity cannot be resolved into individual signals from different analytes that elute from the initial chromatographic step with the same retention time. Instead EI, SIM, and SRM chromatograms are the common forms of LC-MS/MS output used to both identify and quantify a range of analytes in a multi-residue pharmaceutical analysis.

4.6.2 Identification and quantification of an analyte

Both these steps in the overall analytical procedure require the use of reference samples of the target analytes. Fortunately pure samples of all the common pharmaceuticals are available from chemical supply companies and are normally used "as received" without any further purification.

Providing the signal in the EI (or SIM) chromatogram is above the instrumental limit of detection (ILD) (*see* below) determined for that analyte using a given LC–MS/MS instrumental setup, the presence of the analyte in an unknown sample can be confirmed by a peak in the SIM chromatogram measured at the known *m/z* value of the producer ion for that analyte and with the same chromatographic retention time as that measured for a reference sample of pure analyte. Further confirmation is provided by the observation of peaks in the MRM chromatograms occurring for at least two different transitions that are specific to a unique combination of precursor and daughter ions arising from the target analyte and occurring with the same retention time as that of the reference standard.

Once the presence of the target analyte has been confirmed, the next question is: what is its concentration in the unknown sample? The most common method of quantitation involves using a *concentration curve* (plot) in which the peak heights (or peak areas) derived from either SIM or MRM chromatograms are plotted as a function of the known analyte concentration for a series of reference samples that can be prepared in different ways including (*see* Section 3.3.5 in Greaves and Roboz, 2014):

(1) The *external standard* method in which increasingly higher concentrations of a pure sample of the analyte are added (*spiked*) into a series of individual samples of the same matrix subjected to exactly the same extraction, separation, and MS measurement procedures as for the unknown samples but measured separately. The resulting plot of signal intensity (peak height or area) against the known concentration of the analyte (*see* Figure 4.6(a)) is normally linear and hence can be fitted via regression procedures to an equation of the form $y = mx + c$, where y is the dependent variable (measured peak intensity or area) and x is the dependent variable (analyte concentration). Values of m (gradient) and c (intercept) are specific to a given analyte and instrumental settings. Rearrangement of this equation to give $x = (y - c)/m$ enables calculation of the analyte concentration in an unknown sample from measurement of just the signal intensity for that sample.

This external standard method assumes that the efficiency of the initial analyte extraction step is the same for all analyte concentrations. This situation does not always happen and so to compensate for any variable extraction efficiency between different samples another type of calibration procedure is commonly used.

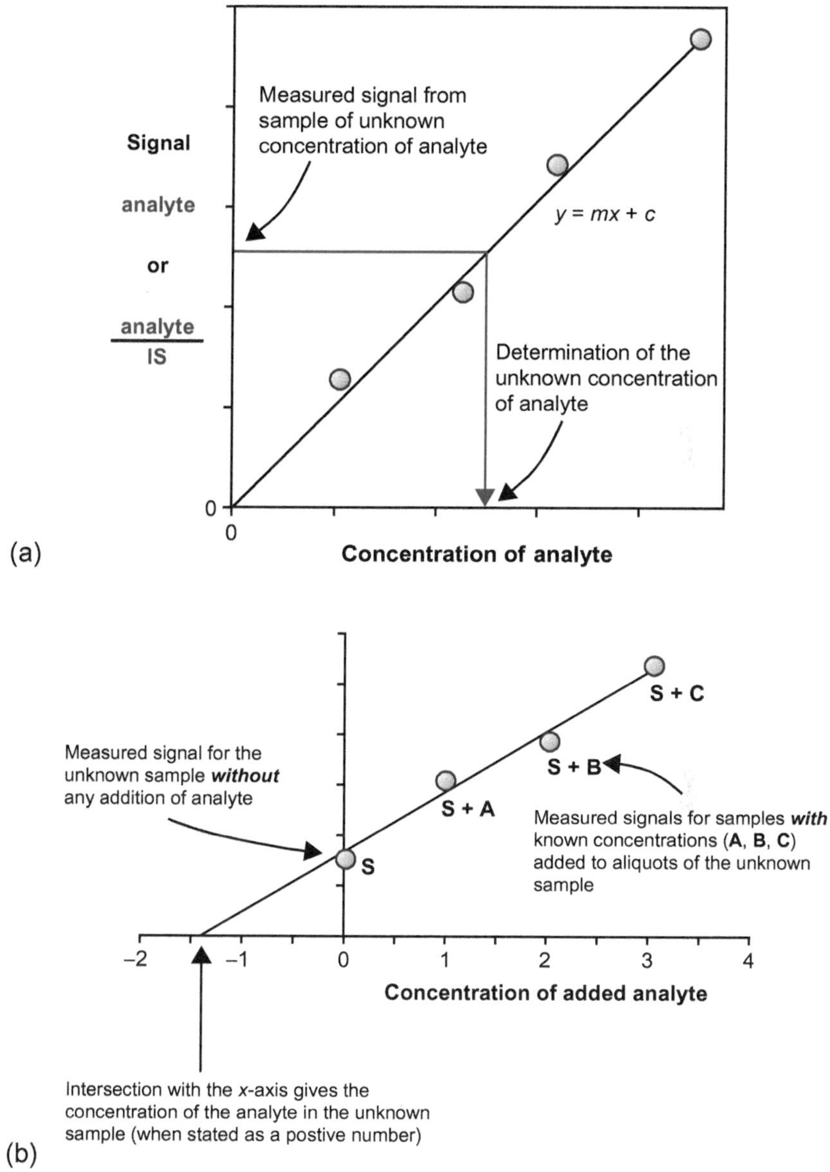

Figure 4.6 Calibration curves determined using: (a) external and internal standards; (b) the standard addition method.

(2) The *internal standard* method involves the addition of a constant amount of an additional compound (known as the *internal standard* (IS)) to samples of increasingly higher concentrations of the analyte as prepared for the external standard method (1) and before the extraction step. The compound used as IS *should* have the same extraction efficiency, chromatographic retention time and mass spectrometric properties as the analyte and these properties can be provided by an IS with as close a chemical structure as possible the analyte. The best choice for an IS is an isotopically-labelled form of the analyte (e.g., diclofencac-d4 with four hydrogen atoms substituted by four deuterium atoms as shown in Figure 4.7). The calibration plot is now prepared by plotting the *ratio* of the signal intensity measured for the analyte to that of the added IS, against the known analyte concentration as also shown in Figure 4.6(a). The same amount of IS is now added to the unknown sample and subjected to the same analytical procedure and the ratio of the analyte signal intensity to that of the IS is determined in the unknown sample. Since the calibration plot is again typically linear, the analyte concentration in the unknown sample can be determined using this ratio in the same manner as just the analyte signal intensity alone is used in the external standard method described above.

By simultaneously measuring *both* the signal of the unknown concentration of analyte and that of the known amount of IS in the same sample, then any variation in extraction efficiency for different samples will be compensated.

A further calibration method to allow for variable extraction efficiencies and matrix effects is:

(3) The *standard addition* method which is even more accurate for the determination of unknown concentrations in a sample and does not require the availability of an appropriate IS. It involves first measuring the signal intensity (S) in the sample without any addition of the analyte and then measurements on aliquots with increasing known amounts of the pure analyte added to the sample. The resulting plot of signal intensity versus concentration of added analyte (*see* Figure 4.7(b)) is again normally linear and can be extrapolated back to the intersection with the x-axis to give the unknown concentration of analyte (expressed as a negative value using the same concentration scale as that for the standard addition aliquots).

Although standard addition is perhaps the most accurate of these three common calibration methods and one that avoids the cost of a suitable IS, it does require

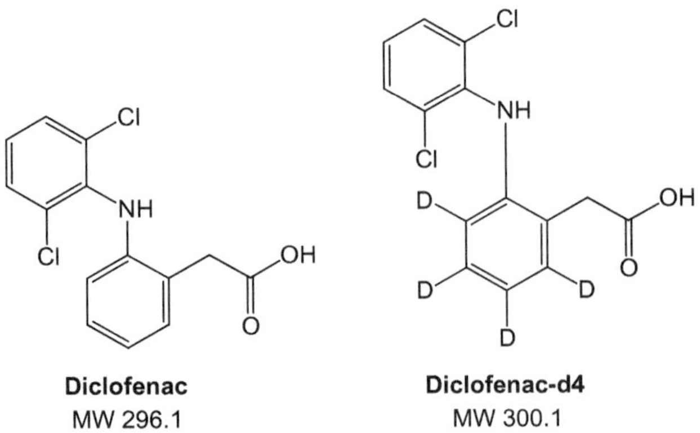

Figure 4.7 Diclofenac and its deuterated internal standard.

measurements of a number of aliquots prepared from the same sample to obtain just one unknown concentration of analyte. Such multiple measurements take extra time (and hence cost) and require an adequate quantity of the raw sample. Hence the internal standard method is a compromise between cost and accuracy and is the most common calibration method currently used for the LC-MS/MS quantitative determination of pharmaceutical levels in environmental samples. An extensive range of mainly deuterium-labeled analogues of all the common pharmaceuticals are now readily available commercially for use as internal standards (although these are often *very* expensive compared to even pure samples of the parent (undeuterated) compounds).

For multiresidue determinations, *mixtures* of the relevant pure standards and their isotopically substituted analogues can used as external or internal standards even although separate EI and MRM chromatograms must still be measured for each target analyte.

4.6.3 Limits of detection

There is a minimum concentration of each analyte below which it is not possible to confirm that the particular analyte species is present in the sample with a specified probability, let alone determine its concentration. In mass spectrometry the signal intensity (expressed as either the peak height or area) in either a mass spectrum or a mass chromatogram is proportional to the number of ions with a particular *m/z* or transition value. As the concentration of a given analyte in a sample decreases, so will its signal intensity until the peak cannot be distinguished from the background electronic noise of the instrumentation. This gives rise to the concept of a minimum concentration (*limit of detection* (LOD) or *instrumental detection limit* (IDL)).

The IDL for MS analysis for a given analyte is determined by measuring the signal intensity observed for peaks by direct injection into the MS ionization source of a series of samples with decreasing concentration of the pure analyte in a specified matrix system. The IDL is defined as the concentration of analyte that gives a ratio of the peak height (*S*) for the analyte to the amplitude of the average noise in the background of the mass spectral trace (*S/N*) that is no less than 3 (e.g., *see* Al-Odaiani et al., 2010). Although a *S/N* ratio of 3 or more is considered adequate to confirm the presence of a given analyte in a sample, the analyte concentration giving a larger *S/N* of 10 is defined as the lower *limit of quantification* (LOQ) (or *instrumental quantification limit* (IQL)). This minimum value must be achieved before any spectral peaks can be attributed to a particular analyte and the peak height (or area) used for quantification purposes.

Values for these two types of detection limit vary with many factors including the type of analyte, the nature of other components in the sample (the sample *matrix*), and the particular mass spectrometer being used and its many different possible instrumental settings which may have been optimized for other purposes than the lowest LOD. So for "analytical thoroughness," any laboratory undertaking such trace analyte measurements needs to determine these parameters for *each* of the proposed target analytes using their particular MS instrument and settings. Table 4.2 lists some IDL and IQL values in the range of 0.05–0.5 and 0.5–1 ng/L respectively, that have been reported by Al-Odaini et al. (2010) for different pharmaceutical analytes using their particular MS/MS instrument.

A third type of detection limit for a given analyte is also often used called the *method limit of detection* (MDL) and this refers to the whole analytical method involving both the initial extraction (using SPE) and then the LC-MS/MS procedures. This parameter has been defined in general terms by the EPA (1996) for any analytical method as:

> *"the minimum concentration of a substance that can be measured and reported with 99% confidence that the analyte concentration is greater than zero."*

To determine the MDL of a given analyte, seven or more replicate samples are prepared in the sample matrix by spiking with a concentration of analyte that results in an instrumental *S/N* ratio between 2.5 and 5. The standard deviation (SD) of these replicate measurements is then calculated and used in the expression.

$$\text{MDL} = t_{(n-1,1-\alpha=0.99)} \times \text{SD}$$

where $t_{(n-1,1-\alpha=0.99)}$ is the students' *t* value appropriate for a 99% confidence level for *n* replicates.

The variation in MDL values with the nature of the sample matrix is illustrated by the values listed in Table 4.2 for a range of pharmaceutical analytes measured in river water below the outfall of a sewage treatment plant, and in the effluent of that plant (Al-Odaini et al., 2010).

In many papers, some of the concentrations of the pharmaceuticals are reported as <LOD but without any indication of what type of detection limit (IDL, LOQ, or MQL) this term is referring to.

4.6.4 Extraction recovery efficiency and matrix effect

As well as the various limit of detection values discussed above, other measures of efficiency for specific steps in the overall analytical procedure used for LC-MS/MS determination of pharmaceuticals in environmental samples include:

(1) *Extraction recovery*: the efficiency of solid phase extraction (SEP) of a given target analyte present in a given matrix depends at least in part on the influence of the chemical nature of the analyte on how it bonds to the solid phase support and how that is affected by the other components of the sample matrix. This extraction efficiency is expressed for each analyte as:

$$\text{Recovery }\% = (c_s / c_r) \times 100$$

where c_s and c_r are the mean concentrations of the analyte determined from LC-MS/MS measurements of a number of replicates of the matrix spiked with a known concentration of analyte either *before* or *after* the SEP step, respectively. An estimate of the precision of this extraction recovery parameter is provided by the calculation of the relative standard deviation of the measurements.

(2) *Matrix effect*: the efficiency of the electron spray ionization (ESI) step in the mass spectrometric part of the overall analytical procedure can either be suppressed or even sometimes enhanced with the nature of the other components in the sample (the *matrix*) (*see* Antignac, 2005). A measure of this matrix effect (ME) for a given analyte and matrix using a particular MS instrumental setup is (Vieno et al., 2006):

$$\text{Matrix effect (ME \%)} = (A_s - (A_{sp} - A_{usp}))/A_s \times 100$$

where A_s and A_{sp} are the peak areas for the analyte in samples of the matrix or ultrapure water respectively, both spiked with the same concentration of analyte reference material *after* extraction and *before* the ionization step. A_{usp} is the peak area of the analyte in a sample of the matrix without spiking and compensates for any background concentration of analyte in the sample. A positive ME% value suggests ionization suppression while a negative value for ME% suggests ionization enhancement.

The ME% parameter as calculated above using the peak area of the analyte signal without any internal standard (IS) correction is called an *absolute* ME% value. If, however, the ratio of peak area to that of the internal standard is used, the ME% calculated with the above expression is a *relative* value. Because the use of an IS already compensates to some extent for matrix effects, relative ME% values are often considerably smaller that the absolute value for the same analyte.

Some values for recovery%, absolute ME% and relative ME% are listed in Table 4.2 as reported by Al-Odaini et al. (2010) for their particular analytical methodology. These values clearly illustrate how both these parameters can vary markedly with the nature of the analyte and the sample matrix.

Finally, although LC-MS/MS is now the most common analytical technique used to detect and quantify pharmaceuticals in a wide range of environmental samples, other techniques that have also been used include gas chromatography (GS) in conjunction with mass spectrometry (GC-MS/MS). However, none of these other methods can generally achieve the selectivity and sensitivity of current LC-MS/MS methods to detect and quantify the low levels of the wide range of pharmaceutical analytes typically present in environmental samples.

4.7 Environmental concentrations of some common pharmaceuticals

The analytical techniques described above have been used to identify and measure the concentrations (levels) of many different pharmaceuticals in different environments in many parts of the world. As analytical techniques become more sensitive, the method and instrumental limits of detection have become progressively lower resulting in more pharmaceuticals being detected and their environmental levels quantified. These results are appearing in an ever-increasing number of reports and other publications. A very small selection of these results is listed in Table 4.3 and illustrate the large variation in the reported levels of the same pharmaceutical in different environments.

Table 4.3 **Environmental concentrations of some common pharmaceuticals**

Pharmaceutical	Type	Type of water	Country	Location	Concentration ng/L	Reference
Paracetamol	Nonspecific	Groundwater	USA		380	Barnes et al. (2008)
	Anti-inflammatory	River water	Serbia	Danube	78,170	Grujic et al. (2009)
			Korea	Han	<5–127	Choi et al. (2008)
		Surface water	UK		<50	Hilton and Thomas (2003)
		River water	Malaysia	Langat	10	Al-Odaini et al. (2010)
Salbutamol	Asthma–β-adrenergic receptor agonist	River water	Italy	Lambro	2.5	Zuccato et al. (2005)
			Italy	Po	1.1	Zuccato et al. (2005)
Simvastatin	Blood lipid lowering	River water	Canada	Otonabee	< LOD[a]	Miao and Metcalfe (2003)
Aspirin	NSAID anti-platelet	River water	Romania	Somes	< 30- 37.2	Moldovan (2006)
Metformin hydrochloride	Anti-hyperglycemic	River water	Malaysia	Langat	293	Al-Odaini et al. (2010)
Omeprazole	Anti-acid proton pump inhibitor	River water	Italy	Lambro, Po	<LOD	Zuccato et al. (2005)
		Surface water	Spain		LOD < 18 ng/L	Gracia et al. (2011)
Metoprolol succinate	Antihypertensive	River water	Finland	Vantaa	<3.8–116	Vieno et al. (2006)
			Finland	Luhtajoki	38	Vieno et al. (2006)
			Sweden	Höje	30–70	Bendz et al. (2005)
			Malaysia	Langat	<MDL	Al-Odaini et al. (2010)
Ibuprofen	NSAID	Groundwater	USA		3110	Barnes et al. (2008)
		Drinking water	South Korea		<1.0	Kim et al. (2007)

		Surface water	South Korea		11–38	Kim et al. (2007)
		Lake water	Germany	Alster	8.7–32	Weigel et al. (2004)
		River water	Romania	Somes	<30–115.2	Moldovan (2006)
			Brazil	Paraiba do Sul	< 10	Stumpf et al. (1999)
			Italy	Po	LOD–9.76	Calamari et al. (2003)
			Germany	Elber	8.7–32	Weigel et al. (2004)
			Luxembourg	Alzette	10–295	Pailler et al. (2009)
			Luxembourg	Mess	9–2383	Pailler et al. (2009)
			South Korea	Mankyung	<5–414	Kim et al. (2007)
			UK	Tyne	144–2370	Roberts and Thomas (2006)
			Sweden	Hoje	10–120	Bendz et al. (2005)
Diclofenac sodium	NSAID	Groundwater	Germany		590	Sacher et al. (2001)
		Surface water	South Korea		1.1–6.8	Kim et al. (2007)
		Surface water	UK		<20–90	Hilton and Thomas (2003)
		River water	Brazil	Sul	20–60	Stumpf et al. (1999)
			Sweden	Hoje	10–120	Bendz et al. (2005)
			Germany	Various	26–72	Hernando et al. (2006)
			Germany	Elber	42–67	Weigel et al. (2004)
			China	Pearl	<1.1–147	Zhao et al. (2009)
			Luxembourg	Mess	0.3–19	Pailler et al. (2009)
			Romania	Somes	<30–115.2	Moldovan (2006)
			Malaysia	Langat	17	Al-Odaini et al. (2010)
Prednisone		Surface water	USA		< 2.2 (LOD)	Batt et al. (2008)

[a] LOD: Limit of detection.

4.8 Summary

The analytical methods described in this chapter for the identification and quantitation of numerous pharmaceuticals in various types of environmental samples are continuing to be refined particularly with advances in LC-MS/MS instrumentation. However, there will always be some method limit of detection value for a given pharmaceutical as determined in a given laboratory and below which it is impossible to decide with an acceptable level of confidence whether that pharmaceutical is present at some even lower concentration in the environmental sample. Furthermore, even if a given pharmaceutical can be positively identified and quantified, it does not necessarily mean that at that level the pharmaceutical poses a risk to humans and/or other biological organisms.

In the next chapter, we discuss the impact and potential harm of pharmaceuticals on wildlife and explore how environmental risk assessments of pharmaceutical contaminants are conducted in laboratory and field studies.

References

Ahrer, W., Buchberger, W., 2001. Analysis of acidic pharmaceutical drug residues in surface water by capillary electrophoresis-electrospray mass spectrometry. Am. Lab. 33 (24), 31.

Ahrer, W., Scherwenk, E., et al., 2001. Determination of drug residues in water by the combination of liquid chromatography or capillary electrophoresis with electrospray mass spectrometry. J. Chromatogr. A 910 (1), 69–78.

Allan, I.J., Vrana, B., et al., 2006. A "toolbox" for biological and chemical monitoring requirements for the European Union's Water Framework Directive. Talanta 69 (2), 302–322.

Al-Odaini, N.A., Zakaria, M.P., et al., 2010. Multi-residue analytical method for human pharmaceuticals and synthetic hormones in river water and sewage effluents by solid-phase extraction and liquid chromatography–tandem mass spectrometry. J. Chromatogr. A 1217 (44), 6791–6806.

Anumol, T., Merel, S., et al., 2013. Ultra high performance liquid chromatography tandem mass spectrometry for rapid analysis of trace organic contaminants in water. Chem. Cent. J. 7 (1), 104.

Antignac, J.-P., De Wasch, K., Monteau, F., De Brabanderd, H., Andre, F., Le Bizec, B., 2005. The ion-suppression phenomenon in liquid chromatography-mass spectrometry and its consequences in the field of residue analysis. Anal. Chim. Acta 529, 129–136.

Baker, D.R., Kasprzyk-Hordern, B., 2011a. Critical evaluation of methodology commonly used in sample collection, storage and preparation for the analysis of pharmaceuticals and illicit drugs in surface water and wastewater by solid phase extraction and liquid chromatography–mass spectrometry. J. Chromatogr. A 1218 (44), 8036–8059.

Baker, D.R., Kasprzyk-Hordern, B., 2011b. Multi-residue analysis of drugs of abuse in wastewater and surface water by solid-phase extraction and liquid chromatography–positive electrospray ionisation tandem mass spectrometry. J. Chromatogr. A 1218 (12), 1620–1631.

Baker, D.R., Kasprzyk-Hordern, B., 2013. Spatial and temporal occurrence of pharmaceuticals and illicit drugs in the aqueous environment and during wastewater treatment: New developments. Sci. Total Environ. 454–455, 442–456.

Baranowska, I., Kowalski, B., 2010. The development of SPE procedures and an UHPLC method for the simultaneous determination of ten drugs in water samples. Water Air Soil Pollut. 211 (1–4), 417–425.

Barnes, K.K., Kolpin, D.W., Furlong, E.T., Zaugg, S.D., Meyer, B.N., Barber, L.B., 2008. A national reconnaissance of pharmaceuticals and other organic wastewater contaminants in the United States—I Groundwater. Sci. Total Environ. 402, 192–200.

Batt, A.L., Kostich, M.S., et al., 2008. Analysis of ecologically relevant pharmaceuticals in wastewater and surface water using selective solid-phase extraction and UPLC-MS-MS. Anal. Chem. 80, 5021–5030.

Bäuerlein, P.S., Mansell, J.E., et al., 2011. Sorption behavior of charged and neutral polar organic compounds on solid phase extraction materials: which functional group governs sorption? Environ. Sci. Technol. 46 (2), 954–961.

Bendz, D., Paxeus, N.A., Ginn, T.R., Loge, F.J., 2005. Occurrence and fate of pharmaceutically active compounds in the environment, a case study: Hoje River in Sweden. J. Hazardous Mater. 122 (3), 195–204.

Calamari, D., Zuccato, E., Castiglioni, S., Bagnati, R., Fanelli, R., 2003. Strategic survey of therapeutic drugs in the Rivers Po and Lambro in Northern Italy. Environ. Sci. Tech. 37, 1241–1248.

Calisto, V., Esteves, V.I., 2009. Psychiatric pharmaceuticals in the environment. Chemosphere 77 (10), 1257–1274.

Camacho-Munoz, D., Martin, J., et al., 2009. An affordable method for the simultaneous determination of the most studied pharmaceutical compounds as wastewater and surface water pollutants. J. Sep. Sci. 32 (18), 3064–3073.

Castiglioni, S., Bagnati, R., et al., 2006. Removal of pharmaceuticals in sewage treatment plants in Italy. Environ. Sci. Technol. 40 (1), 357–363.

Chen, Y., Cao, Q., et al., 2012. Determination of pharmaceuticals from various therapeutic classes in dewatered sludge by pressurized liquid extraction and high performance liquid chromatography and tandem mass spectrometry (HPLC-MS/MS). Int. J. Environ. Anal. Chem. 93 (11), 1159–1173.

Chiaia, A.C., Banta-Green, C., et al., 2008. Eliminating solid phase extraction with large-volume injection LC/MS/MS: analysis of illicit and legal drugs and human urine indicators in US wastewaters. Environ. Sci. Technol. 42 (23), 8841–8848.

Choi, K., Kim, Y., Park, J., Park, C.K., Kim, M., Kim, H.S., Kim, P., 2008. Seasonal variations of several pharmaceutical residues in surface water and sewage treatment plants of the Han River, Korea. Sci. Total Environ. 405, 120–128.

Conley, J.M., Symes, S.J., et al., 2008. Spatial and temporal analysis of pharmaceutical concentrations in the upper Tennessee River basin. Chemosphere 73 (8), 1178–1187.

EPA, 1996. USEPA 1996. Guidelines and Format for Methods to be Proposed at 40 CFR Part 136 or Part 141 (Guidelines and Format Document). U.S. Environmental Protection Agency. Office of Water, Engineering and Analysis Division. Washington, D.C., EPA-821-B-96-003 (National Service Center for Environmental Publications (NSCEP) at http://nepis.epa.gov/Simple.htm (accessed 22.06.15.)).

Fang, Y., Karnjanapiboonwong, A., et al., 2012. Occurrence, fate, and persistence of gemfibrozil in water and soil. Environ. Toxicol. Chem. 31 (3), 550–555.

Gibson, R., Becerril-Bravo, E., et al., 2007. Determination of acidic pharmaceuticals and potential endocrine, disrupting compounds in wastewaters and spring waters by selective elution and analysis by gas chromatography–mass spectrometry. J. Chromatogr. A 1169 (1–2), 31–39.

Gracia-Lor, E., Sancho, J.V., Hernández, F., 2011. Multi-class determination of around 50 pharmaceuticals, including 26 antibiotics, in environmental and wastewater samples by ultra-high performance liquid chromatography-tandem mass spectrometry. J. Chromatogr. A 1218, 2264–2275.

Greaves, J., Roboz, J., 2014. Mass Spectrometry for the Novice. CRC Press, Taylor & Francis Group, Boca Raton, FL, USA. ISBN: 978-1-4200-9418-3.

Grujić, S., Vasiljević, T., Laušević, M., 2009. Determination of multiple pharmaceutical classes in surface and ground waters by liquid chromatography-ion trap-tandem mass spectrometry. J. Chromatogr. A 1216, 4989–5000.

Heath, E., Kosjek, T., et al., 2010. Second interlaboratory exercise on non-steroidal anti-inflammatory drug analysis in environmental aqueous samples. Talanta 81 (4–5), 1189–1196.

Hernando, M.D., Heath, E., Petrovíc, M., Barceló, D., 2006. Trace-level determination of pharmaceuticals residues by LC-MS/MS in natural and treated waters. A pilot-survey study. Anal. Bioanal. Chem. 385, 985–991.

Hilton, M.J., Thomas, K.V., 2003. Determination of selected human pharmaceutical compounds in effluent and surface water samples by high-performance liquid chromatography-electrospray tandem mass spectrometry. J. Chromatogr. A 1015, 129–141.

Jandera, P., 2012. Comprehensive two-dimensional liquid chromatography—practical impacts of theoretical considerations. A review. Cent. Eur. J. Chem. 10 (3), 844–875.

Kim, S.D., Cho, J., Kim, I.S., Vanderford, B.J., Snyder, S.A., 2007. Occurrence and removal of pharmaceuticals and endocrine disruptors in South Korean surface, drinking, and waste waters. Water Res. 41 (5), 1013–1021.

Kleywegt, S., Pileggi, V., et al., 2011. Pharmaceuticals, hormones and bisphenol A in untreated source and finished drinking water in Ontario, Canada—and treatment efficiency. Sci. Total Environ. 409 (8), 1481–1488.

Kolpin, D.W., Furlong, E.T., et al., 2002. Pharmaceuticals, hormones and other organic wastewater contaminants in U.S. streams, 1999–2000: A national reconnaissance. Environ. Sci. Technol. 36, 1202–1211.

Kummerer, K., 2009. Antibiotics in the aquatic environment—A review—Part I. Chemosphere 75 (4), 417–434.

Kurissery, S., Kanavillil, N., et al., 2012. Caffeine as an anthropogenic marker of domestic waste: A study from Lake Simcoe watershed. Ecol. Indic. 23, 501–508.

Lajeunesse, A., Gagnon, C., et al., 2008. Determination of basic antidepressants and their n-desmethyl metabolites in raw sewage and wastewater using solid-phase extraction and liquid chromatography–Tandem mass spectrometry. Anal. Chem. 80 (14), 5325–5333.

Liu, H., Zhang, G.P., et al., 2009. The occurrence of chloramphenicol and tetracyclines in municipal sewage and the Nanming River, Guiyang City, China. J. Environ. Monit. 11 (6), 1199–1205.

Liz, M.V.d., Nagata, N., et al., 2012. Considerações sobre o preparo de amostras contendo micropoluentes estrogênicos. Quim. Nova 35, 1213–1215.

Madureira, T.V., Barreiro, J.C., et al., 2009. Pharmaceutical trace analysis in aqueous environmental matrices by liquid chromatography–ion trap tandem mass spectrometry. J. Chromatogr. A 1216 (42), 7033–7042.

McMaster, M.C., 2005. LC/MS A Practical User's Guide. Wiley, Hoboken New Jersey.

Metcalfe, C.D., Miao, X.S., et al., 2003. Distribution of acidic and neutral drugs in surface waters near sewage treatment plants in the lower Great Lakes, Canada. Environ. Toxicol. Chem. 22 (12), 2881–2889.

Miao, X., Metcalfe, C., 2003. Determination of cholesterol-lowering statin drugs in aqueous samples using liquid chromatography–electrospray ionization tandem mass spectrometry. J. Chromatogr. A 998, 133–141.

Moldovan, Z., 2006. Occurrences of pharmaceutical and personal care products as micropollutants in rivers from Romania. Chemosphere 64, 1808–1817.

Mompelat, S., Jaffrezic, A., et al., 2013. Storage of natural water samples and preservation techniques for pharmaceutical quantification. Talanta 109, 31–45.

Morasch, B., Bonvin, F., et al., 2010. Occurrence and fate of micropollutants in the Vidy Bay of Lake Geneva, Switzerland. Part II: Micropollutant removal between wastewater and raw drinking water. Environ. Toxicol. Chem. 29 (8), 1658–1668.

Murray, K.K., Boyd, R.K., Eberlin, M.N., Langley, G.J., Li, L., Naito, Y., 2013. Definitions of terms relating to mass spectrometry (IUPAC Recommendations 2013). Pure Appl. Chem. 85 (7), 1515–1609.

Na, G., Gu, J., et al., 2011. Detection of 36 antibiotics in coastal waters using high performance liquid chromatography–tandem mass spectrometry. Chin. J. Oceanol. Limnol. 29 (5), 1093–1102.

Nebot, C., Gibb, S.W., et al., 2007. Quantification of human pharmaceuticals in water samples by high performance liquid chromatography–tandem mass spectrometry. Anal. Chim. Acta 598 (1), 87–94.

Okuda, T., Kobayashi, Y., et al., 2008. Removal efficiency of 66 pharmaceuticals during wastewater treatment process in Japan. Water Sci. Technol. 57 (1), 65–71.

Ort, C., Lawrence, M.G., et al., 2010. Sampling for pharmaceuticals and personal care products (PPCPs) and illicit drugs in wastewater systems: are your conclusions valid? A Critical Review. Environ. Sci. Technol. 44, 6024–6035.

Pailler, J.Y., Pfsiter, K.L., Hoffmann, L., Guignard, C., 2009. Solid phase extraction coupled to liquid chromatography-tandem-mass spectrometry analysis of sulphonamides, tetracyclines, analgesics and hormones in surface water and wastewater in Luxembourg. Sci. Total Environ. 407, 4736–4743.

Peysson, W., Vulliet, E., 2013. Determination of 136 pharmaceuticals and hormones in sewage sludge using quick, easy, cheap, effective, rugged and safe extraction followed by analysis with liquid chromatography–time-of-flight-mass spectrometry. J. Chromatogr. A 1290, 46–61.

Pojana, G., Fantinati, A., et al., 2011. Occurrence of environmentally relevant pharmaceuticals in Italian drinking water treatment plants. Int. J. Environ. Anal. Chem. 91 (6), 537–552.

Primel, E.G., Caldas, S.S., et al., 2012. Multi-residue analytical methods for the determination of pesticides and PPCPs in water by LC-MS/MS: a review. Cent. Eur. J. Chem. 10 (3), 876–899.

Rua-Gomez, P.C., Puttmann, W., 2012. Occurrence and removal of lidocaine, tramadol, venlafaxine, and their metabolites in German wastewater treatment plants. Environ. Sci. Pollut. Res. 19 (3), 689–699.

Schlusener, M.P., Bester, K., 2005. Determination of steroid hormones, hormone conjugates and macrolide antibiotics in influents and effluents of sewage treatment plants utilising high-performance liquid chromatography/tandem mass spectrometry with electrospray and atmospheric pressure chemical ionisation. Rapid Commun. Mass Spectrom. 19 (22), 3269–3278.

Schriks, M., van der Linden, S.C., et al., 2013. Occurrence of glucocorticogenic activity in various surface waters in The Netherlands. Chemosphere 93 (2), 450–454.

Segura, P.A., Garcia-Ac, A., et al., 2007. Determination of six anti-infectives in wastewater using tandem solidphase extraction and liquid chromatography–tandem mass spectrometry. J. Environ. Monit. 9 (4), 307–313.

Shelver, W.L., Hakk, H., et al., 2010. Development of an ultra-high-pressure liquid chromatography–tandem mass spectrometry multi-residue sulfonamide method and its application to water, manure slurry, and soils from swine rearing facilities. J. Chromatogr. A 1217 (8), 1273–1282.

Shin, H.S., Oh, J.A., 2012. Simultaneous determination of non-steroidal anti-inflammatory drugs in river water by gas chromatography–mass spectrometry. J. Sep. Sci. 35 (4), 541–547.

Sim, W.J., Lee, J.W., et al., 2011. Occurrence and distribution of pharmaceuticals in wastewater from households, livestock farms, hospitals and pharmaceutical manufactures. Chemosphere 82 (2), 179–186.

Stumpf, M., Ternes, T.A., Wilken, R.D., Rodrigues, S.V., Baumann, W., 1999. Polar drug residues in sewage and natural waters in the state of Rio de Janeiro. Brazil. Sci. Total Environ. 225 (1–2), 135–141.

Sturini, M., Speltini, A., et al., 2009. Solid-phase extraction and HPLC determination of fluoroquinolones in surface waters. J. Sep. Sci. 32 (17), 3020–3028.

Suri, R.S., Singh, T., et al., 2012. Effect of process conditions on the analysis of free and conjugated estrogen hormones by solid-phase extraction–gas chromatography/mass spectrometry (SPE–GC/MS). Environ. Monit. Assess. 184 (3), 1657–1669.

Togola, A., Budzinski, H., 2008. Multi-residue analysis of pharmaceutical compounds in aqueous samples. J. Chromatogr. A 1177 (1), 150–158.

Tong, A.Y.C., 2013. Pharmaceuticals in Sewage: Sources, Levels and Degradation. Ph.D. thesis (Unpublished). University of Otago, Dunedin, New Zealand.

Tran, N.H., Hu, J., et al., 2013. Simultaneous determination of PPCPs, EDCs, and artificial sweeteners in environmental water samples using a single-step SPE coupled with HPLC–MS/MS and isotope dilution. Talanta 113, 82–92.

Uslu, M.O., Jasim, S., et al., 2013. A Survey of Occurrence and Risk Assessment of Pharmaceutical Substances in the Great Lakes Basin. Ozone-Sci. Eng. 35 (4), 249–262.

Van Stempvoort, D.R., Roy, J.W., et al., 2013. An artificial sweetener and pharmaceutical compounds as co-tracers of urban wastewater in groundwater. Sci. Total Environ. 461–462, 348–359.

Vanderford, B.J., Pearson, R.A., et al., 2003. Analysis of endocrine disruptors, pharmaceuticals, and personal care products in water using liquid chromatography/tandem mass spectrometry. Anal. Chem. 75, 6265–6274.

Vanderford, B.J., Mawhinney, D.B., et al., 2011. Assessment of sample preservation techniques for pharmaceuticals, personal care products, and steroids in surface and drinking water. Anal. Bioanal. Chem. 399 (6), 2227–2234.

Vieno, N.M., Tuhkanen, T., Kronberg, L., 2006. Analysis of neutral and basic pharmaceuticals in sewage treatment plants and in recipient rivers using solid phase extraction and liquid chromatography-ion trap tandem mass spectrometry. J. Chromatogr. A 1116, 193–203.

Wang, Z., Huang, Q., et al., 2013. Stereoisomeric profiling of pharmaceuticals ibuprofen and iopromide in wastewater and river water, China. Environ. Geochem. Health 35 (5), 683–691.

Weigel, S., Kallenborn, R., Hühnerfuss, H., 2004. Simultaneous solid-phase extraction of acidic, neutral and basic pharmaceuticals from aqueous samples at ambient (neutral) pH and their determination by gas chromatography-mass-spectrometry. J. Chromatogr. A 1023, 183–195.

Wu, C.X., Spongberg, A.L., et al., 2008. Use of solid phase extraction and liquid chromatography–tandem mass spectrometry for simultaneous determination of various pharmaceuticals in surface water. Int. J. Environ. Anal. Chem. 88 (14), 1033–1048.

Yang, S., Carlson, K.H., 2004. Solid-phase extraction-high-performance liquid chromatography–ion trap mass spectrometry for analysis of trace concentrations of macrolide antibiotics in natural and waste water matrices. J. Chromatogr. A 1038 (1–2), 141–155.

Yu, Y., Wu, L., 2011. Comparison of four extraction methods for the analysis of pharmaceuticals in wastewater. J. Chromatogr. A 1218 (18), 2483–2489.

Yu, K., Li, B., et al., 2012. Direct rapid analysis of multiple PPCPs in municipal wastewater using ultrahigh performance liquid chromatography–tandem mass spectrometry without SPE pre-concentration. Anal. Chim. Acta 738, 59–68.

Zhang, R.J., Zhang, G., et al., 2012. Occurrence and risks of antibiotics in the Laizhou Bay, China: Impacts of river discharge. Ecotoxicol. Environ. Saf. 80, 208–215.

Zhao, J.L., Ying, G.G., et al., 2010. Occurrence and a screening-level risk assessment of human pharmaceuticals in the Pearl River System, South China. Environ. Toxicol. Chem. 29 (6), 1377–1384.

Zhao, J.L., Ying, G.G., Wang, L., Yang, J.F., Yang, X.B., Yang, L.H., Li, X., 2009. Determination of phenolic endocrine disrupting chemicals and acidic pharmaceuticals in surface water of the Pearl Rivers in South China by gas chromatography-negative chemical ionization-mass spectrometry. Sci. Total Environ. 407, 962–974.

Zuccato, E., Castiglioni, S., Fanelli, R., 2005. Identification of the pharmaceuticals for human use contaminating the Italian aquatic environment. J. Hazardous Mater. 122, 205–209.

Impact of pharmaceuticals on the environment

5

5.1 Overview

Although the presence of contaminants such as pharmaceuticals in freshwater systems has been reported as one of the key problems facing humanity (Schwarzenbach et al., 2006), there is still limited evidence to fully substantiate this statement. In most instances, it remains unknown whether they contribute to detrimental effects that adversely affect organisms in the receiving aquatic or terrestrial environment (Daughton and Ruhoy, 2009). The previous chapters of this book have focused on the issues associated with how pharmaceuticals come to be discharged into the environment (Chapters 2 and 3) and how they can be identified and quantified (Chapter 4). However, these aspects often do not provide enough information to fully assess the potential risk that the pharmaceuticals might pose to the environment (Boxall et al., 2012).

As noted in Chapter 1, *risk* can be quantified in terms of *hazard* and *exposure*. In brief, a pharmaceutical poses no risk to the environment if it remains unused and safely stored somewhere or if unused portions of a dispensed prescription are disposed of or destroyed by environmentally acceptable methods (Chapters 2 and 3). However, it is apparent from their detection that pharmaceuticals are discharged into the environment (Chapter 4). The key to determine whether a particular pharmaceutical poses an environmental risk is to characterize the effects or harms that could arise in organisms exposed to it.

5.2 Environmental risk assessment methodologies

The environmental risk assessment process for a particular pharmaceutical is complex and based on overall usage, its *persistence*, *fate* and *transport*, and *ecotoxicity* in the environment. There is often insufficient information to assess this risk. For example, prescribing data may be incomplete in certain countries (Chapter 2), and the concentrations of selected pharmaceuticals detected in a few samples of sewage may not be representative of the actual diurnal and seasonal discharge patterns.

Two concepts used in the assessment of environmental risk are:

(1) *Predicted environmental concentration* (PEC): the calculated concentration of a chemical in the environment. This is based on modeling and third-party data derived from known environmental *persistence* (ability to be degraded), *transport* (ability to penetrate biological membranes and accumulate inside flora and fauna), and *fate* (the ability for organisms in the environment to metabolize and detoxify the pharmaceutical, nullifying any detrimental

The Life-Cycle of Pharmaceuticals in the Environment. http://dx.doi.org/10.1016/B978-1-907568-25-1.00005-0

effects arising from the bioaccumulation of the pharmaceutical). In addition, a comparison can be made between the *measured environmental concentrations* (MECs) derived from the detection and quantitation of specific pharmaceutical residues in water and sediment and the calculated PEC values in order to verify the accuracy of models used (Roos et al., 2012).

(2) *Predicted no-effect concentration* (PNEC): the concentration of a pharmaceutical at which no pharmacological effect is expected to occur for a specific organism. A PNEC value is usually derived using ecotoxicity testing data (*see* ecotoxicology below).

A PEC/PNEC ratio of <1 suggests that there is no significant risk associated with the environmental presence of the pharmaceutical. A ratio ≥ 1 suggests that further environmental risk characterization or management of the pharmaceutical is necessary (*see* risk quotients below).

Values for PEC and PNEC can be incorporated into a traditional approach used to quantitate environmental risk posed by a given pharmaceutical based on the available usage data, disposal, and environmental behavior (Roos et al., 2012) as summarized in Figure 5.1.

The variables contributing to the PEC value depicted in Figure 5.1 are factors which *increase* or *decrease* the *exposure* of an aquatic or terrestrial organism to a pharmaceutical. However, these factors do not address whether the pharmaceutical

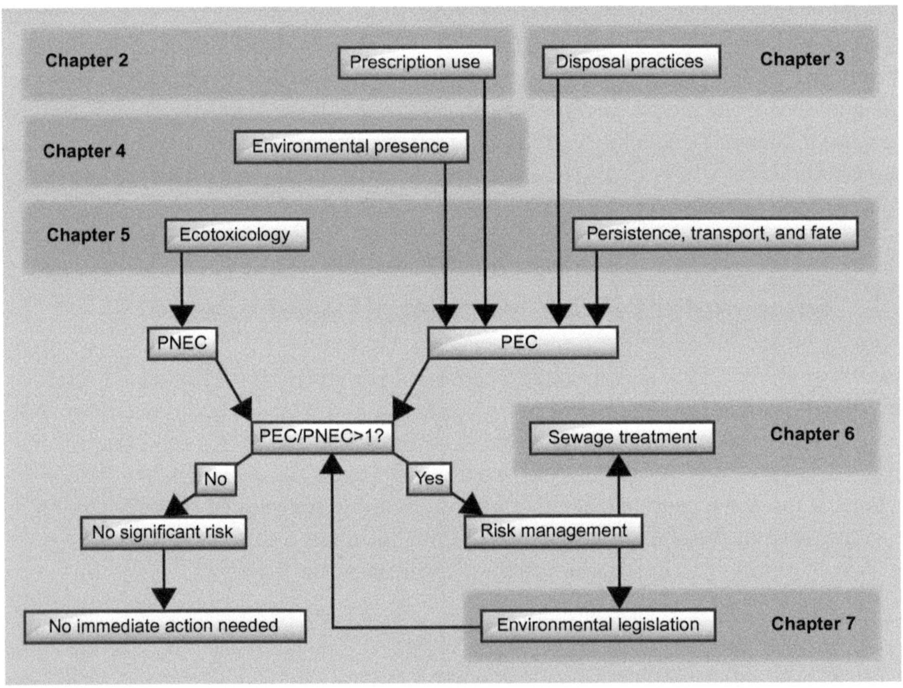

Figure 5.1 Aspects of risk assessment in the environmental life-cycle of pharmaceuticals. PNEC, predicted no-effect concentration; PEC, predicted environmental concentration. Chapter references refer to the present text.
Adapted from concepts by Backhaus and Faust (2012).

is inherently *toxic* or *hazardous* to the exposed organism. The *toxicity* of a pharmaceutical to aquatic or terrestrial organisms must be determined to assess potential effects on wider ecosystems. In a similar manner to understanding drug effects or the pharmacodynamics of human body systems (Chapter 2), ecotoxicology determines the *toxicodynamics* or the underlying pharmacological response/effect arising from an exposure to a particular amount of drug. Laboratory and field toxicology experiments are designed to determine the concentration of drug that will lead to a detrimental effect or *endpoint* in the test organism. A collation of data obtained from these experiments can be used to predict the concentration below which *no effects* from the pharmaceutical are observed (PNEC).

The environmental risk of a pharmaceutical calculated from the PEC/PNEC ratio can be expressed as a *risk quotient* (RQ). The magnitude of exposure can be considered proportional to the PEC but inversely proportional to the PNEC, and so the risk equation:

$$\text{Risk} = \text{Exposure} \times \text{Hazard}$$

can be rewritten as

$$\text{Risk}(\text{RQ}) = \text{PEC} / \text{PNEC}$$

In other words, the environmental risk of a particular pharmaceutical can be quantified by the ratio of its calculated PEC to its PNEC. If environmental concentration data are available, they can be used to replace PEC to give the ratio MEC/PNEC as a measure of RQ. The environmental risk of a pharmaceutical can be classified into categories of significance according to the calculated RQ (Table 5.1).

In general, a chemical with an $RQ < 1$ is considered to have a low or insignificant environmental risk that does not require immediate management. However, a chemical with an $RQ \geq 1$ suggests there is at least the *potential* for that chemical to be detrimental to wildlife.

Other risk assessment methodologies commonly used include the *fish plasma model* (FPM) (Huggett et al., 2003), and the method of *aquatic environment ranking* (Cooper et al., 2008). In brief, the FPM uses fish as a model organism and calculates a ratio between two concentrations—the *human therapeutic plasma concentration* (HTPC) for a pharmaceutical (data for which can be obtained from clinical websites such as www.micromedex.com, accessed 4-9-14) and an estimated *fish*

Table 5.1 Interpretation of environmental risk derived from the Risk Quotient (RQ) for a chemical

Environmental risk level	Risk quotient (RQ)
Insignificant	<0.1
Low	0.1–1.0
Moderate	1.0–10
High	>10

Source of data: Wennmalm and Gunnarsson (2009).

steady-state plasma concentration (FssPC) obtained from the PEC and the lipophilicity (i.e. the ability for a pharmaceutical to penetrate the biological membranes of a fish and accumulate in tissue) as a function of log *P* (see *bioaccumulation* below). Subsequently, a concentration ratio or quotient (CR) obtained by dividing the HTPC value by the FssPC value, provides an estimate of the environmental risk. In this model, a lower CR implies a higher risk of pharmacological activity in the fish model (which is not necessarily adverse). The aquatic environment ranking method incorporates environmental persistence, surface water concentration, and crustacean toxicity for quantitation (Roos et al., 2012).

These risk assessment methodologies often rely heavily on usage data in addition to the persistence, transport, and fate of a pharmaceutical. As explained in Chapter 2, usage data patterns differ between countries and therefore the environmental risks posed by a pharmaceutical can also vary with country.

A more global approach in assessing the potential dangers posed by pharmaceuticals in the environment is to use a *hazard assessment* methodology. For pharmaceutical compounds, one such strategy is the *Persistence*, *Bioaccumulation*, and *Toxicity* (PBT) scheme devised by the Stockholm County Council in Sweden (Castensson et al., 2009). The PBT classification is independent of prescription or usage data and categorizes pharmaceuticals in terms of three variable measures: *persistence (P)*, *bioaccumulation (B)*, and *toxicity (T)*. A pharmaceutical is given a ranking score of 0–3 under each of the *P*, *B*, and *T* categories. The score derived from the sum of these categories provides an overall measure of the environmental hazard of the pharmaceutical (Deblonde and Hartemann, 2013). While this categorization is useful, the ranking system may overestimate or underestimate the true extent of the potential environmental hazard posed by a pharmaceutical. For example, the synthetic estrogen hormones 17α-ethinylestradiol and levonorgestrel are both assigned a *T* score of 3. However, their respective PNEC values according to the Stockholm County Council differ by a factor of 50,000 (Roos et al., 2012). It can be concluded that while such categorization is more universal, the actual hazard of a pharmaceutical is more accurately assessed through the risk which incorporates the exposure of environmental organisms to the pharmaceutical (Roos et al., 2012).

Another hazard assessment scheme is the *critical environmental concentration* (CEC) assessment method described by Fick et al. (2010). For the CEC method, the concentration of a pharmaceutical that is predicted from the available clinical data to cause a pharmacological effect in fish (irrespective of whether this effect is positive or negative) is used in place of the PEC along with log *P* (*see* bioaccumulation below) to derive a value for the FssPC. The use of the CEC parameter rather than the PEC means that the assessment is independent of exposure and can be used to assess potentially harmful pharmaceuticals when their usage or sales data in a particular country are not available.

Finally, the quantitative *structure–activity relationship* (QSAR) assessment method can be used for ranking the environmental hazard of pharmaceuticals (Sanderson et al., 2004; Sanderson and Thomsen, 2007). In this method, a number of physicochemical properties of a molecule, such as its size, shape, and ability to dissolve into oils (i.e., lipophilicity), are mathematically related to the minimum concentration (*C*) necessary to induce a pharmacological effect in a particular organism. For a pharmaceutical to penetrate organisms, the log *P* term which is a measure of oil solubility (*see* bioaccumulation below) is usually the primary variable considered in the mathematic model as expressed in the linear equation:

$$\log(1/C) = k_1 \log P + k_2$$

Here, k_1 and k_2 are constants that are specific to a given organism and are derived from the linear fit of $\log(1/C)$ versus $\log P$ values. Once the equation adequately fits a data set, the minimum effect concentration (and hence PNEC) of a pharmaceutical with a known $\log P$ can be calculated from the model equation. A QSAR model is usually only useful for a set of pharmaceutical compounds with similar structural and physicochemical properties (e.g. an amine substituent group) and for a specific organism. The ability of a QSAR to predict the effect of a given pharmaceutical on a particular organism is dependent on mathematical equations modeled on actual data sets.

The US EPA has developed a software programme known as ECOSAR (ECOlogical Structure–Activity Relationships) which can be freely downloaded from the US EPA website (available http://www.epa.gov/oppt/newchems/tools/21ecosar.htm, accessed 19-8-14). This programme can be used to estimate acute (short-term) and chronic (long-term or delayed) toxicity of a chemical to aquatic organisms such as fish, aquatic invertebrates, and aquatic plants by using computerized structure activity relationships (SARs). Details on the molecular structure of the pharmaceutical are entered using the SMILES (Simplified Molecule Input Line Entry System)-based structural formula. The accuracy in predicting the true toxicity of a pharmaceutical generated by a QSAR model is, however, sometimes questionable, given that experimental data sets for some classes of pharmaceuticals are often limited (Reuschenbach et al., 2008).

There is no substitute for high-quality data on the persistence, transport, fate, and ecotoxicity of a given pharmaceutical to conduct an environmental risk assessment using any of the above models. In the next section, we discuss how experimental data relevant to the determination of the environmental risk of pharmaceuticals is collected.

5.3 Experimental variables involved in risk assessment

5.3.1 Environmental persistence

The environmental persistence of a pharmaceutical is evaluated from its half-life ($t_{1/2}$) defined as the time taken for the parent molecule to decay to half its original concentration. The magnitude of this parameter is based upon the overall rate of decomposition of a contaminant (i.e., the chemical conversion of a pharmaceutical by the sum of the various physicochemical factors in the environment to form particular by-product(s)).

The most simplistic decomposition processes in the environment are usually modeled assuming *first-order* degradation kinetics in which the concentration of the pharmaceutical ($[\mathrm{P}]_t$) at time t is related to the initial concentration ($[\mathrm{P}]_0$) via the exponential relationship:

$$[\mathrm{P}]_t = [\mathrm{P}]_0 \, \mathrm{e}^{-k_{\mathrm{app}} t}$$

where k_{app} is the *apparent first-order degradation* rate constant. Mathematical integration of this equation gives the linear equation:

$$\ln[\mathrm{P}]_t = \ln[\mathrm{P}]_0 - k_{app}t$$

from which an expression for $t_{½}$ can be derived:

$$t_{½} = \frac{\ln 2}{k_{app}}$$

where ln 2 is the natural logarithm of 2 (approximately 0.693). A large (rapid) k_{app} (decay rate) implies a shorter value for the half-life ($t_{½}$) and less environmental persistence.

The rate of decomposition is largely governed by k_{app} which can be expressed as the sum of the rate constants for a number of environmentally-relevant decay processes:

$$k_{app} = k_{hydro} + k_{microbial} + k_{photo} + k_{other}$$

The physicochemical and microbiological processes that affect k_{app} include:

Hydrolysis (k_{hydro}): This is a type of chemical reaction which can occur when a pharmaceutical comes into contact with water and leads to the formation of decomposition products that typically dissolve in the water. For example, hydrolysis may contribute to the decomposition rate of a pharmaceutical in a sewer leading to its partitioning into the liquid phase in a dissolved form. Hydrolysis commonly occurs for drugs that have water-reactive functional groups, such as *ester* bonds. Hydrolysis is a purely chemical process that is not normally facilitated by the action of microbes (i.e., it is an *abiotic* degradation process).

Microbial degradation or *biodegradation* ($k_{microbial}$): Microcosms of anaerobic or aerobic bacteria can metabolise some pharmaceuticals by using them as a carbon source (*biotic* degradation). For example, the decomposition rate of a pharmaceutical in sewage water can be accelerated through activated biological sludge treatment (see Chapter 6). Dissolved organic matter (DOM) and sediments in the environment rich in microbial fauna can also contribute to the overall decomposition of pharmaceuticals.

Photolysis (k_{photo}): The exposure of pharmaceuticals to sunlight and artificial light sources such as the use of ultraviolet (UV) radiation in advanced oxidation processes (see Chapter 6) may induce the abiotic photochemical decomposition processes. There are two types of photolysis: (1) *direct photolysis* occuring when energy (typically arising from UV-B and UV-A radiation (290–400 nm)) is directly transferred from the light source to specific chemical bonds in the pharmaceutical molecule leading to their cleavage; and (2) *indirect photolysis* occuring when the chemical bonds of the pharmaceutical are cleaved by a reactive chemical species that has been initially generated photochemically by irradiation. For example, hydroxyl radicals (•OH) formed from the direct photolysis of nitrate ions or DOM in natural water samples may facilitate degradation of pharmaceuticals.

Apart from hydrolysis, biodegradation, and photolysis, there are other physicochemical processes whose rate constants are collected in the single k_{other} term. These can include the transfer of a pharmaceutical from one environment compartment to another, leading to its apparent dissipation. For example, certain pharmaceuticals may be *adsorbed* from sewage water by particulates during the activated sludge process (see Chapter 6) leading to the apparent disappearance of the pharmaceutical.

In practice, the overall rate constant (k_{app}) pertaining to the total environmental persistence of a pharmaceutical cannot be readily determined from a single set of experiments. However, each of the contributing rate constants (k_{hydro}, $k_{microbial}$, k_{photo}, k_{other}) can be determined by experimentation involving a specific set of laboratory or field conditions. Such experiments can provide information on whether a particular degradation mechanism is an important contributor to the overall environmental degradation of the pharmaceutical. To facilitate the comparison of kinetic data between studies with different experimental conditions, it is desirable to use standardized test methods. The Organization for Economic Cooperation and Development (OECD) provides such standardized guidelines for biotic and abiotic tests to determine the degradation kinetics of a chemical (see www.oecd.org, accessed 6-9-14). For example, the potential environmental biodegradability of a chemical can be extrapolated from experiments conforming to OECD ready biodegrability (OECD 301), inherent (OECD 302), or simulation (OECD 303, 307, 308, or 309) test specifications. Similarly, the potential for a particular chemical to undergo environmental hydrolysis or photolysis can be determined under OECD 111 and 316 specifications, respectively.

The persistence of pharmaceuticals is also dependent on the environmental compartments where it will partition (i.e., water, soil, sediment) as they each have various degradation mechanisms. Examples of the varying environmental persistence of pharmaceuticals will be presented in the next section of this chapter.

5.3.1.1 Environmental persistence of pharmaceuticals in water

Various environmental factors affect the $t_{½}$ of pharmaceuticals in water. These include the amount of incident sunlight received by a particular aquatic environment (Verma et al., 2007), water turbidity and water flow (Robinson et al., 2007), and the presence of reactive radical species (Luo et al., 2012). An additional factor that can affect environmental persistence in the aqueous phase is the adsorption of pharmaceuticals onto the suspended colloidal organic matter present in natural water, which leads to apparent decreases in their concentrations (Maskaoui and Zhou, 2010).

In surface waters, abiotic degradation processes such as hydrolysis and direct photolysis predominate over biodegradation due to the abundance of sunlight (De Laurentiis et al., 2013). Jiang et al. (2010) investigated the degradation of four cephalosporin antibiotics in surface water and observed degradation $t_{½}$ values of 2.7–18.7 days through hydrolysis. Ceftriaxone was observed to degrade by photolysis with a $t_{½}$ of 4.1 days. However, biodegradation was observed to be predominant in sediment (*see* below). The availability of sunlight in surface water also promoted the direct and indirect photolysis of the hallucinogenic pharmaceutical ketamine while no microbial or hydrolytic degradation was observed (Lin et al., 2014). The degradation of estrone (E1) in surface water was more pronounced under sunlight with higher UV-B intensities suggesting that photolysis is an important mechanism in reducing the $t_{1/2}$ of this steroid (Atkinson et al., 2011).

Indirect photolysis leading to the formation of •OH has also been observed to enhance the rate of degradation of lidocaine, tramadol, venlafaxine, *O*-desmethyltramadol, and *O*-desmethylvenlafaxine in surface water (Rua-Gomez and Puttmann, 2013). Reactive oxygen species (ROS) such as singlet oxygen ($^1O_2$) and excited states of DOM may facilitate pharmaceutical removal in addition to direct photolytic degradation (Passananti et al., 2014). For example, the predominant degradation processes of ibuprofen in surface water were observed to be both direct photolysis and indirect photolysis by reaction with •OH radicals and excited states of DOM formed from direct photolysis (Ruggeri et al., 2013). The environmental removal of nicotine in surface water has been attributed predominantly to the action of •OH radicals, with a minor contribution from $^1O_2$ and reactive species involving nitrogen dioxide (Passananti et al., 2014). The degradation of mefenamic acid in water is dependent on both direct and indirect photolysis mechanisms (Werner et al., 2005).

Some pharmaceuticals do not degrade via direct photolysis in sunlight because they either do not absorb UV-A or UV-B radiation and/or they are not efficiently converted to photoproducts (Boreen et al., 2003). In such cases, biodegradation processes may become significant. For example, Bartels and von Tumpling (2008) observed that oseltamivir carboxylate did not degrade by direct photolysis in shallow water but rather from a combination of indirect photolysis and biodegradation. This lack of direct photolysis was attributed to the poor light absorption of oseltamivir carboxylate in the UV-A and UV-B region of sunlight. The role of indirect photolysis may also be important for pharmaceuticals that do not undergo direct photolysis. For example, the direct photolysis of trimethoprim has been observed to be poor but was significantly enhanced by indirect photolysis in the presence of DOM (Luo et al., 2012). Enhanced photolytic degradation was also observed for sulfonamides in natural water compared to laboratory solutions and this was attributed to the abundance of excited DOM formed under natural sunlight conditions (Batchu et al., 2014). In addition, the production of HO· radicals upon irradiation of Suwannee River Fulvic Acid (SRFA), a model compound for DOM, has been observed (Jacobs et al., 2012). The formation of this type of radical improved caffeine degradation in natural water compared to direct photolysis alone. Further studies have confirmed the improved degradability of pharmaceuticals in natural water conditions (Zeng et al., 2012; Leech et al., 2009).

Degradation of pharmaceuticals may also be *inhibited* in surface waters by DOM as a result of the turbidity reducing the penetration of sunlight. For example, modeling of experimental data from the simulated sunlight irradiation of propranolol hydrochloride showed that PEC values in UK and US river catchments can be significantly reduced by sunlight photolysis (Robinson et al., 2007). Environmental concentrations of propranolol were therefore predicted to be lower in the more intense sunlight conditions of summer than in winter and the degradation enhanced in long rivers with low turbidity and flow conditions.

The transformation of pharmaceuticals in surface water can also be modulated by the aqueous oxygen content which can influence the formation of ROS. Dissolved oxygen has been observed to inhibit the direct photolytic degradation of naproxen but enhance the indirect photolysis of this pharmaceutical (Ma et al., 2014). Enhanced concentrations of singlet oxygen ($^1O_2$) have been observed in the presence of *photosensitizers* such as DOM which absorbs sunlight (Ter Halle and Richard, 2006). Singlet oxygen was observed to be primarily responsible for the degradation of atorvastatin in simulated natural waters (Razavi et al., 2011).

Although photolysis may be the predominant process in the degradation of a pharmaceutical, there is often limited information on the toxicity of the resulting photolytic by-product chemicals compared to the parent compound. These transformation products may be as environmentally persistent and have similar transportation, fate, and effects as their parent pharmaceuticals. Wang and Lin (2014) have noted that pharmaceuticals are expected to be found in mixtures (*see* mixture toxicity below) which may affect their phototransformation rates in surface water. In a mixture, one pharmaceutical may act as a DOM component that increases or decreases the rate of photodegradation of another pharmaceutical. This process raises questions as to the level of direct photolysis actually occurring in aquatic systems. It is challenging to assess the risk of the plethora of resulting phototransformation products on exposed ecosystems. As such, it cannot be concluded with any certainty that environmental phototransformation is a significant purification process leading to a reduction in the environmental risk of the parent pharmaceuticals.

The presence of colloidal carriers or suspended particulate organic matter in surface water may affect the apparent environmental persistence of a pharmaceutical that is adsorbed to these matrices. Maskaoui and Zhou (2010) found up to 84% of the total amount of pharmaceutical material in a water-suspended solid system may be adsorbed onto the solid, leading to a potential underestimation of pharmaceutical concentrations when only the water phase is analyzed. Adsorption of a pharmaceutical onto suspended organic matter reduces its mobility and bioavailability (i.e., its ability to penetrate biological cell membranes of environmental biota). It may also increase the long-term environmental persistence of the pharmaceutical by reducing water-based degradation processes such as those induced by direct sunlight. However, such adsorbed pharmaceuticals can then be exposed to soil or sediment degradation processes.

5.3.1.2 Environmental persistence in soils and sediment

Pharmaceuticals partitioned onto sediment may persist quiescently for decades. Tamtam et al. (2011) collected up to 50-year old sediment samples from a portion of the Seine River downstream of urban Paris (France) and detected traces of quinolone and sulphonamides including norfloxacin, flumequine, oxolinic acid, nalidixic acid, and sulfamethoxazole. The use of radioactive lead- and cesium-dating methods allowed the authors to conclude that the initial presence of the pharmaceuticals in the sediment coincided with the time when they were first marketed. Therefore, it is plausible to suggest that such sediment deposits may provide a long-term reservoir of pharmaceuticals which may ultimately leach from the sediment as runoff and become bioavailable once again to environmental biota.

However, the environmental persistence of pharmaceuticals that are biodegradable may be reduced in sediments (Perelo, 2013). Negligible degradation was observed for oseltamivir carboxylate (OC) in surface water samples from the River Po and the Venice lagoon (Italy) (Sacca et al., 2009). However the addition of 5% of sediment content increased the degradability of OC in the water and a $t_{½}$ of 15 days was determined under these conditions. These results suggest that microbial degradation processes associated with the sediment predominated in the removal of OC under these conditions. The biodegradability of oseltamivir in aqueous solution with 5% sediment content was also observed in laboratory studies involving two Japanese surface water samples (Accinelli et al., 2010).

Sediment or soil biodegradation may be modulated by oxygen. Radke and Maier (2014) observed increased transformation rates induced by the addition of oxygen to anoxic (i.e., oxygen-lacking) river water-sediment systems. Lin and Gan (2011) investigated the sorption and degradation of some NSAIDs and antibiotics in two soil samples collected from arid regions. Biodegradation of the pharmaceuticals naproxen, trimethoprim, diclofenac, ibuprofen, and sulfamethoxazole was more effective under aerobic conditions ($t_{1/2}$=4.8–69.3 days) while all compounds apart from sulfamethoxazole were moderately persistent (most $t_{1/2}$>50 days) under anaerobic conditions. Fang et al. (2012) estimated the level of biodegradation of gemfibrozil under aerobic conditions in sandy and silt loam soils at 25.4% and 11.3%, respectively over 14 days. However, anaerobic conditions were not investigated in the study. Furthermore, while biodegradation of four antidiabetic drugs was slow in soil under aerobic conditions ($t_{1/2}$=5–120 days), the degradation rate was observed to be significantly slower under anaerobic conditions (Mrozik and Stefanska, 2014).

Microbial activity utilising the organic chemical content of pharmaceuticals as a source of carbon is primarily responsible for their degradation in soils. Yu et al. (2013a) reported that the sterilization of soil, which eliminates all microbial activity, leads to an increase in $t_{½}$ of pharmaceuticals resident in the soil. Pal et al. (2011) investigated the persistence of several illicit drugs, including methamphetamine and pseudoephedrine, in South Australian soils under sterile and nonsterile conditions. Biotic degradation processes were observed to predominate over abiotic processes. However, the degree of biotic and abiotic degradation observed for different soil types and incubation periods varied. Nevertheless, there was no significant degradation of methamphetamine by abiotic degradation processes over a one-year period.

Biodegradability of pharmaceuticals may be dependent on the amount and types of microbes and the organic carbon material in the soil, whereby soils with a higher organic content may improve biodegradability (Yu et al., 2013a). However, the composition of soil microbial ecosystems may change over time in response to the presence of pharmaceuticals. For example, Cui et al. (2013) observed that the addition of the antibiotic ciprofloxacin affected the ability of the microcosm to metabolize certain nutrients (i.e., D-glucose, lactose, D-mannose) after a 40-day incubation. Indeed, antibiotics in soils may induce the selection of antibiotic-resistant strains of microbes (Thiele-Bruhn, 2003) especially when the antibiotics become adsorbed onto soil particles (Wegst-Uhrich et al., 2014).

The biodegradation efficiency of pharmaceuticals in soils may also be affected by the ambient temperature. Al-Rajab et al. (2010) observed that the rate of mineralization of the anti-HIV agent tenofovir to form harmless inorganic compounds was increased with increasing temperature up to 40 °C. In addition, the biological transformation rates of estradiol-3-sulfate and estrone-3-sulfate as reported by Scherr et al. (2008, 2009) were more efficient at an ambient soil temperature of 25 °C, compared to

7.5 or 15 °C. The disappearance of clotrimazole from loam soil was also observed to be slower at a lower temperature (Sabourin et al., 2011).

As for the situation involving phototransformation of pharmaceuticals in surface water, the biodegradation of a pharmaceutical in soil may not necessarily nullify a pharmaceutical's biological activity. In fact, some biodegradation processes may make pharmaceuticals more bioavailable and even potentially more harmful in the environment. Shrestha et al. (2012) have documented the transformation of the estrogen conjugate 17β-estradiol-3-glucuronide back to its parent compound, 17β-estradiol (E2) which was subsequently oxidized to a variable extent in both unfertilized cultivated field topsoil and subsoil to estrone (E1). This transformation can be attributed to the ubiquitous presence of β-glucuronidases in soil bacteria. In addition, these enzymes may also be introduced from the application of biosolids, the nutrient-rich organic materials resulting from the treatment of sewage sludge, onto agricultural land. A similar degradation process has also been observed with the estrogen conjugate, 17β-estradiol-3-sulfate, under the action of arylsulfatase enzymes and the degradation of estrone-3-sulfate to E1 in soil (Scherr et al., 2008, 2009). Furthermore, <2% of the carbamazepine spiked in soil by Li et al. (2013) under aerobic conditions was mineralized to CO_2 and other harmless inorganic substances. The remaining >98% was converted into a large number of transformation products, including acridones and acridine, both of which are potentially harmful by-products that are resistant to further environmental degradation.

Microbial biodegradation efficiency may vary for different *stereoisomers* of the same pharmaceutical. In chemistry, stereoisomers are two distinct chemicals with the same molecular formula (number of atoms of each element) but differing in their bonding arrangement in three-dimensional space. A special form of stereoisomer is the *enantiomer*, a molecule that can present itself as a pair of non-superimposable mirror images in three-dimensional space in the manner of a "left-hand" and a "right-hand." In biological systems, sometimes only the "left-handed" or the "right-handed" forms of these molecules can be metabolized or degraded by microbial enzymes. Therefore, the persistence of the medication may be underestimated in cases where the poorly metabolized enantiomer is the active ingredient in the form of the drug that is marketed while degradation studies are undertaken using a mixture of both enantiomers. Bagnall et al. (2013) studied the fate of methamphetamine and amphetamine in river microcosms. Degradation was attributed solely to biotic processes, and a change in the ratio of the enantiomers present in the biological system was observed over a course of 15 days. These results suggested that one enantiomeric form of either methamphetamine or amphetamine was preferentially metabolized by the river microcosms compared to the other form.

Current studies on the fate and persistence of pharmaceuticals in soil and sediment are primarily based on analysis involving a single pharmaceutical species. However, the rate of degradation of one type of pharmaceutical may be modulated by another when they are present as a mixture in the soil. Monteiro and Boxall (2009) reported that the degradation rate of naproxen was decreased in soil spiked with a mixture of other pharmaceuticals. This suggests that the presence of other pharmaceuticals or organic compounds may reduce their degradability in soil or sediment which can lead to an underestimation of the environmental risk of a pharmaceutical.

5.3.1.3 Environmental persistence in marine and coastal wetland environments

Pharmaceuticals have been detected in the water and sediments of estuarine and coastal areas (e.g., Richardson et al., 2005; Gulkowska et al., 2007; Minh et al., 2009; Wille et al., 2010; Hoang et al., 2011). However, few studies have investigated their environmental fate in these environmental systems (Laane et al., 2013). Some research on the distribution of antibiotics in water-sediment systems of the Pearl River estuary system has been undertaken by Liang et al. (2013). They found that the distribution of pharmaceutical content between the water and sediment phases was dependent on the organic carbon content of the sediment. In addition, Ying and Kookana (2003) studied the degradation of several emerging contaminants, including pharmaceuticals, in coastal sediment and water collected from the Adelaide coast (Australia). The degradation of 17β-estradiol (E2) and 17α-ethinylestradiol (EE2) in seawater was found to occur within 56 days. In addition, E2 and EE2 degradation in sediment under aerobic conditions was relatively fast, with $t_{½}$ = 4.4 and ~20 days, respectively. The degradation behavior of pharmaceuticals in marine sediment was observed to be similar to that of freshwater soil and sediment with no significant degradation being observed under anaerobic/anoxic conditions.

Thuy and Loan (2014) investigated the degradation of the antibiotics ciprofloxacin, griseofulvin, and rifampicin in marine water and mixed water-sediment systems. The degradation behavior of pharmaceuticals was observed to be comparable to that observed in freshwater and freshwater sediment. In the water-only system, photolysis was confirmed as the predominant degradation process whereas biodegradation was more important in marine sediment. In addition, adsorption to sediment decreased the environmental degradation of the three antibiotics. These results are consistent with the findings of a previous study on the recalcitrance of antibiotics in freshwater river sediment (Tamtam et al., 2011).

5.3.1.4 Categorization of environmental persistence of pharmaceuticals

Stockholm County Council has developed a ranking framework to assess the persistence of a pharmaceutical using the criteria listed in Table 5.2.

From the examples in the previous section, there are clearly some cases that satisfy the criteria for being persistent or very persistent, such as the striking persistence of 50-year-old antibiotic residues in river sediments (Tamtam et al., 2011). There are also examples where pharmaceuticals with relatively short degradation $t_{½}$ may pose low risks to exposed biota. This could be the case for ceftriaxone that has a degradation $t_{½}$ of 4.1 days in surface water (Jiang et al., 2010).

Overall, there is limited information on the environmental degradation kinetics of pharmaceuticals. Little is known of their behavior in mixtures or multicomponent systems and how this will affect environmental degradation processes (i.e., hydrolysis, microbial degradation, and photolysis). In particular, more studies on environmental variables and ambient conditions to modulate the efficiency of degradation processes are warranted to better assess the persistence of pharmaceuticals. There has been much research on the

Table 5.2 Environmental persistence of pharmaceuticals as classified by Stockholm County Council

Environmental system	$t_{1/2}$ (abiotic or biotic degradation) (days)	Categorization
Freshwater	>40	Persistent
Freshwater	>60	Very persistent
Marine water	>60	Persistent
Freshwater sediment	>120	Persistent
Freshwater sediment	>180	Very persistent
Marine sediment	>180	Persistent

Data from Table 9 in http://www.fass.se/pdf/Environmental_classification_of_pharmaceuticals-120816.pdf (accessed 9-9-15)

environmental fate of pharmaceuticals in water and the interaction of sunlight, radical species, and DOM on their degradation. However, there is a dearth of information on their degradation in sediment or soil or their behavior in marine or coastal environments. It is therefore pertinent that future research into the environmental fate and persistence of pharmaceuticals should address a diverse array of environmental systems and microcosms to provide a broader perspective on their potential environmental risk.

5.3.2.1 Bioconcentration factors of pharmaceuticals

Bioconcentration is the ability of an organism to accumulate a chemical from the ambient environment. It is quantitatively expressed in terms of the *bioconcentration factor* (BCF) which is the ratio of the concentration of the chemical in the organism to its concentration in the ambient environment. A range BCF values for the bioaccumulation of pharmaceuticals have been reported. For example, BCF values of 2.2 and 12.6 have been reported for the bioaccumulation of carbamazepine by the green algae *Pseudokirchneriella subcapitata* and the crustacean *Thamnocephalus platyurus*, respectively (Vernouillet et al., 2010). Wang and Gardinali (2012) studied the bioaccumulation of caffeine, diphenhydramine, and carbamazepine in mosquitofish (*Gambusia holbrooki*) and reported BCF values of 29 ± 26, 821 ± 422, and 108 ± 144, respectively, for the three pharmaceuticals which were comparable to the results of another mosquitofish study (Wang and Gardinali, 2013). However, the magnitude of BCF values were in the order of diphenhydramine > caffeine > carbamazepine after correcting for steady-state concentrations and uptake and elimination kinetics observed in the fish.

BCF values may be concentration-dependent for a drug in an organism. The bioconcentration of moxifloxacin and rosuvastatin in the mussel (*Lampsilis siliquoidea*) has been reported by Gilroy et al. (2014) to give BCF values between 0.03–70 and 0–0.05, respectively, when mussels were exposed to the drugs at concentrations up to 100 mg/L. The inability of the mussels to filter drugs was suspected to have increased the BCF at the higher drug concentrations, and these higher concentrations were correlated with decreased mussel-feeding activity. However, these sublethal effects were observed at concentrations much higher than the correspoding PEC vlaues for these two pharmaceuticals, suggesting a low risk to at least these two mussel species from the environmental presence of moxifloxacin and rosuvastatin. Indeed, the ability for drugs to be eliminated efficiently by the organism may reduce apparent BCF values.

Once a drug is distributed within an organism from the surrounding ambient environment, it may preferentially accumulate in some tissues or organs which can lead to different BCF values for each tissue-type or organ. Liu et al. (2014) explored the bioconcentration of roxithromycin in goldfish and found that BCF values were higher in the liver (2.15–38.0) and bile (0.950–20.7) compared to the gill (0.0506–19.7) and muscle (0.0439–13.8) tissues. Tissue BCF values were regularly observed to be slightly higher than those in plasma during bioaccumulation monitoring of carbamazepine in the fathead minnow *Pimephales notatus* and channel catfish *Ictalurus punctatus* and ranged from 1.8 to 7.1 (Garcia et al., 2012). Upon exposure of these two fish species to the oral contraceptive norethisterone, Nallani et al. (2012) reported that bioaccumulation was greatest in the liver followed by plasma, gill, brain, and muscle with BCF values ranging from 2.6 to 40.8. In summary, the greatest magnitude of bioaccumulation usually occurs in liver for many higher organisms such as fish, as it is this organ which is responsible for the metabolism of many pharmaceuticals and plays a role in lipoprotein utilization and storage.

The pH of the ambient water surrounding the organism may affect drug uptake. Anskjaer et al. (2013) examined the bioaccumulation of sulfadiazine in *Daphnia magna* from surrounding water with a pH of 6.0 and 8.5 and reported BCF values of 50 and 36 respectively. However, while the bioaccumulation of sulfadiazine was 1.5 times greater at pH 6.0 compared to 8.5, the toxicity observed was over 10 times greater at pH 6.0 compared to 8.5. These results suggest that drug toxicity to environmental organisms may not be linearly proportional to the amount of drug accumulated in the organism.

Finally, estimation of BCF values for a pharmaceutical can vary depending on experimental conditions used. Cuklev et al. (2012) studied the bioaccumulation of ketoprofen and diclofenac by rainbow trout (*Oncorhynchus mykiss*). They observed that ketoprofen bioaccumulated less from surface water than diclofenac (BCF values of 0.05 and. 4 respectively). However, in undiluted sewage effluent, ketoprofen was observed to bioaccumulate into fish tissue more than diclofenac. This illustrates that there are limitations in using BCFs to assess risk. If BCF values are to be used to accurately assess risk, one must also consider various other parameters including metabolism and organ-specific bioconcentration in relation to ecotoxicological endpoints (Franke, 1996).

5.3.3 Fate of pharmaceuticals in environmental organisms (toxicokinetics)

After bioaccumulation in an organism, the subsequent fate of a pharmaceutical may largely depend on the ability of the organism to metabolize it. As such, BCF values may need to be refined to take into account such metabolic capabilities in an organism (Gomez et al., 2010). In Chapter 2, we discussed patient variability in pharmacokinetics which may influence the therapeutic drug concentration in humans. These same concepts can be similarly applied to the behaviour of a pharmaceutical in other (non-human) organisms but clinical data for a pharmaceutical derived from human studies may not be necessarily representative of the drug's behavior in these other biological species. For example, isolated liver fractions (S9) from rainbow trout (*O. mykiss*) contain cytochrome P450 (CYP) CYP1 and CYP3 isoenzymes that are genetically similar to their human variants. However, no metabolism of drugs that are specifically metabolized by CYP3A4 in humans, such as diltiazem and carbamazepine, was observed in these liver fractions of trout (Connors et al., 2013b). On the other hand, drugs which can be metabolized by CYP1A2 in humans, such as propranolol and diclofenac, were observed to be also metabolized by trout liver fractions. Another study showed that fluoxetine was not metabolized by liver S9 fractions in trout (Connors et al., 2013a). This is consistent with fluoxetine being metabolized by CYP2D6, CYP2C9, and CYP3A4 in humans while fish lack both CYP2C and 2D homologs (Smith et al., 2010).

Cytochrome P450 enzymes with similar gene sequences in both humans and fish, but may not possess the same types of metabolic activity at least in rainbow trout (Connors et al., 2013b). In contrast to rainbow trout liver fractions, the human metabolism of phenacetin to paracetamol by CYP1A2 and dextromethorphan to dextrorphan by CYP2D6 was observed in zebrafish larvae (*Danio rerio*) (Alderton et al., 2010). In addition, various human Phase II metabolism pathways were also observed in the fish. These results further demonstrate different metabolic pathways of the same pharmaceutical between humans and fish and between different fish species. Cytochrome P450 enzymes are only responsible for Phase I metabolism (*see* Chapter 2), and homogenized trout liver fractions contain a combination of both Phase I and Phase II conjugation enzymes (i.e., various glucuronyltransferases and sulfotransferases) and so these *in vitro* liver fraction studies do not provide enough information to determine the exact enzymes responsible for the metabolism.

As observed in bacterial microcosms, the metabolism of a pharmaceutical may be dependent on its stereoisomerism (*see* discussion above). In rainbow trout liver fractions, the *R*(+) enantiomer of propranolol is cleared by liver fractions at a much slower rate than the *S*(−) enantiomer while the opposite trend was found in humans (Connors et al., 2013a). There may also be other factors modulating the clearing of pharmaceuticals as found with the metabolism of propranolol that occur at different rates depending on the species of trout (Gomez et al., 2010; Connors et al., 2013a).

5.4 Ecotoxicology

In general terms, ecotoxicology is the study of the effects of natural and synthetic chemicals (stressors) in the biosphere considered within a context of ecological and human health (Newman, 2010). It is a multidisciplinary field incorporating concepts arising from many scientific disciplines including toxicology, biology, analytical, environmental and organic chemistry, physiology, ecology, genetics, microbiology, biochemistry, immunology, molecular biology, soil, water and air sciences, and economics.

The aim of ecotoxicology is to assess the acute and chronic lethality of chemicals using a range of test methodologies derived from mammalian toxicology. *Acute lethality* refers to the death of an organism after a brief and intense exposure to a chemical or complex mixture while *chronic lethality* results from a more prolonged exposure (Newman, 2010). Acute and chronic toxicity tests are designed to assess the short- and long-term effects of chemicals on key biological parameters, including survival, reproduction, and physiological and biochemical responses. The approach used for the testing involves exposures of selected organisms to increasing amounts of the test substance. The aim is to derive a standard measure of toxicity such as the *median lethal concentration* (LC50). The LC50 parameter is the predicted dose resulting in the death of 50% of the exposed individuals over a set exposure period.

Other important ecotoxicological tools include the use of *biomarkers* which are measurable indicators of some biological condition in an organism. These can be developed with a better knowledge of the mechanisms of toxicity and the causality of effects, and can provide a functional measure of organism response to single and complex mixtures of bioavailable chemicals (Adams et al., 2001). The challenge with assessing the risk of pharmaceuticals is that some have totally new and often multiple mechanisms of action (MOAs) that are rarely fully understood even for the targeted therapeutic effects. The continuous commercial release of new pharmaceuticals in addition to other anthropogenic chemicals, poses significant challenges to the development and design of robust monitoring frameworks based on a chemical-by-chemical approach for regulating the occurrence of pollutants in the environment (Daughton, 2004). The characterization of hazards resulting from exposures to chemicals commonly used in environmental risk assessment is derived from ecotoxicological data. The basis for assessing the ecotoxicology of pharmaceuticals in environmental biota (*toxicodynamics*) will be discussed below.

5.4.1 Acute versus chronic ecotoxicity

The ecotoxicity of a chemical can usually be described in terms of the length of exposure to the hazard, which in the context of the current text is one (or more) pharmaceutical(s) (Chapter 1). *Acute toxicity* describes any detrimental effects arising from the exposure of an organism to a pharmaceutical hazard usually over a span of no more than 15 days. In contrast, *chronic toxicity* describes detrimental effects resulting from the long-term exposure (≥15 days to years) of an organism to the pharmaceutical hazard (Fent et al., 2006). The definition of chronic toxicity is satisfied in a situation where regular assaults of the pharmaceutical on the organism occur as discrete events but on a regular basis (i.e., the *pseudopersistent* exposure of an organism to a particular pharmaceutical with a short environmental $t_{1/2}$ such as might result from regular sewage discharge events) or when the organism is exposed continuously to the pharmaceutical hazard (i.e., a rarely used pharmaceutical discharged occasionally into the environment but having a long environmental $t_{1/2}$).

The pharmacology associated with acute and chronic toxicity also differs. Acute toxicity usually arises from a nonspecific mode of action on tissues independent of particular receptors or target molecules able to modulate a toxic response. Often, acute toxicity results from the interaction of the chemical hazard (the pharmaceutical) with the cellular membranes of organisms, leading to cellular or tissue damage or even death (Fent et al., 2006). The disruption of cell membranes may also lead to a decrease in cellular biochemical functions and to an alteration of the usual physiological responses or behavior of an organism to environmental stimuli. For instance, putative molecular interactions between eight antidepressants and phosphatidylcholine, a common cellular phospholipid in cell membranes, have been demonstrated by computer modeling *in silico* (Minguez et al., 2014). The authors suggest that the acute toxicity of these antidepressants on cells is more dependent on their physicochemical structure than actual drug–receptor responses in organisms. These antidepressant molecules have both hydrophilic and hydrophobic parts to their chemical structure (i.e. they are *amphipathic*) and this is a property which easily allows them to pass through and lodge inside the cell membrane where the ester functional groups can potentially be hydrolysed. These nonspecific cellular disruption mechanisms are commonly referred to as *narcosis*. Sanderson and Thomsen (2009) reviewed 569 acute effects data that was available for a selection of 275 pharmaceuticals. Ninety-two percent of this data suggested that the acute effects were the result of narcosis. In addition, 68% of the acute effects observed for the selected pharmaceuticals suggest that most of the observations were based on toxicity associated with a narcotic mode of action.

In acute toxicity, usually a very high concentration of a test chemical is necessary for a short deleterious assault on cell membranes or to cause cell death. As discussed in Chapter 4 and elsewhere in this book, the levels of pharmaceuticals in the environment rarely exceed μg/L concentrations. However, most of the acute toxicity associated with pharmaceuticals leading to the death of organisms occurs for concentrations a thousand or more times higher i.e. in the mg/L range. For example, a 96 h acute toxicity test of the verapamil heart treatment drug on early life-stage larvae of common carp (*Cyprinus carpio*) that induced a 50% death in the tested population gave rise to an estimate of the LC50 value for this drug of 4.8 ± 0.2 mg/L (Steinbach et al., 2013). However, no mortality was observed when the same larvae were exposed to environmentally relevant (μg/L) concentrations of this drug. Kim and Lee (2009) reported the death of 50% of *Thamnocephalus platyurus* when exposed to a 3.95 mg/L concentration of mefenamic acid. A similar LC50 value of 8.04 mg/L was determined from 96 h acute exposure of *Oryzias latipes* to mefenamic acid. In some instances, exposures of organisms to pharmaceuticals at concentrations much higher than their real or calculated environmental concentrations still does not result in acute toxicity. Tong et al. (2011) investigated the 96 h acute toxicity of oseltamivir phosphate (2.55 mg/L) on the marine harpacticoid copepod *Quinquelaophonte* sp., and no lethality or changes in organism behavior were observed for this drug concentration. Such studies suggest that it is very unlikely that any acute effects will occur in organisms arising from the environmental presence or persistence of pharmaceuticals.

Instead the presence of pharmaceuticals in the environment is most likely to result in more subtle chronic toxicity effects. Chronic toxicity is usually associated with particular drug–receptor actions that initiate a particular pharmacological response in an aquatic or terrestrial organism (Sanderson and Thomsen, 2009). Therefore, chronic toxicity endpoints are rarely based on lethality but rather on subtle physiological, behavioral or biochemical changes induced by exposure of an organism to the pharmaceutical. For example, Pascoe et al. (2003) investigated the exposure of the freshwater polyp *Hydra vulgaris* to a selection of pharmaceuticals each at a concentration of 10 μg/L over a period of 17 days. *Hydra vulgaris* is an animal with prominent feeding tentacles and so measured toxicological endpoints from the exposure included the ability of the test organisms to feed and to produce tentacle buds. The ability of dissected organisms to regenerate feeding tentacles was observed to be inhibited by chronic exposure to diazepam, digoxin, and amlodipine. However, there was no significant change to feeding behavior compared to organisms not exposed to the pharmaceuticals. Bergheim et al. (2012) observed a low-level of bacterial chronic toxicity in a mixture of tramadol and its phototransformation products but not from a similar mixture of ranitidine.

Lamichhane et al. (2013) examined the toxicity of carbamazepine (100–300 μg/L) to three successive generations of the water flea *Ceriodaphnia dubia*. The authors found reduced reproductive rates, slower maturation, and reduced weight in some of the generations over a test period > 2 weeks. In another study, chronic exposure to mefenamic acid at mg/L levels resulted in reduced reproduction of the crustacea *Daphnia magna* and *Moina macrocopa* and vitellogenin biosynthesis (*see* below) in the zebrafish *Danio rerio* (Collard et al., 2013). The results of these studies were observed at concentrations generally above the PEC of most pharmaceuticals in the environment, but they provide insights into potential MOAs. Feito et al. (2013) examined the survival of spores from the fern *Polystichum setiferum* after exposure to venlafaxine. Acute toxicity was observed for the lowest observed effect concentrations (LOEC) of 1 μg/L and 0.1 μg/L after 24 h and 48 h of exposure, respectively. As an aside, it is worth noting that the test spores are representative of critical and sensitive life stage of development.

Ecosystems are composed of multiple organism types which means that our knowledge of chronic toxicity in the environment is unlikely to be complete, as chronic toxicity measurements require an understanding of the form and function of tissues specfic to an organism of interest (Brausch et al., 2012). Usually, chronic toxicity tests are based on standardized protocols involving selected organisms and there is a level of uncertainty around the validity of the results of these tests in order to assess the risk to other organisms. However, they can still provide a useful insight into the long-term exposure of organisms to a particular chemical.

Stockholm County Council (available www.fass.se/environment, accessed 23-8-14) considers ecotoxicological data used for risk assessment acceptable if they are based on test organisms from three different *trophic* levels (i.e., the food chain stratification from predator to prey). A pharmaceutical is considered toxic by definition if a no observed effect concentration (NOEC) obtained from at least a chronic toxicity test is <0.01 mg/L. Some pharmaceuticals discussed above fit into this category. For example, the 48-h LOEC for venlafaxine in *Polystichum setiferum* was estimated at 0.1 μg/L (Feito et al., 2013). Overall, there is limited ecotoxicological data available to environmental managers to conduct robust risk assessments and a strategy to further understand the potential effects of pharmaceuticals on environmental biota is required. In the following section, we discuss some test organisms and the toxicity endpoints commonly used for environmental risk assessment of environmental pollutants including pharmaceuticals.

5.4.2 Ecotoxicology testing in fish

Fish are generally the uppermost trophic level commonly used in ecotoxicological testing and are the most similar to the structure and function of humans. They are the most commonly used vertebrates in ecotoxicology and share many physiological and endocrinological similarities with humans. Their complex anatomy makes them suitable for testing a wide range of toxicological endpoints which may be too subtle to be objectively determined in organisms at lower trophic levels. In addition, some human receptor types are also present in fish. As such, pharmaceuticals possessing pharmacological activity for these human receptor types can be predicted to be also bioactive in fish. For example, two different estrogen receptors (ER-α, ER-β) have been characterized in mammals and these same two receptors along with ER-β2 have been found to be present in teleost fish (Hawkins et al., 2005). In humans, estrogens bind to the ER to modulate ovulation and other reproductive functions and fish are also controlled by similar physiological processes (Crisp et al., 1998). Specifically, estrogens are responsible for the regulation of gonadal functions and development as well as oogenesis (i.e., the production and development of eggs). However, there are genetic differences between human and fish ERs. While ER-α is similar in form and function across all invertebrates, including humans and fish, humans possess only a single form of ER-β while fish have both ER-β1 and ER-β2 receptors (Nelson and Habibi, 2013). In addition, the ER mode of action is very complicated. ERs are intracellular (i.e., located in the inside of the cell) and estrogens must penetrate the cell membrane before they can bind to the receptor and exert effects on development, growth, and reproduction, in addition to other bodily functions. When an estrogen-activates an ER, the complex relocates to the cell nucleus with a host of coregulation proteins and modulates the estrogen-regulated reproductive processes (see Nelson and Habibi, 2013 for a comprehensive overview). There are over 300 different estrogen-coregulated proteins in mammals alone.

Because of the diversity of these coregulator proteins and the different number of ER types in humans and fish, the pharmacology of estrogens in humans may not be necessarily predictive of their effects in fish. In humans, the ability of ER-α to bind estradiol (E2) is much greater than that for ER-β, whereas the opposite is true of fish (i.e., the receptor types between species show different *binding affinities*) (Nelson and Habibi, 2013). In addition, fish ERs generally have a lower affinity for E2 than human ERs and may also show temperature-dependent binding (Nelson and Habibi, 2013).

One particular biomarker that can be used to determine the effects of estrogenic compounds on fish is the induction of vitellogenin, a group of calcium- and zinc-rich proteins responsible for the development of embryos in egg-laying (*oviparous*) animals. Vitellogenins assist in the formation of vitellins, which are major constituents of egg-yolk protein and are present at elevated levels in reproductive females (Jones et al., 2000). It is unlikely that vitellogenin is detected in male fish plasma. However, vitellogenin can be biosynthesized in male fish exposed to environmental estrogens (Marin and Matozzo, 2004). Various protein and molecular-based methods are used to assess vitellogenesis, or the production of vitellogenin, in fish (Wheeler et al., 2005). The assessment of the estrogenic potential of a chemical can be undertaken by measuring vitellogenin production in hepatocytes isolated from fish liver that can be maintained in culture (Navas and Segner, 2006).

However, the pharmacological nature of vitellogenin induction in fish is not completely defined. While the ER-α induces vitellogenesis in *in vitro* fish liver cells, there is some evidence that ER-β subtypes are responsible for the production of vitellogenin in male fish. In addition, the amount of ER-α in females may vary seasonally (Nelson and Habibi, 2013). Thus, vitellogenin and the effect of estrogens on the reproductive behavior of fish may be sex-, receptor-, or seasonally-dependent (Nelson and Habibi, 2013). However, irrespective of the pharmacological nature of vitellogenin production in fish, its presence may represent a significant health risk. For example, elevated levels of vitellogenin have been correlated with the development of a female reproductive tract or eggs (oocytes) in the testis of the male roach (*Rutilus rutilus*) (Tyler and Jobling, 2008). These changes to the reproductive tract of male fish may ultimately lead to population crashes and abnormal breeding behaviors.

There are a number of toxicity tests for fish and some common ones used in ecotoxicology processes include the fish acute toxicity test (OECD 203) and the fish early life stage test (OECD 210)) (available at http://www.oecd-ilibrary.org/environment/test-no-210-fish-early-life-stage-toxicity-test_9789264203785-en, accessed 28-8-14). The acute toxicity test endpoint is usually mortality whereas a variety of endpoints, such as hatching, survival, deformities to fish and behavior, can be assessed in chronic toxicity tests. It should be noted that there are serious ethical issues with the use of animals in toxicity testing, and alternative test models such as the zebrafish fish embryo toxicity (FET) provide a more ethically acceptable alternative to whole organism models (Strahle et al., 2012).

5.4.3 Ecotoxicology testing with Daphnia

Daphnia sp. are water flea crustacea commonly used for ecotoxicological testing. They are easily cultured in the laboratory and because these animals are small, 20 or 30 of them can be cultured at one time in a vessel, as long as food and aeration is sufficient (Adema, 1978). These properties allow this organism to be routinely used in ecotoxicological testing. Various OECD standard tests have been published that use *Daphnia* sp. to determine the toxicity of environmental pollutants. For example, the *D. magna* reproduction test (OECD 211) determines the LOEC from a chronic 21-day exposure test (available http://www.oecd-ilibrary.org/environment/test-no-211-daphnia-magna-reproduction-test_9789264070127-en, accessed 26-8-14). A single female *D. magna* is placed in a temperature, oxygen, and pH-controlled medium of 50–100 mL volume. Each crustacean is exposed to five different concentrations of the pollutant. The total number of living offspring produced per parent animal that do not die accidentally during the test and the number of living offspring produced per surviving parent animal at the end of the test are reported. After 21 days, the number of offspring produced by the exposed parent *Daphnia* is then compared to that of the control *Daphnia* not exposed to the pollutant. The *Daphnia* reproduction test is commonly used by ecotoxicologists because female Daphnia can reproduce at age 5–10 days and additionally produce every 3–4 days (Ebert, 2005). This fast rate of reproduction (*fecundity*) allows any adverse effects resulting from hazard exposure to be readily observed.

For accurate ecotoxicological assessment using the *Daphnia* reproduction test, it is important that the ambient conditions, such as temperature, are carefully controlled to avoid confounded results. Martins et al. (2013) measured the chronic toxicity of the antibacterial florfenicol using the *Daphnia* reproduction test and found that *effective concentrations* (EC_x; *see* discussion note on *mixture toxicity* below) were influenced by temperature. At 20 °C ambient temperature, population growth was inhibited at a florfenicol concentration of 12.6 mg/L but this was lowered to 3.1 mg/L at 25 °C. This study demonstrates that environmental factors such as temperature can potentially significantly modulate the ecotoxicity of a pharmaceutical.

Unlike fish, *Daphnia* are lower-order plankton-like creatures with limited anatomical complexities. However, the daphnid is a widely used toxicity assessment model with an extensive endocrine system that can provide much information about the endocrine-disrupting potential of pharmaceuticals (Hutchinson, 2007). Indeed, the reproduction test (OECD 211) can provide information about the putative mode of toxicity. Berninger et al. (2011) examined the toxicity of the pharmaceutical diphenhydramine and estimated NOECs of 0.8 μg/L for *Daphnia* and 2.8 μg/L for the fathead minnow *Pimephales promelas*. Both these concentrations are potentially environmentally relevant. The lower threshold toxicity observed with *Daphnia* suggests that diphenylhydramine can interact with reproduction processes, possibly through acetylcholine and histamine receptors which are antagonized in humans. The relevance of these results to human pharmacology remains to be investigated.

Many ecotoxicological tests involving whole organisms are based on easily measurable endpoints such as lethality, growth, or behavior. Such tests include the *Daphnia* immobilization test (OECD 202) (available http://www.oecd-ilibrary.org/environment/test-no-202-daphnia-sp-acute-immobilisationtest_9789264069947-en, accessed 26-8-14). This is a short-term or acute toxicity test that involves testing the chemical at five different concentrations. *Daphnia* of no more than 24 h of age are used in the tests and the number of immobilized organisms is recorded after exposure to the pollutant after 24 and 48 h.

Concentrations above environmentally relevant levels are usually required for immobilization of *Daphnia* which is consistent with what has been reported using other acute toxicity tests with pharmaceuticals. For example, the median 48 h lethal concentrations (LC_{50-48}) for the exposure of *Daphnia* to caffeine, norfloxacin, and sulfamethoxazole were 177.49, 187.48, and 234.18 mg/L, respectively (Lu et al., 2013). Such results can be useful for immediately determining whether a pharmaceutical is hazardous to environmental biota as well as whether further testing is required to identify the particular mode of toxicity and confirm the risk.

5.4.4 Ecotoxicology testing with plants and algae

The determination of effects arising from environmental pollutants at the lowest trophic level usually involves the exposure of freshwater algae, cyanobacteria, or higher plants, such as *Lemna gibba* (duckweed) to the pollutant of interest. These tests incorporate the algal growth inhibition tests (OECD 201) (available http://www.oecd-ilibrary.org/environment/test-no-201-alga-growth-inhibition-test_9789264069923-en, accessed 28-8-14) or the *L. gibba* assay (OECD 221) (available http://www.oecd-ilibrary.org/environment/test-no-221-lemna-sp-growth-inhabition-test_9789264016194-en, accessed 28-8-14).

In the OECD 201 algal growth inhibition tests, a fast-growing species is preferable. Such algal species may include but are not limited to *Pseudokirchneriella subcapitata* and *Desmodesmus subspicatus.* Samples of these algae are exposed to the pollutant of interest over 72 h at five different concentrations, and the resulting algal growth over the 72 h timeframe is measured by the amount of light that is absorbed by the sample and compared with a control. In the *L. gibba* assay, 9–12 organisms are placed in test vessels and exposure to the pollutant of interest for no more than seven days. For estimating LOEC and NOEC with the *L. gibba* assay, the lowest test concentration chosen for the pollutant of interest should not yield significant decreases in growth, whereas a significant decrease in growth should be clearly expected with the highest test concentration. To determine estimates of LOEC or NOEC, the total growth, the surface area of each organism, and the wet and dry weights are taken into account and compared with the same parameters for a control sample. The tests always require the use of a reference toxicant to assess the constancy of response of a given species of test organisms to that chemical. It is assumed that any change in sensitivity to the reference substance will indicate the existence of some similar change in degree of sensitivity to other chemicals whose toxicity is to be determined.

A study investigating some fluoroquinolone antibiotics demonstrated very limited toxicity to three of the five species tested including *Pseudokirchneriella subcapitata* (algal species), *Daphnia magna* (waterflea) and *Pimephales promelas* (fathead minnow) while *Microcystis aeruginosa* (algal species) was the most sensitive (Robinson et al. 2005). However, effective concentrations in the 100–200 μg/L range have been reported for the exposure of *P. subcapitata* to anti-cancer pharmaceuticals (Brezovsek et al., 2014). The toxicity values obtained by the exposure of *P. subcapitata* to ciprofloxacin were in the mg/L range (Martins et al., 2012). The results for *Desmodesmus subspicatus* are also similar, with effective concentrations > 8 mg/L for ciprofloxacin (Ebert et al., 2011) and a LOEC of 40 mg/L was observed for 5-fluorouracil (Zounkova et al., 2010). No effects were observed on *L. gibba* upon exposure to diphenhydramine in acute toxicity tests at levels up to 10 mg/L (Berninger et al., 2011).

Interestingly, *L. gibba* may be more susceptible to the antibiotic sulfonamide. Brain et al. (2008) identified an increase of *para*-aminobenzoic acid (PABA) in *L. gibba* samples following exposure to sulfamethoxazole at the 3–150 μg/L range. An increase of PABA suggests that sulfonamides may act with a specific mode of action as competitive inhibitors of dihydropteroate synthase (DHPS) in *L. gibba* with a similar mechanism in bacteria. Indeed, DHPS is important for the growth and repair of cells as it is responsible for the utilization of PABA for the biosynthesis of folic acid in bacteria. However, it is not present in human cells as humans can scavenge folic acid from the diet.

In summary, while many of these algae and plants are not susceptible to pharmaceuticals, further research may be necessary to identify those pharmaceuticals which exert alternative modes of action in these organisms.

5.5 Mixture toxicity

Pharmaceuticals are generally not released into the environment as isolated, single components but rather as complex mixtures with other pollutants, especially in aquatic systems (as already noted earlier in this and previous chapters). As a consequence, organisms (including humans) are rarely exposed to isolated pharmaceuticals in the environment, but instead to these complex chemical mixtures, the individual components of which might be present at concentrations too low to raise concern on their own. There is limited information to estimate whether additive or even synergistic effects can render these mixtures more potent than when the components are on their own (Schwarzenbach et al., 2006). If a pharmaceutical is excreted from the body after human use, there is the possibility that it is also excreted with metabolites which may be equally or even more bioactive than the parent pharmaceutical itself. Studies to date suggest that current sewage treatment technologies (see Chapter 6) have different levels of removal efficacy depending on the how the chemical characteristics of the pharmaceutical can influence degradation mechanisms such as hydrolysis, microbial degradation, or photolysis that may partially transform the parent pharmaceutical into bioactive products (Luo et al., 2014).

The widespread occurrence of mixtures of pharmaceuticals means that current frameworks for determining the environmental risk of individual pharmaceuticals may not satisfactorily reflect their actual environmental risk when they are present in mixtures. For example, the use of the risk quotient (RQ) (*see* Section 5.2) as a measure of risk assessment estimates environmental risk from an individual pharmaceutical's predicted concentration and available ecotoxicological data. Important limitations with the use of RQs include that they do not consider the environmental risk of metabolites or transformation products in the same assessment. In addition, the hazard categorization schemes developed for assessing the intrinsic environmental harm of pharmaceuticals (such as the Stockholm PBT classification discussed above) and many ecotoxicology studies involving test organisms, often summarize data from hazards of individual pharmaceuticals rather than the complex mixtures of them that undoubtedly exist in practice.

Ecotoxicologists have long known about *synergism* in environmental contexts, where the expected effect or response from an organism's exposure to a pollutant is enhanced because the organism has been exposed to another pollutant. In addition, certain ambient environmental conditions may contribute to enhance toxicological responses observed in an organism. As discussed above, florfenicol was determined to be more toxic to *Daphnia* at an ambient temperature of 25 °C compared to 20 °C (Martins et al., 2013). These *multiple stressors* within an environmental system can often affect the accuracy of collected ecotoxicological data, given that it is usually difficult to retrospectively assess these stressors once the data has been collected.

While some stressors may be difficult to quantify, there are two general concepts involved in understanding toxicity arising from the multiple chemical components in a mixture and these are known as *concentration addition* (CA) and *independent action* (IA) (Altenburger et al., 2003). In CA, the expected pharmacological response from each pharmaceutical component, when normalized for their potencies, is the sum of their parts in the mixture (i.e., classic *additive toxicity*) whereas this is not the case in IA (Backhaus and Faust, 2012). In CA, the same level of toxicity to an organism occurs if any pharmaceutical in the mixture is substituted for another pharmaceutical at the same potency level.[1] Mathematically, this situation can be expressed as (Altenburger et al., 2000):

$$EC_{x,\text{mix}} = \left[\left(p_1 / EC_{x,1}\right) + \left(p_2 / EC_{x,2}\right) + \left(p_3 / EC_{x,3}\right) + \ldots + \left(p_n / EC_{x,n}\right)\right]$$

$$EC_{x,\text{mix}} = \sum_{i=1}^{n} \frac{p_i}{EC_{x,i}}$$

Here, $EC_{x,\text{mix}}$ is the effective concentration at an x level of pharmacological response for the entire mixture of n pharmaceutical components, p_i is the *fraction* of concentration of a particular pharmaceutical component ($i = 1, 2, 3, \ldots n$) in the total mixture and the $EC_{x,i}$ is the effective concentration at an x level of pharmacological response for an individual pharmaceutical component (i). Note that the sum of pharmaceutical component fractions in the mixture must equate to unity ($\Sigma p_i = 1$). For example, the p_i values for aspirin and ibuprofen would be 0.8 and 0.2, respectively for 1 L of a two-component mixture consisting of 0.8 mg of aspirin at 0.2 mg of ibuprofen to give a total concentration of pharmaceuticals in the mixture of 1 mg/L.

Concentration addition would be expected in a mixture of pharmaceuticals from the same therapeutic class, as the overall response most likely arises from a common pharmacological mode of action for each pharmaceutical (Altenburger et al., 2004). Even if a pharmaceutical is present in the environment at sub-therapeutic levels, its effects may be additive to those of any other pharmaceutical present and trigger a pharmacological response in an exposed organism. For example, Cleuvers (2004) assessed the mixture toxicity of diclofenac, ibuprofen, naproxen, and acetylsalicylic acid (aspirin) to *D. magna* and found that the sum toxicity of the concentrations tested could be accounted for through concentration addition. Indeed, all four pharmaceuticals are NSAIDs and the pharmacological mode of action for each drug involves the same COX inhibition. While toxic effects on the test organisms for these four pharmaceuticals in individual single-component systems were only observed at

[1] It is important to note that, *potency* describes the amount of a pharmaceutical needed to achieve a particular effect or response at the receptor, tissue, organ, or organism level. Pharmaceuticals that achieve a pharmacological response at lower doses are therefore more potent. Because different pharmaceuticals exhibit different potencies, the same concentrations of different pharmaceuticals, such as aspirin (1 mg/L) and ibuprofen (1 mg/L) for example, are not equivalent toxicological substituents in any consideration of concentration addition. Instead, the concentration must be weighted against its *effective concentration* (EC_x) at a particular level (x) of pharmacological response. This particular level is commonly 50% (EC_{50}) and is the concentration at the inflection point of a *dose* (independent variable) *vs response* (dependent variable) curve. To use CA to predict mixture toxicity, the EC_x values of all pharmaceuticals in the mixture must therefore be known. It should also be noted that potency is independent of a pharmaceutical's *efficacy*. Efficacy describes the magnitude of the pharmacological response that can be achieved irrespective of dose. A potent drug is not an efficacious drug if dose increases do not proportionally increase the pharmacological response observed at low doses in a subject or organism (i.e. there is a *low ceiling* to the efficacy in such an instance). CA models fit data only for a particular magnitude of a pharmacological response and do not determine whether a particular mixture is more toxic/efficacious than another.

concentrations much higher than environmentally relevant levels, mixture toxicities were observed at levels of the individual pharmaceuticals that were considerably lower.

Concentration addition can occur even in situations where a mixture may contain pharmaceuticals from various therapeutic classes. For example, Bain and Kumar (2014) studied cellular changes to a rainbow trout tissue-culture cell line (cytotoxicity) from exposure to binary mixtures of fluoxetine (a selective serotonin reuptake inhibitor (SSRI), *see* below), propranolol (a β-blocker), and diclofenac (an NSAID). The observed cytotoxicity arising from all three possible binary mixtures of these pharmaceuticals could be adequately predicted by CA. This suggests that the observed cytotoxicity arises from an unknown pharmacological action but one that is common to all three pharmaceuticals and independent of their known therapeutic modes of action which are clearly very different. Indeed, CA may sometimes be the result of non-specific narcosis (*see* above) for pharmaceuticals with limited similarities in their pharmacological modes of action. This was observed by Cleuvers (2003) in his study of clofibric acid and carbamazepine in *Daphnia* and algae. Alsop and Wood (2013) observed additive toxicities in a mixture of metal ions and pharmaceuticals and also accounted for these toxicities by CA.

However, toxicities observed for some mixtures are sometimes better modeled by the independent action (IA) concept rather than CA. IA assumes that mixture components act in a *dissimilar* mode of action (i.e., same receptor, different tissue/organ or different receptor, same tissue/organ). The IA concept also assumes that the specific pharmacological effect or response observed is equivalent to the probability of the particular components in a mixture also exerting a pharmacological response on the organism in a statistically *independent* and *random* nature. It can be mathematically described as:

$$E(c_{\text{mix}}) = 1 - \left[(1 - Ec_1)(1 - Ec_2)(1 - Ec_3)\ldots(1 - Ec_n)\right]$$

$$E(c_{\text{mix}}) = 1 - \prod_{i=1}^{n}(1 - Ec_i)$$

Here, $E(c_{\text{mix}})$ is the expected pharmacological effect or response of a mixture at a concentration of c_{mix}, and Ec_i the pharmacological effect or response achieved by a particular pharmaceutical component ($i = 1, 2, 3 \ldots n$) at a concentration of c_i. Note that $E(c_{\text{mix}})$ and Ec_i are arbitrary and describe the magnitude of an expected pharmacological effect. Also, the IA equation does not take into account any consideration of effective concentrations as EC_x values for each pharmaceutical component are not required for the prediction. As such, an EC_x for an x level of pharmacological response cannot be calculated using this IA concept. Furthermore, the expression $(1 - Ec_i)$ describes the probability that a pharmaceutical component in a mixture is pharmacologically *nonresponsive*. Multiplication of all the nonresponding probabilities estimates the total nonresponse probability of the mixture. Subtracting this total from unity allows the probability and hence the predicted pharmacological effect of the mixture to be calculated (Faust et al., 2003).

The use of the above IA equation to calculate risk is illustrated by the hypothetical mixture of drugs A and B that induce individual pharmacological responses of 0.5 and 0.25, respectively (Goolsby et al., 2013):

$$\begin{aligned} E(c_{\text{mix}}) &= 1 - \left[(1 - Ec_{\text{A}})(1 - Ec_{\text{B}})\right] \\ &= 1 - \left[(1 - 0.5)(1 - 0.25)\right] \\ &= 0.625 \end{aligned}$$

In other words, if IA was the predictive model used to estimate the toxicity of this binary mixture, the mixture toxicity would be expected to be 1.25 times (0.625/0.5) and 2.5 times (0.625/0.25), respectively, the individual toxicities of drug A and drug B in solution alone.

IA has been used to account for the mixture toxicity of clofibric acid and carbamazepine on *Desmodesmus subspicatus* and *Lemna minor* (Cleuvers, 2003). However, mixtures which purely exhibit combined toxicity via an IA mechanism rarely occur. For example, Escher et al. (2010) observed that, comparing the CA and IA models, the IA model was closest in predicting the mixture toxicity of both oseltamivir phosphate and its active metabolite, oseltamivir carboxylate, in algae. However, neither model was a very good fit to the data. The apparent IA observed may have been caused by toxicokinetic effects, whereby oseltamivir phosphate may be preferentially bioaccumulated by algae compared to oseltamivir carboxylate. Similar toxicokinetic effects which may skew mixture toxicity modeling have also been reported by Brezovsek et al. (2014). In other instances, the predictive capacity of both CA and IA models has been shown to be unreliable when used with acute toxicity data at low concentrations (Nieto et al., 2013). Toxicodynamics and mode of action may also influence the predictive ability of both the toxicity models. However, one drug may nullify (i.e., antagonize) the pharmacological response of the other and in such instances, the observed toxicity would be less than that predicted by CA.

Obviously, further research is necessary to better characterize the toxicity of mixtures as both CA and IA models are limited in their utility to account for all situations of mixture toxicity (Ginebreda et al., 2014). The inclusion of better bioaccumulation, toxicokinetic, toxicodynamic, and ecotoxicological data in toxicity models may potentially improve mixture toxicity predictions (Backhaus and Karlsson, 2014). A study investigating the effects of combinations of 17α-ethinylestradiol with the pharmaceuticals carbamazepine, diclofenac, and metoprolol demonstrated that these drugs induce multigenerational effects in the cladoceran *D. magna* (Dietrich et al., 2010). The environmental significance and risk of these chemicals such as their ecotoxicology remains poorly understood and there is too little information on their fate in the environment to characterize the long-term risk of EDCs on their own or as mixtures (Fent et al., 2006; Pal et al., 2010).

While the real or potential environmental impacts of *all* pharmaceuticals should be measured or at least calculated, we will now illustrate the current state of knowledge concerning the risk posed by just three very common pharmaceuticals. Paracetamol is without doubt the most commonly used analgesic alone or in combination with other APIs in dosage forms (Chapter 2). The subtle endocrine disruption observed with fluoxetine on many aquatic organisms has raised public health concerns; however, data on its detrimental effects to wildlife is still not firmly established (Stanley et al., 2007; Sanchez-Arguello et al., 2009). In addition, the environmental presence of diclofenac has been linked to the only reported ecological disaster to date arising from pharmaceutical usage albeit for non-human purposes (Chapter 1). The different physicochemical properties of these three pharmaceuticals also make their environmental behavior worthy of comparison. Furthermore the diverse environmental behavior of these three drugs confirms the need for a detailed environmental risk assessment of individual pharmaceuticals alone and in a mixture on a case-by-case basis (see Chapter 9).

5.6 Case study: Paracetamol (acetaminophen) in the environment

Paracetamol (*N*-(4-hydroxyphenyl)ethanamide, acetaminophen) is a widely used analgesic and antipyretic agent. This pharmaceutical and its respective Phase II and I metabolites paracetamol glucuronide and *p*-aminophenol (see Chapter 2 for a discussion on pharmacokinetics) have been confirmed in Portuguese river and surface water at concentrations between 1.63 and 3.67 μg/L (Modi et al., 2012). The measurable presence of paracetamol and its human metabolites is somewhat surprising, as many studies report its very efficient removal by conventional sewage treatment technologies (see Chapter 6) as ≥90% (Law et al., 2004; Lin et al., 2010; de Graaff et al., 2011; Matsuo et al., 2011) and even more efficient removal using tertiary treatment or advanced oxidation processes (Yang et al., 2008; Tambosi et al., 2010; de Luna et al., 2012; Quesada-Penate et al., 2012). Furthermore, paracetamol is readily biodegradable and can be used as a carbon source by the *Pseudomonas aeruginosa* strain HJ1012 (Hu et al., 2013).

Paracetamol biodegradation may contribute to its removal during activated sludge sewage treatment, as it is readily metabolized to acetamide or *p*-aminophenol by *Pseudomonas aeruginosa* strain HJ1012 (Shankar et al., 2002). Paracetamol was also observed to be efficiently adsorbed onto sediment (Kulkarni et al., 2008) where further microbial degradation can occur. However, paracetamol is still recognized as a *pseudo-persistent* contaminant as it is one of the most commonly prescribed pharmaceuticals in the world, with an estimated consumption of over 3.2 billion tablets each year in the UK alone (Maclean et al., 2009). *Pseudo-persistence* is defined as the continued environmental presence of a pharmaceutical irrespective of its ability for physicochemical or microbial removal, because it is continuously introduced into the environment (Kummerer, 2009). The detection of paracetamol and its metabolites in river water suggests that the extent of continuous human use and its disposal in sewage overwhelms its effective removal by conventional sewage treatment (Modi et al., 2012). In addition, while paracetamol is effectively removed by conventional sewage treatments, it is not considered to be readily biodegradable in OECD 301F tests, as only 57% biodegradation has been observed after 28 days (*LGC Standards*, http://www.lgcstandards.com/WebRoot/Store/Shops/LGC/FilePathPartDocuments/ST-WB-MSDS-1319727-1-1-1.PDF, accessed 1-8-14).

There is still uncertainty about the environmental persistence and fate of paracetamol. Dusing et al. (2009) investigated the role of DOM (Chapter 6) and nitrate ions in paracetamol degradation under natural sunlight irradiation (UV-B and UV-A). The degradation half-life ($t_{1/2}$) of paracetamol under irradiation but in the absence of DOM and nitrate ions was estimated to be 72 days (1,732 h). The addition of nitrate ions alone improved paracetamol degradability, shortening the $t_{1/2}$ of paracetamol to 44 days (1,065 h). The $t_{1/2}$ of paracetamol was further reduced to 34 days (837 h) and 25 days (620 h) under irradiation with DOM alone or in combination with nitrate ions, respectively. These results suggest that paracetamol is resistant to direct photolysis under the weak irradiation intensities of sunlight, but its removal is somewhat enhanced through indirect photolysis in the environment and involving components such as DOM and nitrate ions (*see* Section 5.3.1 above). However, this is contradicted by a recent study which suggests that direct photolysis of paracetamol could be as important as microbial degradation in surface water (De Laurentiis et al., 2014). Consistent with these results is the previous observation of rapid paracetamol direct photolysis to form an unidentified photoproduct and CO_2 in 14 days (Kulkarni et al., 2008).

It is hard to compare the above experimental results as the experimental designs varied greatly between the laboratory and field studies. For example, it is difficult to compare degradation kinetics obtained from the reaction of paracetamol in water containing a range of DOM and nitrate ion concentrations. Furthermore, DOM is a very heterogeneous material and different components may promote or inhibit photolysis depending on the extent to which they generate reactive chemical species (ROS) and/or absorb the incident irradiation before it can interact with the paracetamol. Natural water-sediment systems may also have a wide variety of microbes, DOM, and chemical species such as nitrate ions which are virtually impossible to standardise for comparison of the photolytic degradation results.

The paracetamol degradation results are however consistent with those obtained from current environmental risk and hazard assessments. The Swedish Stockholm County Council has established a classification system to estimate the environmental risk and hazard (Chapter 1) of pharmaceuticals (available www.janusinfo.se/Global/Miljo_och_lakemedel/miljobroschyr_engelsk_2012_uppslag.pdf, accessed 31-7-13). While paracetamol is considered environmentally persistent using this particular classification system, it is not bioaccumulative or toxic to aquatic organisms. The environmental risk to the ecological health of aquatic organisms posed by this widespread pharmaceutical is also considered to be very low. However, this environmental risk assessment is based upon Swedish data and such information on the use of this common drug in many other countries is limited. It is therefore premature to firmly conclude that the global risk of paracetamol to humans and other flora or fauna is negligible.

Given the predicted RQ values that have been reported in the peer-reviewed literature, the environmental risk of paracetamol is deemed to be low to moderate (Table 5.3).

However while this reported range of RQ values suggests that adverse effects of the parent paracetamol on ecosystems are unlikely, there is a lack of information about the environmental behavior of its metabolites or transformation products. These include human metabolites and by-products formed from the physicochemical and microbial removal of paracetamol in sewage treatment facilities. The production of two known toxic

Table 5.3 Risk estimates for paracetamol

Country of study	Risk quotient (RQ) PEC/PNEC	Reference
Denmark	0.5	Stuer-Lauridsen et al. (2000)
UK	0.09	Jones et al. (2002)
Switzerland	0.1	Lienert et al. (2007)
Norway	0.048–2.0*	Grung et al. (2008)
Greece	0.14	Stamatis et al. (2013)
US	0.1	Yu et al. (2013b)
France	1.6	Bouissou-Schurtz et al. (2014)
Greece	0.01	Kosma et al. (2014)

*Calculated from PNECs for a variety of test organisms.

intermediates arising from the human metabolism of paracetamol, 1,4-benzoquinone (BQ) and *N*-acetyl-p-benzoquinone imine, have been identified as by-products following chlorination (Mangione-Smith et al., 2005). However, both are rapidly removed by treatment with sulfites which are commonly added along with the chlorine-generating hypochlorite in tertiary treatment. It is therefore unknown whether these transformation products have a significant impact on the environment, particularly where only secondary treatment is used.

5.7 Case study: Fluoxetine in the environment

Fluoxetine (also commonly known as Prozac) ((RS)-*N*-methyl-3-phenyl-3-[4-(trifluoromethyl)phenoxy]propan-1-amine) is a widely used SSRI antidepressant that is commonly detected in surface waters. For example, fluoxetine concentrations of up to 0.012 μg/L were measured in a 1999–2000 reconnaissance study of US lakes, streams, and rivers (Matthes and Albus, 2014). Schultz et al. (2010) also reported fluoxetine concentrations of 20 ng/L in a river downstream of a wastewater reclamation plant in the US.

Despite its frequent occurrence in the environment, the removal of fluoxetine by activated sludge processes (Chapter 6) has been reported to be >80% (Vasskog et al., 2006). However, the biodegradability of a 1-mg/L fluoxetine aqueous solution was negligible when exposed to bacterial inoculums for 28 days in an OECD 301 test (Kwon and Armbrust, 2006). The adsorption and dissipation of fluoxetine from water into sediments was observed to be rapid (between 1.3 and 2.7 days). In contrast, samples of sediment-free fluoxetine solution gave rise to $t_{1/2}$ values of 112–133 days when irradiated with UV-A and B lamps. These results suggest that the environmental partitioning of fluoxetine into sediment rather than water is an important factor in assessing its fate and hence risk in the environment. Adsorption onto solid matrices may also account for the apparently high sewage removal efficacy observed previously (Vasskog et al., 2006) and this has been confirmed in studies where fluoxetine was found to effectively adsorb onto biosolids (Sanchez-Arguello et al., 2009; Arney et al., 2013) which are typically sewage sludges often mixed with other materials that have been treated and/or stabilized to the extent that they can be safely applied to land.

Fluoxetine undergoes slow direct photodegradation ($t_{1/2}$ = 55.2 h, Chapter 6) but this is not the main degradation process (Lam et al., 2005). This $t_{1/2}$ value was reduced to 21 days when indirect photolysis of fluoxetine was undertaken in the presence of synthetic humic acids which form ROS upon irradiation (Kwon and Armbrust, 2006). However, this enhanced photolytic degradation was not observed for fluoxetine in natural lake water samples compared to buffered laboratory solutions containing the humic material (Kwon and Armbrust, 2006). This suggests that studies carried out with synthetic humic acids may not fully reflect the degradation kinetics determined using natural DOM. However, Lam et al. (2005) found that fluoxetine was photoreactive and degraded in deionized water when exposed to simulated sunlight. Other radical species present in the natural water, such as HO·, singlet oxygen ($^1O_2$) and the presence of nitrate, may have been responsible for the apparently increased rate of degradation.

Fluoxetine demonstrated no detectable degradation in a biosolids-amended soil mesocosm experiment after three years of monitoring (Walters et al., 2010). It also dissipates and persists in soils after application in biosolid form onto agricultural lands and other sludge-amended soils with little to no significant degradation (Chaar and Lee, 2012; Sarkar et al., 2014). However, such an observation may not be specific to fluoxetine. Biosolids can reduce the degradation of contaminants such as pharmaceuticals as this complex biosolid matrix can act as an alternative source of carbon for soil microbes and in competition with that provided by trace levels of any pharmaceuticals present in the matrix (Schwartz and Wagner, 2000).

Fluoxtene has been shown to accumulate in the fatty tissue in freshwater fish, specifically in the brain, and cause effects such as anxiety behavior in adult male fathead minnow (*P. promelas*). A recent hazard assessment process concluded that antidepressant drugs like fluoxetine would be expected to reach therapeutic levels in fish exposed to water at or below a common regulatory environmental trigger concentration for pharmaceuticals of 1 μg/L (Brooks, 2014). Despite the observation that fluoxetine can penetrate fish tissue (Brooks et al., 2005) and has endocrine-disrupting activity (Sanchez-Arguello et al., 2009; El-Bassat et al., 2012), there is still insufficient evidence to show that it is a significant environmental risk as RQ values are from 0.05 to 0.09 (Lienert et al., 2007; Besse et al., 2008).

Fluoxetine is metabolized by humans to norfluoxetine by CYP450 *N*-demethylation (Chapter 2) and thus any complete assessment of environmental fate must include analysis for this metabolite (Bucher et al., 2003). As for fluoxetine, norfluoxetine may be potentially dissipated into biosolids. Concentrations of fluoxetine and norfluoxetine in two sewage influent samples were 11.0 and 11.5 ng/L, respectively. However, both compounds were below the quantitation limit of 7 ng/L in the final treated effluents. Norfluoxetine was measured in a sewage effluent outflow in Montreal, Canada at concentrations of 12–13 ng/L during two different sampling campaigns (Lajeunesse et al., 2008) and in surface river water of the Upper Tennessee River Basin (Conley et al., 2008).

To conclude, fluoxetine is considered moderately hazardous by Stockholm County Council as it is classified as highly persistent and toxic but not particularly bioaccumulative (available www.janusinfo.se/Global/Miljo_och_lakemedel/miljobroschyr_engelsk_2012_uppslag.pdf, accessed 31-7-13). However, the current literature suggests that fluoxetine and norfluoxetine may persist in environmental sediments and both these compounds are poorly degraded by hydrolysis, biodegradation, and photolytic processes. There is still uncertainty around the environmental risks posed by both these chemical species.

5.8 Case study: Diclofenac in the environment

Diclofenac (also commonly known in the form of the sodium salt as Voltaren) (2-(2-(2,6-dichlorophenylamino)phenyl)acetic acid) is a widely used NSAID which is often detected in the environment (e.g., Comeau et al., 2008; Vystavan et al., 2012; Kozisek et al., 2013) as it is poorly removed in conventional sewage treatment (Stulten et al., 2008). Estimates of the removal efficacy for diclofenac using conventional sewage treatment vary from 39% (Paxeus, 2004; Kosjek et al., 2007) to 60–70% (Huang et al., 2011). Only 30–35% diclofenac was removed from activated sludge in an observation by Ribeiro et al. (2012). Diclofenac adsorption to the activated sludge formed during sewage treatment processes rather than biodegradation, may also contribute to its partial sewage treatment removal (Kabak and Basibuyuk, 2012). Certain functional groups present in the chemical structure of diclofenac (see Table 4.2), such as –Cl and –N–H, can inhibit bacterial growth and as a consequence the efficacy of secondary treatment (Ziylan and Ince, 2011). The removal efficacy of diclofenac depends also on a multitude of sewage treatment parameters, such as the solids retention time (SRT)[2] (Plósz et al., 2012).

Diclofenac is excreted in the form of glucuronides and so likely deconjugation processes during sewage treatment may lead to an apparent increase detected in treated effluents compared to influent streams (Lee et al., 2012). For instance, a diclofenac concentration of 1,032 ng/L was detected in treated effluent while influent concentrations were in the range of 100–200 ng/L (Pedrouzo et al., 2011). This is consistent with the results from a study of three wastewater treatment plants in Ireland (Lacey et al., 2012) where the drug was detected in effluent but not in influent. The Phase II glucuronides of diclofenac (Chapter 2) have been shown to biotransform back to parent diclofenac in sewage treatment facilities (Lee et al., 2012). Diclofenac did not undergo significant biodegradation with activated sludge inoculums during a 28-day OECD 301C ready test. However, the human phase I diclofenac metabolite, 4-hydroxydiclofenac, degraded slowly (Lee et al., 2012). In contrast, the phase II diclofenac metabolite, β-*O*-acyl glucuronide, deconjugated to the parent diclofenac within seven days of the biodegradability test.

Diclofenac and its metabolites have been found to accumulate in aquatic organisms. For example, the acyl glucuronides of hydroxydiclofenac have been detected in

[2] The length of time that activated sludge material is resident in the sewage treatment process involving aerobic biodegradation.

the bile of rainbow trout (*O. mykiss*) (Kallio et al., 2010). In a follow-up field study, diclofenac was also found in the bile of two wild fish species sampled downstream of a Finnish wastewater treatment facility (Brozinski et al., 2013).

The persistence of diclofenac after passage through common sewage treatment processes was not reflected in the results of land application studies. A study on soils with a range of texture (sandy loam, loam, clay loam) showed rapid mineralization of diclofenac with an estimated half-life of <5 days (Al-Rajab et al., 2010). These results suggest that diclofenac is readily biodegradable in agricultural soils.

Diclofenac is also unlikely to persist in surface water mainly as a result of photolytic degradation. The concentration of diclofenac was observed to drop from 370 to 12 ng/L in major inflows and outflows of a Swiss lake, respectively (Buser et al., 1998). Experiments on sediment adsorption, biodegradation, photolysis, and hydrolysis of diclofenac in the lake water determined that photolysis was the predominant mechanism of diclofenac removal. Diclofenac has been reported to be transformed into carbazole-1-acetic acid and 8-chlorocarbazole-acetic acid by the action of UV-A and B radiation (Moore et al., 1990; Boreen et al., 2003; Perez-Estrada et al., 2005; Peuravuori, 2012). The formation and decay of this 8-chlorocarbazole-acetic acid have been reported in lake water (Poiger et al., 2001). Diclofenac is also transformed by sunlight to known and unidentified photoproducts in the presence of DOM and nitrate ions with photocyclization shown to be the main phototransformation pathway (Peuravuori, 2012).

As noted above, diclofenac has been linked with a major ecological disaster leading to the decline of a vulture population (Oaks et al., 2004), but further research is warranted to fully assess its environmental risk in other situations. There appears to be currently insufficient information on the environmental fate and behavior of diclofenac and its photoproducts and metabolites to conduct any full risk assessment process. This is confirmed by Stockholm County Council that classifies diclofenac as a highly persistent substance but considers there is insufficient bioaccumulation and toxicity data to assess risks (available http://www.janusinfo.se/Global/Miljo_och_lakemedel/Miljobroschyr_2014_engelsk_webb.pdf, accessed 31-7-13). Other risk assessments (Table 5.4) generally consider the risk of diclofenac to be negligible as a PEC/PNEC ratio ≤ 1 is considered low.

Table 5.4 Risk assessment for diclofenac

Country of study	Risk quotient (RQ) PEC/PNEC	Reference
UK	0.01	Jones et al. (2002)
Switzerland	0.3	Lienert et al. (2007)
Norway	0.34–3.8*	Grung et al. (2008)
France	0.25	Besse et al. (2008)
Spain	0.151	Nieto et al. (2013)
US	0.1	Yu et al. (2013b)

*Calculated from PNECs determined for a variety of test organisms.

Table 5.5 Predominant environmental degradation mechanisms for paracetamol, fluoxetine, and diclofenac

Pharmaceutical	Major environmental degradation mechanism	Reference
Paracetamol	Biodegradation	Hu et al. (2013)
Fluoxetine	Sorption onto sediments and solids	Redshaw et al. (2008)
Diclofenac	Photodegradation	Buser et al. (1998)

5.9 Summary

Given the number of pharmaceuticals on the global market, each with their own physicochemical characteristics, it is challenging to determine whether any of them can lead to adverse effects on the receiving ecosystems. The environmental risk assessment for each pharmaceutical needs to be evaluated independently under all possible potential environmental conditions before any robust conclusions can be drawn.

The pharmaceuticals presented as case studies in this chapter clearly illustrate this diversity in environmental behavior (Table 5.5).

All three pharmaceuticals exhibit acute and chronic toxicity in various ways, given their respective toxicokinetics and toxicodynamics. However, there is currently a lack of data to conduct extensive risk assessment for these and other pharmaceuticals. The risk assessments conducted with current best information suggest there is no evidence for immediate alarm arising from the environmental presence and persistence of at least these three pharmaceuticals. To fully characterize the risk of pharmaceuticals will require new approaches and frameworks that ensure a cohesive use of resources and expertise. For instance, a new process has recently been designed to engage individuals from academia, industry, and government agencies to identify the key questions required to assess the risk of pharmaceuticals (Boxall et al., 2012). This was followed by a workshop where experts were invited to prioritize the questions with the top-ranked question being: *How important are PPCPs relative to other chemicals and nonchemical stressors in terms of biological impacts in the natural environment*? (Boxall et al., 2012). It is a certainty that coordination of resources and expertise from across the global community is required to address such questions and overcome the challenge of fully assessing the risk pharmaceuticals may pose to the environment.

References

Accinelli, C., Sacca, M.L., Fick, J., Mencarelli, M., Lindberg, R., Olsen, B., 2010. Dissipation and removal of oseltamivir (Tamiflu) in different aquatic environments. Chemosphere 79, 891–897.

Adams, S.M., Giesy, J.P., Tremblay, L.A., Eason, C.T., 2001. The use of biomarkers in ecological risk assessment: recommendations from the Christchurch conference on Biomarkers in Ecotoxicology. Biomarkers 6, 1–6.

Adema, D.M.M., 1978. Daphnia magna as a test animal in acute and chronic toxicity tests. Hydrobiologia 59, 125–134.
Alderton, W., Berghmans, S., Butler, P., Chassaing, H., Fleming, A., Golder, Z., Richards, F., Gardner, I., 2010. Accumulation and metabolism of drugs and CYP probe substrates in zebrafish larvae. Xenobiotica 40, 547–557.
Al-Rajab, A.J., Sabourin, L., Chapman, R., Lapen, D.R., Topp, E., 2010. Fate of the antiretroviral drug tenofovir in agricultural soil. Sci. Total Environ. 408, 5559–5564.
Alsop, D., Wood, C.M., 2013. Metal and pharmaceutical mixtures: is ion loss the mechanism underlying acute toxicity and widespread additive toxicity in zebrafish? Aquat. Toxicol. 140, 257–267.
Altenburger, R., Backhaus, T., Boedeker, W., Faust, M., Scholze, M., Grimme, L.H., 2000. Predictability of the toxicity of multiple chemical mixtures to Vibrio fischeri: mixtures composed of similarly acting chemicals. Environ. Toxicol. Chem. 19, 2341–2347.
Altenburger, R., Nendza, M., Schuurmann, G., 2003. Mixture toxicity and its modeling by quantitative structure-activity relationships. Environ. Toxicol. Chem. 22, 1900–1915.
Altenburger, R., Walter, H., Grote, M., 2004. What contributes to the combined effect of a complex mixture? Environ. Sci. Technol. 38, 6353–6362.
Anskjaer, G.G., Rendal, C., Kusk, K.O., 2013. Effect of pH on the toxicity and bioconcentration of sulfadiazine on Daphnia magna. Chemosphere 91, 1183–1188.
Arney, J., Street, R.L., Naik, A.D., 2013. Consumers' various and surprising responses to direct-to-consumer advertisements in magazine print. Patient Prefer Adherence 7, 95–102.
Atkinson, S.K., Marlatt, V.L., Kimpe, L.E., Lean, D.R.S., Trudeau, V.L., Blais, J.M., 2011. Environmental factors affecting ultraviolet photodegradation rates and estrogenicity of estrone and ethinylestradiol in natural waters. Arch. Environ. Contam. Toxicol. 60, 1–7.
Backhaus, T., Faust, M., 2012. Predictive environmental risk assessment of chemical mixtures: a conceptual framework. Environ. Sci. Technol. 46, 2564–2573.
Backhaus, T., Karlsson, M., 2014. Screening level mixture risk assessment of pharmaceuticals in STP effluents. Water Res. 49, 157–165.
Bagnall, J., Malia, L., Lubben, A., Kasprzyk-Hordern, B., 2013. Stereoselective biodegradation of amphetamine and methamphetamine in river microcosms. Water Res. 47, 5708–5718.
Bain, P.A., Kumar, A., 2014. Cytotoxicity of binary mixtures of human pharmaceuticals in a fish cell line: approaches for non-monotonic concentration–response relationships. Chemosphere 108, 334–342.
Bartels, P., von Tumpling, W., 2008. The environmental fate of the antiviral drug oseltamivir carboxylate in different waters. Sci. Total Environ. 405, 215–225.
Batchu, S.R., Panditi, V.R., Gardinali, P.R., 2014. Photodegradation of sulfonamide antibiotics in simulated and natural sunlight: implications for their environmental fate. J. Environ. Sci. Health B 49, 200–211.
Bergheim, M., Giere, R., Kummerer, K., 2012. Biodegradability and ecotoxicitiy of tramadol, ranitidine, and their photoderivatives in the aquatic environment. Environ. Sci. Pollut. Res. 19, 72–85.
Berninger, J.P., Du, B.W., Connors, K.A., Eytcheson, S.A., Kolkmeier, M.A., Prosser, K.N., Valenti, T.W., Chambliss, C.K., Brooks, B.W., 2011. Effects of the antihistamine diphenhydramine on selected aquatic organisms. Environ. Toxicol. Chem. 30, 2065–2072.
Besse, J.P., Kausch-Barreto, C., Garric, J., 2008. Exposure assessment of pharmaceuticals and their metabolites in the aquatic environment: application to the French situation and preliminary prioritization. Hum. Ecol. Risk. Assess. 14, 665–695.
Boreen, A.L., Arnold, W.A., McNeill, K., 2003. Photodegradation of pharmaceuticals in the aquatic environment: a review. Aquat. Sci. 65, 320–341.

Bouissou-Schurtz, C., Houeto, P., Guerbet, M., Bachelot, M., Casellas, C., Mauclaire, A.C., Panetier, P., Delval, C., Masset, D., 2014. Ecological risk assessment of the presence of pharmaceutical residues in a French national water survey. Regul. Toxicol. Pharmacol. 69, 296–303.

Boxall, A.B.A., Rudd, M.A., Brooks, B.W., Caldwell, D.J., Choi, K., Hickmann, S., Innes, E., Ostapyk, K., Staveley, J.P., Verslycke, T., Ankley, G.T., Beazley, K.F., Belanger, S.E., Berninger, J.P., Carriquiriborde, P., Coors, A., DeLeo, P.C., Dyer, S.D., Ericson, J.F., Gagne, F., Giesy, J.P., Gouin, T., Hallstrom, L., Karlsson, M.V., Larsson, D.G.J., Lazorchak, J.M., Mastrocco, F., McLaughlin, A., McMaster, M.E., Meyerhoff, R.D., Moore, R., Parrott, J.L., Snape, J.R., Murray-Smith, R., Servos, M.R., Sibley, P.K., Straub, J.O., Szabo, N.D., Topp, E., Tetreault, G.R., Trudeau, V.L., Van Der Kraak, G., 2012. Pharmaceuticals and personal care products in the environment: what are the big questions? Environ. Health Perspect. 120, 1221–1229.

Brain, R.A., Ramirez, A.J., Fulton, B.A., Chambliss, C.K., Brooks, B.W., 2008. Herbicidal effects of sulfamethoxazole in Lemna gibba: using p-aminobenzoic acid as a biomarker of effect. Environ. Sci. Technol. 42, 8965–8970.

Brausch, J.M., Connors, K.A., Brooks, B.W., Rand, G.M., 2012. Human pharmaceuticals in the aquatic environment: a review of recent toxicological studies and considerations for toxicity testing. Rev. Environ. Contam. Toxicol. 218, 1–99.

Brezovsek, P., Elersek, T., Filipic, M., 2014. Toxicities of four anti-neoplastic drugs and their binary mixtures tested on the green alga Pseudokirchneriella subcapitata and the cyanobacterium Synechococcus leopoliensis. Water Res. 52, 168–177.

Brooks, B.W., 2014. Fish on Prozac (and Zoloft): ten years later. Aquat. Toxicol. 151, 61–67.

Brooks, B.W., Chambliss, C.K., Stanley, J.K., Ramirez, A., Banks, K.E., Johnson, R.D., Lewis, R.J., 2005. Determination of select antidepressants in fish from an effluent-dominated stream. Environ. Toxicol. Chem. 24, 464–469.

Brozinski, J.M., Lahti, M., Meierjohann, A., Oikari, A., Kronberg, L., 2013. The anti-inflammatory drugs Diclofenac, Naproxen and Ibuprofen are found in the bile of wild fish caught downstream of a wastewater treatment plant. Environ. Environ. Sci. Technol. 47, 342–348.

Bucher, H.C., Tschudi, P., Young, J., Periat, P., Welge-Lussen, A., Zust, H., Schindler, C., 2003. Effect of amoxicillin-clavulanate in clinically diagnosed acute rhinosinusitis: a placebo-controlled, double-blind, randomized trial in general practice. Arch. Intern. Med. 163, 1793–1798.

Buser, H.R., Poiger, T., Muller, M.D., 1998. Occurrence and fate of the pharmaceutical drug diclofenac in surface waters: Rapid photodegradation in a lake. Environ. Sci. Technol. 32, 3449–3456.

Castensson, S., Eriksson, V., Lindborg, K., Wettermark, B., 2009. A method to include the environmental hazard in drug prescribing. Pharm. World Sci. 31, 24–31.

Cleuvers, M., 2003. Aquatic ecotoxicity of pharmaceuticals including the assessment of combination effects. Toxicol. Lett. 142, 185–194.

Cleuvers, M., 2004. Mixture toxicity of the anti-inflammatory drugs diclofenac, ibuprofen, naproxen, and acetylsalicylic acid. Ecotoxicol. Environ. Saf. 59, 309–315.

Collard, H.R.J., Ji, K., Lee, S., Liu, X., Kang, S., Kho, Y., Ahn, B., Ryu, J., Lee, J., Choi, K., 2013. Toxicity and endocrine disruption in zebrafish (Danio rerio) and two freshwater invertebrates (Daphnia magna and Moina macrocopa) after chronic exposure to mefenamic acid. Ecotoxicol. Environ. Saf. 94, 80–86.

Comeau, F., Surette, C., Brun, G.L., Losier, R., 2008. The occurrence of acidic drugs and caffeine in sewage effluents and receiving waters from three coastal watersheds in Atlantic Canada. Sci. Total Environ. 396, 132–146.

Conley, J.M., Symes, S.J., Schorr, M.S., Richards, S.M., 2008. Spatial and temporal analysis of pharmaceutical concentrations in the upper Tennessee River basin. Chemosphere 73, 1178–1187.

Connors, K.A., Du, B.W., Fitzsimmons, P.N., Chambliss, C.K., Nichols, J.W., Brooks, B.W., 2013a. Enantiomer-specific in vitro biotransformation of select pharmaceuticals in rainbow trout (Oncorhynchus mykiss). Chirality 25, 763–767.

Connors, K.A., Du, B.W., Fitzsimmons, P.N., Hoffman, A.D., Chambliss, C.K., Nichols, J.W., Brooks, B.W., 2013b. Comparative pharmaceutical metabolism by rainbow trout (Oncorhynchus mykiss) liver S9 fractions. Environ. Toxicol. Chem. 32, 1810–1818.

Cooper, E.R., Siewicki, T.C., Phillips, K., 2008. Preliminary risk assessment database and risk ranking of pharmaceuticals in the environment. Sci. Total Environ. 398, 26–33.

Crisp, T.M., Clegg, E.D., Cooper, R.L., Wood, W.P., Anderson, D.G., Baetcke, K.P., Hoffmann, J.L., Morrow, M.S., Rodier, D.J.., Schaeffer, J.E., Touart, L.W., Zeeman, M.G., Patel, Y.M., 1998. Environmental endocrine disruption: an effects assessment and analysis. Environ. Health Perspect. 106(Suppl. 1), 11–56.

Cui, H., Wang, S.P., Jia, S.G., Zhang, N., Zhou, Z.Q., 2013. Influence of ciprofloxacin on the microbial catabolic diversity in soil. J. Environ. Sci. Health B 48, 869–877.

Cuklev, F., Fick, J., Cvijovic, M., Kristiansson, E., Forlin, L., Larsson, D.G.J., 2012. Does ketoprofen or diclofenac pose the lowest risk to fish? J. Hazard. Mater. 229, 100–106.

Daughton, C.G., 2004. Non-regulated water contaminants: emerging research. Environ. Impact Assess. Rev. 24, 711–732.

Daughton, C.G., Ruhoy, I.S., 2009. Environmental footprint of pharmaceuticals: the significance of factors beyond direct excretion to sewers. Environ. Toxicol. Chem. 28, 2495–2521.

de Graaff, M.S., Vieno, N.M., Kujawa-Roeleveld, K., Zeeman, G., Temmink, H., Buisman, C.J.N., 2011. Fate of hormones and pharmaceuticals during combined anaerobic treatment and nitrogen removal by partial nitritation-anammox in vacuum collected black water. Water Res. 45, 375–383.

De Laurentiis, E., Minella, M., Sarakha, M., Marrese, A., Minero, C., Mailhot, G., Brigante, M., Vione, D., 2013. Photochemical processes involving the UV absorber benzophenone-4 (2-hydroxy-4-methoxybenzophenone-5-sulphonic acid) in aqueous solution: reaction pathways and implications for surface waters. Water Res. 47, 5943–5953.

De Laurentiis, E., Prasse, C., Ternes, T.A., Minella, M., Maurino, V., Minero, C., Sarakha, M., Brigante, M., Vione, D., 2014. Assessing the photochemical transformation pathways of acetaminophen relevant to surface waters: transformation kinetics, intermediates, and modelling. Water Res. 53, 235–248.

de Luna, M.D.G., Veciana, M.L., Su, C.-C., Lu, M.-C., 2012. Acetaminophen degradation by electro-Fenton and photoelectro-Fenton using a double cathode electrochemical cell. J. Hazard. Mater. 217, 200–207.

Deblonde, T., Hartemann, P., 2013. Environmental impact of medical prescriptions: assessing the risks and hazards of persistence, bioaccumulation and toxicity of pharmaceuticals. Public Health 127, 312–317.

Dietrich, S., Ploessl, F., Bracher, F., Laforsch, C., 2010. Single and combined toxicity of pharmaceuticals at environmentally relevant concentrations in Daphnia magna—a multigenerational study. Chemosphere 79, 60–66.

Dusing, R., Handrock, R., Klebs, S., Tousset, E., Vrijens, B., 2009. Impact of supportive measures on drug adherence in patients with essential hypertension treated with valsartan: the randomized, open-label, parallel group study VALIDATE. J. Hypertens. 27, 894–901.

Ebert, D., 2005. Introduction to Daphnia biology. Ecology, Epidemiology, and Evolution of Parasitism in Daphnia. National Center for Biotechnology Information, US.

Ebert, I., Bachmann, J., Kuhnen, U., Kuster, A., Kussatz, C., Maletzki, D., Schluter, C., 2011. Toxicity of the fluoroquinolone antibiotics enrofloxacin and ciprofloxacin to photoautotrophic aquatic organisms. Environ. Toxicol. Chem. 30, 2786–2792.

El-Bassat, R.A., Touliabah, H.E., Harisa, G.I., 2012. Toxicity of four pharmaceuticals from different classes to isolated plankton species. Afr. J. Aquat. Sci. 37, 71–80.

Escher, B.I., Bramaz, N., Lienert, J., Neuwoehner, J., Straub, J.O., 2010. Mixture toxicity of the antiviral drug Tamiflu (R) (oseltamivir ethylester) and its active metabolite oseltamivir acid. Aquat. Toxicol. 96, 194–202.

Fang, Y., Karnjanapiboonwong, A., Chase, D.A., Wang, J.F., Morse, A.N., Anderson, T.A., 2012. Occurrence, fate, and persistence of gemfibrozil in water and soil. Environ. Toxicol. Chem. 31, 550–555.

Faust, M., Altenburger, R., Backhaus, T., Blanck, H., Boedeker, W., Gramatica, P., Hamer, V., Scholze, M., Vighi, M., Grimme, L.H., 2003. Joint algal toxicity of 16 dissimilarly acting chemicals is predictable by the concept of independent action. Aquat. Toxicol. 63, 43–63.

Feito, R., Valcarcel, Y., Catala, M., 2013. Preliminary data suggest that venlafaxine environmental concentrations could be toxic to plants. Chemosphere 90, 2065–2069.

Fent, K., Weston, A.A., Caminda, D., 2006. Ecotoxicology of pharmaceuticals. Aquat. Toxicol. 76, 122–159.

Fick, J., Lindberg, R.H., Tysklind, M., Larsson, D.G.J., 2010. Predicted critical environmental concentrations for 500 pharmaceuticals. Regul. Toxicol. Pharmacol. 58, 516–523.

Franke, C., 1996. How meaningful is the bioconcentration factor for risk assessment? Chemosphere 32, 1897–1905.

Garcia, S.N., Foster, M., Constantine, L.A., Huggett, D.B., 2012. Field and laboratory fish tissue accumulation of the anti-convulsant drug carbamazepine. Ecotoxicol. Environ. Saf. 84, 207–211.

Gilroy, E.A.M., Klinck, J.S., Campbell, S.D., McInnis, R., Gillis, P.L., de Solla, S.R., 2014. Toxicity and bioconcentration of the pharmaceuticals moxifloxacin, rosuvastatin, and drospirenone to the unionid mussel Lampsilis siliquoidea. Sci. Total Environ. 487, 537–544.

Ginebreda, A., Kuzmanovic, M., Guasch, H., de Alda, M.L., Lopez-Doval, J.C., Munoz, I., Ricart, M., Romani, A.M., Sabater, S., Barcelo, D., 2014. Assessment of multi-chemical pollution in aquatic ecosystems using toxic units: compound prioritization, mixture characterization and relationships with biological descriptors. Sci. Total Environ. 468, 715–723.

Gomez, C.F., Constantine, L., Huggett, D.B., 2010. The influence of gill and liver metabolism on the predicted bioconcentration of three pharmaceuticals in fish. Chemosphere 81, 1189–1195.

Goolsby, E.W., Mason, C.M., Wojcik, J.T., Jordan, A.M., Black, M.C., 2013. Acute and chronic effects of diphenhydramine and sertraline mixtures in ceriodaphnia dubia. Environ. Toxicol. Chem. 32, 2866–2869.

Grung, M., Kallqvist, T., Sakshaug, S., Skurtveit, S., Thomas, K.V., 2008. Environmental assessment of Norwegian priority pharmaceuticals based on the EMEA guideline. Ecotoxicol. Environ. Saf. 71, 328–340.

Gulkowska, A., He, Y.H., So, M.K., Yeung, L.W.Y., Leung, H.W., Giesy, J.P., Lam, P.K.S., Martin, M., Richardson, B.J., 2007. The occurrence of selected antibiotics in Hong Kong coastal waters. Mar. Pollut. Bull. 54, 1287–1293.

Hawkins, M.B., Godwin, J., Crews, D., Thomas, P., 2005. The distributions of the duplicate oestrogen receptors ER-beta a and ER-beta b in the forebrain of the Atlantic croaker (Micropogonias undulatus): Evidence for subfunctionalization after gene duplication. Proceedings 272, 633–641.

Hoang, T.T.T., Nga, L.P., Tu, T.C.L., 2011. Antibiotic contaminants in coastal wetlands from Vietnamese shrimp farming. Environ. Sci. Pollut. Res. 18, 835–841.

Hu, J., Zhang, L.L., Chen, J.M., Liu, Y., 2013. Degradation of paracetamol by Pseudomonas aeruginosa strain HJ1012. J. Environ. Sci. Health A Tox. Hazard Subst. Environ. Eng. 48, 791–799.

Huggett, D.B., Cook, J.C., Ericson, J.F., Williams, R.T., 2003. A theoretical model for utilizing mammalian pharmacology and safety data to prioritize potential impacts of human pharmaceuticals to fish. Hum. Ecol. Risk. Assess. 9, 1789–1799.

Hutchinson, T.H., 2007. Small is useful in endocrine disrupter assessment—four key recommendations for aquatic invertebrate research. Ecotoxicology 16, 231–238.

Jacobs, L.E., Weavers, L.K., Houtz, E.F., Chin, Y.P., 2012. Photosensitized degradation of caffeine: role of fulvic acids and nitrate. Chemosphere 86, 124–129.

Jiang, M.X., Wang, L.H., Ji, R., 2010. Biotic and abiotic degradation of four cephalosporin antibiotics in a lake surface water and sediment. Chemosphere 80, 1399–1405.

Jones, P.D., De Coen, W.M., Tremblay, L., Giesy, J.P., 2000. Vitellogenin as a biomarker for environmental estrogens. Water Sci. Technol. 42(7-8), 1–14.

Jones, O.A., Voulvoulis, N., Lester, J.N., 2002. Aquatic environmental assessment of the top 25 English prescription pharmaceuticals. Water Res. 36, 5013–5022.

Kabak, H., Basibuyuk, M., 2012. Adsorption of Diclofenac and Paracetamol by activated sludge. Ekoloji. 21, 41–48.

Kallio, J.M., Lahti, M., Oikari, A., Kronberg, L., 2010. Metabolites of the aquatic pollutant diclofenac in fish bile. Environ. Sci. Technol. 44, 7213–7219.

Kim, J.-S., Lee, Y.-S., 2009. Antioxidant activity of Maillard reaction products derived from aqueous glucose/glycine, diglycine, and triglycine model systems as a function of heating time. Food Chem. 116, 227–232.

Kosma, C.I., Lambropoulou, D.A., Albanis, T.A., 2014. Investigation of PPCPs in wastewater treatment plants in Greece: occurrence, removal and environmental risk assessment. Sci. Total Environ. 466, 421–438.

Kozisek, F., Pomykacova, I., Jeligova, H., Cadek, V., Svobodova, V., 2013. Survey of human pharmaceuticals in drinking water in the Czech Republic. J. Water Health 11, 84–97.

Kulkarni, A.S., Balkrishnan, R., Anderson, R.T., Edin, H.M., Kirsch, J., Stacy, M.A., 2008. Medication adherence and associated outcomes in Medicare Health Maintenance Organization—enrolled older adults with Parkinson's disease. Mov. Disord. 23, 359–365.

Kummerer, K., 2009. The presence of pharmaceuticals in the environment due to human use—present knowledge and future challenges. J. Environ. Manage. 90, 2354–2366.

Kwon, J.W., Armbrust, K.L., 2006. Laboratory persistence and fate of fluoxetine in aquatic environments. Environ. Toxicol. Chem. 25, 2561–2568.

Laane, R., Vethaak, A.D., Gandrass, J., Vorkamp, K., Kohler, A., Larsen, M.M., Strand, J., 2013. Chemical contaminants in the Wadden Sea: sources, transport, fate and effects. J. Sea Res. 82, 10–53.

Lacey, C., Basha, S., Morrissey, A., Tobin, J.M., 2012. Occurrence of pharmaceutical compounds in wastewater process streams in Dublin, Ireland. Environ. Monit. Assess. 184, 1049–1062.

Lajeunesse, A., Gagnon, C., Sauve, S., 2008. Determination of basic antidepressants and their n-desmethyl metabolites in raw sewage and wastewater using solid-phase extraction and liquid chromatography—tandem mass spectrometry. Anal. Chem. 80, 5325–5333.

Lam, M.W., Young, C.J., Mabury, S.A., 2005. Aqueous photochemical reaction kinetics and transformations of fluoxetine. Environ. Sci. Technol. 39, 513–522.

Lamichhane, K., Garcia, S.N., Huggett, D.B., DeAngelis, D.L., La Point, T.W., 2013. Chronic effects of carbamazepine on life-history strategies of ceriodaphnia dubia in three successive generations. Arch. Environ. Contam. Toxicol. 64, 427–438.

Law, K.S., Wong, C.S., Song, L.J., 2004. The construct and criterion validity of emotional intelligence and its potential utility for management studies. J. Appl. Psychol. 89, 483–496.

Lee, H.-J., Lee, E., Yoon, S.-H., Chang, H.-R., Kim, K., Kwon, J.-H., 2012. Enzymatic and microbial transformation assays for the evaluation of the environmental fate of diclofenac and its metabolites. Chemosphere. 87, 969–974.

Leech, D.M., Snyder, M.T., Wetzel, R.G., 2009. Natural organic matter and sunlight accelerate the degradation of 17 beta-estradiol in water. Sci. Total Environ. 407, 2087–2092.

Li, J.Y., Dodgen, L., Ye, Q.F., Gan, J., 2013. Degradation kinetics and metabolites of carbamazepine in soil. Environ. Sci. Technol. 47, 3678–3684.

Liang, X.M., Chen, B.W., Nie, X.P., Shi, Z., Huang, X.P., Li, X.D., 2013. The distribution and partitioning of common antibiotics in water and sediment of the Pearl River Estuary, South China. Chemosphere 92, 1410–1416.

Lienert, J., Gudel, K., Escher, B.I., 2007. Screening method for ecotoxicological hazard assessment of 42 pharmaceuticals considering human metabolism and excretory routes. Environ. Sci. Technol. 41, 4471–4478.

Lin, K.D., Gan, J., 2011. Sorption and degradation of wastewater-associated non-steroidal anti-inflammatory drugs and antibiotics in soils. Chemosphere 83, 240–246.

Lin, A.Y.C., Lin, C.F., Tsai, Y.T., Lin, H.H.H., Chen, J., Wang, X.H., Yu, T.H., 2010. Fate of selected pharmaceuticals and personal care products after secondary wastewater treatment processes in Taiwan. Water Sci. Technol. 62, 2450–2458.

Lin, A.Y.C., Lee, W.N., Wang, X.H., 2014. Ketamine and the metabolite norketamine: persistence and phototransformation toxicity in hospital wastewater and surface water. Water Res. 53, 351–360.

Liu, J.C., Lu, G.H., Wang, Y.H., Yan, Z.H., Yang, X.F., Ding, J.N., Jiang, Z., 2014. Bioconcentration, metabolism, and biomarker responses in freshwater fish Carassius auratus exposed to roxithromycin. Chemosphere 99, 102–108.

Lu, G.H., Li, Z.H., Liu, J.C., 2013. Effects of selected pharmaceuticals on growth, reproduction and feeding of Daphnia magna. Fresenius Environ. Bull. 22, 2583–2589.

Luo, X.Z., Zheng, Z., Greaves, J., Cooper, W.J., Song, W.H., 2012. Trimethoprim: kinetic and mechanistic considerations in photochemical environmental fate and AOP treatment. Water Res. 46, 1327–1336.

Luo, Y.L., Guo, W.S., Ngo, H.H., Nghiem, L.D., Hai, F.I., Zhang, J., Liang, S., Wang, X.C.C., 2014. A review on the occurrence of micropollutants in the aquatic environment and their fate and removal during wastewater treatment. Sci. Total Environ. 473, 619–641.

Ma, D.J., Liu, G.G., Lv, W.Y., Yao, K., Zhang, X.D., Xiao, H.H., 2014. Photodegradation of naproxen in water under simulated solar radiation: mechanism, kinetics, and toxicity variation. Environ. Sci. Pollut. Res. 21, 7797–7804.

Maclean, J.R., Chapman, R.H., Ferrufino, C.P., Krishnarajah, G., 2009. Drug titration patterns and HbA(1c) levels in type 2 diabetes. Int. J. Clin. Pract. 63, 1008–1016.

Mangione-Smith, R., Wong, L., Elliott, M.N., McDonald, L., Roski, J., 2005. Measuring the quality of antibiotic prescribing for upper respiratory infections and bronchitis in 5 US health plans. Arch. Pediatr. Adolesc. Med. 159, 751–757.

Marin, M.G., Matozzo, V., 2004. Vitellogenin induction as a biomarker of exposure to estrogenic compounds in aquatic environments. Mar. Pollut. Bull. 48, 835–839.

Martins, N., Pereira, R., Abrantes, N., Pereira, J., Goncalves, F., Marques, C.R., 2012. Ecotoxicological effects of ciprofloxacin on freshwater species: data integration and derivation of toxicity thresholds for risk assessment. Ecotoxicology 21, 1167–1176.

Martins, A., Guimaraes, L., Guilhermino, L., 2013. Chronic toxicity of the veterinary antibiotic florfenicol to Daphnia magna assessed at two temperatures. Environ. Toxicol. Pharmacol. 36, 1022–1032.

Maskaoui, K., Zhou, J.L., 2010. Colloids as a sink for certain pharmaceuticals in the aquatic environment. Environ. Sci. Pollut. Res. 17, 898–907.

Matsuo, H., Sakamoto, H., Arizono, K., Shinohara, R., 2011. Behavior of pharmaceuticals in waste water treatment plant in Japan. Bull. Environ. Contam. Toxicol. 87, 31–35.

Matthes, J., Albus, C., 2014. Improving adherence with medication a selective literature review based on the example of hypertension treatment. Dtsch. Arztebl. Int. 111, 41, U13.

Minguez, L., Farcy, E., Ballandonne, C., Lepailleur, A., Serpentini, A., Lebel, J.M., Bureau, R., Halm-Lemeille, M.P., 2014. Acute toxicity of 8 antidepressants: what are their modes of action? Chemosphere 108, 314–319.

Minh, T.B., Leung, H.W., Loi, I.H., Chan, W.H., So, M.K., Mao, J.Q., Choi, D., Lam, J.C.W., Zheng, G., Martin, M., Lee, J.H.W., Lam, P.K.S., Richardson, B.J., 2009. Antibiotics in the Hong Kong metropolitan area: ubiquitous distribution and fate in Victoria Harbour. Mar. Pollut. Bull. 58, 1052–1062.

Modi, A.C., Ingerski, L.M., Rausch, J.R., Glauser, T.A., Drotar, D., 2012. White coat adherence over the first year of therapy in pediatric epilepsy. J. Pediatr. 161, 695.

Monteiro, S.C., Boxall, A.B.A., 2009. Factors affecting the degradation of pharmaceuticals in agricultural soils. Environ. Toxicol. Chem. 28, 2546–2554.

Moore, B.E., Thomson, S.R., Zhen, D., Duke, C.C., 1990. Photochemical studies of the anti-inflammatory drug diclofenac. Photochem. Photobiol. 52, 685–690.

Mrozik, W., Stefanska, J., 2014. Adsorption and biodegradation of antidiabetic pharmaceuticals in soils. Chemosphere 95, 281–288.

Nallani, G.C., Paulos, P.M., Venables, B.J., Edziyie, R.E., Constantine, L.A., Huggett, D.B., 2012. Tissue-specific uptake and bioconcentration of the oral contraceptive norethindrone in two freshwater fishes. Arch. Environ. Contam. Toxicol. 62, 306–313.

Navas, J.M., Segner, H., 2006. Vitellogenin synthesis in primary cultures of fish liver cells as endpoint for in vitro screening of the (anti)estrogenic activity of chemical substances. Aquat. Toxicol. 80, 1–22.

Nelson, E.R., Habibi, H.R., 2013. Estrogen receptor function and regulation in fish and other vertebrates. Gen. Comp. Endocrinol. 192, 15–24.

Newman, M.C., 2010. Fundamentals of Ecotoxicology, third ed. CRC Press, Boca Raton, United States.

Nieto, E., Blasco, J., Gonzalez-Ortegon, E., Drake, P., Hampel, M., 2013. Is Atyaephyra desmarestii a useful candidate for lethal and sub-lethal toxicity tests on pharmaceutical compounds? J. Hazard. Mater. 263, 256–265.

Oaks, J.L., Gilbert, M., Virani, M.Z., Watson, R.T., Meteyer, C.U., Rideout, B.A., Shivaprasad, H.L., Ahmed, S., Chaudry, M.J.I., Arshad, M., Mahmood, S., Ali, A., Khan, A.A., 2004. Diclofenac residues as the cause of vulture population decline in Pakistan. Nature 427, 630–633.

Pal, A., Gin, K.Y.H., Lin, A.Y.C., Reinhard, M., 2010. Impacts of emerging organic contaminants on freshwater resources: review of recent occurrences, sources, fate and effects. Sci. Total Environ. 408, 6062–6069.

Pal, R., Megharaj, M., Kirkbride, K.P., Heinrich, T., Naidu, R., 2011. Biotic and abiotic degradation of illicit drugs, their precursor, and by-products in soil. Chemosphere 85, 1002–1009.

Pascoe, D., Karntanut, W., Muller, C.T., 2003. Do pharmaceuticals affect freshwater invertebrates? A study with the cnidarian Hydra vulgaris. Chemosphere 51, 521–528.

Passananti, M., Temussi, F., Iesce, M.R., Previtera, L., Mailhot, G., Vione, D., Brigante, M., 2014. Photoenhanced transformation of nicotine in aquatic environments: involvement of naturally occurring radical sources. Water Res. 55, 106–114.

Pedrouzo, M., Borrull, F., Pocurull, E., Marce, R.M., 2011. Presence of pharmaceuticals and hormones in waters from sewage treatment plants. Water Air Soil Pollut. 217, 267–281.

Perelo, L.W., 2013. Review: in situ and bioremediation of organic pollutants in aquatic sediments. J. Hazard. Mater. 177, 81–89.

Perez-Estrada, L.A., Malato, S., Gernjak, W., Aguera, A., Thurman, E.M., Ferrer, I., Fernandez-Alba, A.R., 2005. Photo-fenton degradation of diclofenac: Identification of main intermediates and degradation pathway. Environ. Sci. Technol. 39, 8300–8306.

Peuravuori, J., 2012. Aquatic photochemistry of diclofenac in the presence of natural dissolved organic chromophoric material and nitrate. Int. J. Environ. Anal. Chem. 92, 1470–1492.

Poiger, T., Buser, H.R., Muller, M.D., 2001. Photodegradation of the pharmaceutical drug diclofenac in a lake: pathway, field measurements, and mathematical modeling. Environ. Toxicol. Chem. 20, 256–263.

Plosz, B.G., Langford, K.H., Thomas, K.V., 2012. An activated sludge modeling framework for xenobiotic trace chemicals (ASM-X): Assessment of diclofenac and carbamazepine. Biotechnol. Bioeng. 109, 2757–2769.

Quesada-Penate, I., Julcour-Lebigue, C., Jauregui-Haza, U.J., Wilhelm, A.M., Delmas, H., 2012. Degradation of paracetamol by catalytic wet air oxidation and sequential adsorption—catalytic wet air oxidation on activated carbons. J. Hazard. Mater. 221, 131–138.

Radke, M., Maier, M.P., 2014. Lessons learned from water/sediment-testing of pharmaceuticals. Water Res. 55, 63–73.

Razavi, B., Ben Abdelmelek, S., Song, W.H., O'Shea, K.E., Cooper, W.J., 2011. Photochemical fate of atorvastatin (lipitor) in simulated natural waters. Water Res. 45, 625–631.

Redshaw, C.H., Cooke, M.P., Talbot, H.M, McGrath, S., Rowland, S.J., 2008. Low biodegradability of fluoxetine HCl, diazepam and their human metabolites in sewage sludge-amended soil. J. Soils Sediments. 8, 217–230.

Reuschenbach, P., Silvani, M., Dammann, M., Warnecke, D., Knacker, T., 2008. ECOSAR model performance with a large test set of industrial chemicals. Chemosphere 71, 1986–1995.

Ribeiro, A.R., Castro, P.M.L., Tiritan, M.E., 2012. Chiral pharmaceuticals in the environment. Environ. Chem. Lett. 10, 239–253.

Richardson, B.J., Larn, P.K.S., Martin, M., 2005. Emerging chemicals of concern: pharmaceuticals and personal care products (PPCPs) in Asia, with particular reference to Southern China. Mar. Pollut. Bull. 50, 913–920.

Robinson, A.A., Belden, J.B., Lydy, M.J., 2005. Toxicity of fluoroquinolone antibiotics to aquatic organisms. Environ. Toxicol. Chem. 24, 423–430.

Robinson, P.F., Liu, Q.T., Riddle, A.M., Murray-Smith, R., 2007. Modeling the impact of direct phototransformation on predicted environmental concentrations (PECs) of propranolol hydrochloride in UK and US rivers. Chemosphere 66, 757–766.

Roos, V., Gunnarsson, L., Fick, J., Larsson, D.G.J., Ruden, C., 2012. Prioritising pharmaceuticals for environmental risk assessment: towards adequate and feasible first-tier selection. Sci. Total Environ. 421, 102–110.

Rua-Gomez, P.C., Puttmann, W., 2013. Degradation of lidocaine, tramadol, venlafaxine and the metabolites O-desmethyltramadol and O-desmethylvenlafaxine in surface waters. Chemosphere 90, 1952–1959.

Ruggeri, G., Ghigo, G., Maurino, V., Minero, C., Vione, D., 2013. Photochemical transformation of ibuprofen into harmful 4-isobutylacetophenone: pathways, kinetics, and significance for surface waters. Water Res. 47, 6109–6121.

Sabourin, L., Al-Rajab, A.J., Chapman, R., Lapen, D.R., Topp, E., 2011. Fate of the antifungal drug clotrimazole in agricultural soil. Environ. Toxicol. Chem. 30, 582–587.

Sacca, M.L., Accinelli, C., Fick, J., Lindberg, R., Olsen, B., 2009. Environmental fate of the antiviral drug Tamiflu in two aquatic ecosystems. Chemosphere 75, 28–33.

Sanchez-Arguello, P., Fernandez, C., Tarazona, J.V., 2009. Assessing the effects of fluoxetine on Physa acuta (Gastropoda, Pulmonata) and Chironomus riparius (Insecta, Diptera) using a two-species water-sediment test. Sci. Total Environ. 407, 1937–1946.

Sanderson, H., Thomsen, M., 2007. Ecotoxicological quantitative structure-activity relationships for pharmaceuticals. Bull. Environ. Contam. Toxicol. 79, 331–335.

Sanderson, H., Thomsen, M., 2009. Comparative analysis of pharmaceuticals versus industrial chemicals acute aquatic toxicity classification according to the United Nations classification system for chemicals. Assessment of the (Q)SAR predictability of pharmaceuticals acute aquatic toxicity and their predominant acute toxic mode-of-action. Toxicol. Lett. 187, 84–93.

Sanderson, H., Johnson, D.J., Reitsma, T., Brain, R.A., Wilson, C.J., Solomon, K.R., 2004. Ranking and prioritization of environmental risks of pharmaceuticals in surface waters. Regul. Toxicol. Pharmacol. 39, 158–183.

Scherr, F.F., Sarmah, A.K., Di, H.J., Cameron, K.C., 2008. Modeling degradation and metabolite formation kinetics of estrone-3-sulfate in agricultural soils. Environ. Sci. Technol. 42, 8388–8394.

Scherr, F.F., Sarmah, A.K., Di, H.J., Cameron, K.C., 2009. Degradation and metabolite formation of 17 beta-estradiol-3-sulphate in New Zealand pasture soils. Environ. Int. 35, 291–297.

Schultz, M.M., Furlong, E.T., Kolpin, D.W., Werner, S.L., Schoenfuss, H.L., Barber, L.B., Blazer, V.S., Norris, D.O., Vajda, A.M., 2010. Antidepressant pharmaceuticals in two US effluent-impacted streams: occurrence and fate in water and sediment, and selective uptake in fish neural tissue. Environ. Sci. Technol. 44, 1918–1925.

Schwartz, M.P., Wagner, P.J., 2000. Which medicines do our patients want from us? J. Fam. Pract. 49, 339–341.

Schwarzenbach, R.P., Escher, B.I., Fenner, K., Hofstetter, T.B., Johnson, C.A., von Gunten, U., Wehrli, B., 2006. The challenge of micropollutants in aquatic systems. Science 313, 1072–1077.

Shankar, P.R., Partha, P., Nagesh, S., 2002. Prescribing patterns in medical outpatients. Int. J. Clin. Pract. 56, 549–551.

Shrestha, S.L., Casey, F.X.M., Hakk, H., Smith, D.J., Padmanabhan, G., 2012. Fate and transformation of an estrogen conjugate and its metabolites in agricultural soils. Environ. Sci. Technol. 46, 11047–11053.

Smith, E.M., Chu, S.G., Paterson, G., Metcalfe, C.D., Wilson, J.Y., 2010. Cross-species comparison of fluoxetine metabolism with fish liver microsomes. Chemosphere 79, 26–32.

Stamatis, N., Triantafyllidis, V., Hela, D., Konstantinou, I., 2013. Occurrence and distribution of selected pharmaceutical compounds on sewage-impacted section of River Acheloos, Western Greece. Int. J. Environ. Anal. Chem. 93, 1602–1619.

Stanley, J.K., Ramirez, A.J., Chambliss, C.K., Brooks, B.W., 2007. Enantiospecific sublethal effects of the antidepressant fluoxetine to a model aquatic vertebrate and invertebrate. Chemosphere 69, 9–16.

Steinbach, C., Fedorova, G., Prokes, M., Grabicova, K., Machova, J., Grabic, R., Valentova, O., Kroupova, H.K., 2013. Toxic effects, bioconcentration and depuration of verapamil in the early life stages of common carp (Cyprinus carpio L.). Sci. Total Environ. 461, 198–206.

Strahle, U., Scholz, S., Geisler, R., Greiner, P., Hollert, H., Rastegar, S., Schumacher, A., Selderslaghs, I., Weiss, C., Witters, H., Braunbeck, T., 2012. Zebrafish embryos as an alternative to animal experiments-A commentary on the definition of the onset of protected life stages in animal welfare regulations. Reprod. Toxicol. 33, 128–132.

Stuer-Lauridsen, F., Birkved, M., Hansen, L.P., Lutzhoft, H.C.H., Halling-Sorensen, B., 2000. Environmental risk assessment of human pharmaceuticals in Denmark after normal therapeutic use. Chemosphere 40, 783–793.

Stulten, D., Zuhlke, S., Lamshoft, M., Spiteller, M., 2008. Occurrence of diclofenac and selected metabolites in sewage effluents. Sci. Total Environ. 405, 310–316.

Tambosi, J.L., de Sena, R.F., Favier, M., Gebhardt, W., Jose, H.J., Schroder, H.F., Moreira, R., 2010. Removal of pharmaceutical compounds in membrane bioreactors (MBR) applying submerged membranes. Desalination 261, 148–156.

Tamtam, F., Le Bot, B., Dinh, T., Mompelat, S., Eurin, J., Chevreuil, M., Bonte, P., Mouchel, J.M., Ayrault, S., 2011. A 50-year record of quinolone and sulphonamide antimicrobial agents in Seine River sediments. J. Soils Sediments 11, 852–859.

Ter Halle, A., Richard, C., 2006. Simulated solar light irradiation of mesotrione in natural waters. Environ. Sci. Technol. 40, 3842–3847.

Thiele-Bruhn, S., 2003. Pharmaceutical antibiotic compounds in soils—a review. J. Plant Nutr. Soil Sci. 166, 145–167.

Thuy, H.T.T., Loan, T.T.C., 2014. Degradation of selected pharmaceuticals in coastal wetland water and sediments. Water Air Soil Pollut. 225, 9.

Tong, A.Y.C., Braund, R., Tan, E.W., Tremblay, L.A., Stringer, T., Trought, K., Peake, B.M., 2011. UV-induced photodegradation of oseltamivir (Tamiflu) in water. Environ. Chem. 8, 182–189.

Tyler, C.R., Jobling, S., 2008. Roach, sex, and gender-bending chemicals: the feminization of wild fish in english rivers. Bioscience 58, 1051–1059.

Vasskog, T., Berger, U., Samuelsen, P.J., Kallenborn, R., Jensen, E., 2006. Selective serotonin reuptake inhibitors in sewage influents and effluents from Tromso, Norway. J. Chromatogr. A 1115, 187–195.

Verma, B., Headley, J.V., Robarts, R.D., 2007. Behaviour and fate of tetracycline in river and wetland waters on the Canadian Northern Great Plains. J. Environ. Sci. Health A Tox. Hazard. Subst. Environ. Eng. 42, 109–117.

Vernouillet, G., Eullaffroy, P., Lajeunesse, A., Blaise, C., Gagne, F., Juneau, P., 2010. Toxic effects and bioaccumulation of carbamazepine evaluated by biomarkers measured in organisms of different trophic levels. Chemosphere 80, 1062–1068.

Vystavna, Y., Huneau, F., Grynenko, V., Vergeles, Y., Celle-Jeanton, H., Tapie, N., Budzinski, H., Le Coustumer, P., 2012. Pharmaceuticals in rivers of two regions with contrasted socio-economic conditions: occurrence, accumulation, and comparison for Ukraine and France. Water Air Soil Pollut. 223, 2111–2124.

Walters, E., McClellan, K., Halden, R.U., 2010. Occurrence and loss over three years of 72 pharmaceuticals and personal care products from biosolids-soil mixtures in outdoor mesocosms. Water Res. 44, 6011–6020.

Wang, J., Gardinali, P.R., 2012. Analysis of selected pharmaceuticals in fish and the fresh water bodies directly affected by reclaimed water using liquid chromatography-tandem mass spectrometry. Anal. Bioanal. Chem. 404, 2711–2720.

Wang, J., Gardinali, P.R., 2013. Uptake and depuration of pharmaceuticals in reclaimed water by mosquito fish (Gambusia holbrooki): a worst-case, multiple-exposure scenario. Environ. Toxicol. Chem. 32, 1752–1758.

Wang, X.H., Lin, A.Y.C., 2014. Is the phototransformation of pharmaceuticals a natural purification process that decreases ecological and human health risks? Environ. Pollut. 186, 203–215.

Wegst-Uhrich, S.R., Navarro, D.A.G., Zimmerman, L., Aga, D.S., 2014. Assessing antibiotic sorption in soil: a literature review and new case studies on sulfonamides and macrolides. Chem. Cent. J. 8, 12.

Wennmalm, A., Gunnarsson, B., 2009. Pharmaceutical management through environmental product labeling in Sweden. Environ. Int. 35, 775–777.

Werner, J.J., McNeill, K., Arnold, W.A., 2005. Environmental photodegradation of mefenamic acid. Chemosphere 58, 1339–1346.

Wheeler, J.R., Gimeno, S., Crane, M., Lopez-Juez, E., Morritt, D., 2005. Vitellogenin: a review of analytical methods to detect (anti) estrogenic activity in fish. Toxicol. Mech. Methods 15, 293–306.

Wille, K., Noppe, H., Verheyden, K., Vanden Bussche, J., De Wulf, E., Van Caeter, P., Janssen, C.R., De Brabander, H.F., Vanhaecke, L., 2010. Validation and application of an LC-MS/MS method for the simultaneous quantification of 13 pharmaceuticals in seawater. Anal. Bioanal. Chem. 397, 1797–1808.

Yang, L., Yu, L.E., Ray, M.B., 2008. Degradation of paracetamol in aqueous solutions by TiO2 photocatalysis. Water Res. 42, 3480–3488.

Ying, G.G., Kookana, R.S., 2003. Degradation of five selected endocrine-disrupting chemicals in seawater and marine sediment. Environ. Sci. Technol. 37, 1256–1260.

Yu, Y., Liu, Y., Wu, L.S., 2013a. Sorption and degradation of pharmaceuticals and personal care products (PPCPs) in soils. Environ. Sci. Pollut. Res. 20, 4261–4267.

Yu, Y., Wu, L.S., Chang, A.C., 2013b. Seasonal variation of endocrine disrupting compounds, pharmaceuticals and personal care products in wastewater treatment plants. Sci. Total Environ. 442, 310–316.

Zeng, C., Ji, Y.F., Zhou, L., Zhang, Y., Yang, X., 2012. The role of dissolved organic matters in the aquatic photodegradation of atenolol. J. Hazard. Mater. 239, 340–347.

Ziylan, A., Ince, N.H., 2011. The occurrence and fate of anti-inflammatory and analgesic pharmaceuticals in sewage and fresh water: Treatability by conventional and non-conventional processes. J. Hazard. Mater. 187, 24–36.

Zorita, S., Martensson, L., Mathiasson, L., 2009. Occurrence and removal of pharmaceuticals in a municipal sewage treatment system in the south of Sweden. Sci. Total Environ. 407, 2760–2770.

Zounkova, R., Kovalova, L., Blaha, L., Dott, W., 2010. Ecotoxicity and genotoxicity assessment of cytotoxic antineoplastic drugs and their metabolites. Chemosphere 81, 253–260.

Degradation of pharmaceuticals in wastewater

6

The detectable environmental levels that have been reported for many human pharmaceuticals arise in large part from their excretion or disposal into wastewater (WW) (sewage) where they can undergo various levels of treatment leading to their breakdown (degradation) before any remaining pharmaceutical content is discharged into the receiving environment. There are many different types of wastewater treatment (WWT) and so in this chapter we introduce the main ones and illustrate their performances in removing a range of common pharmaceuticals from the sewage influent stream.

6.1 Basic steps in wastewater treatment

In a typical WWT plant (WWTP) wastewater undergoes a series of treatment steps, which fall into three broad categories (EPA, 2004; MfE, 2003):

Physical: Solids are removed using physical screens and settling out of the heavier particles under the influence of gravity. Other lighter solid particles are entrapped in air and float to the surface where they can be skimmed off. Passing the wastewater through filters and membranes or adsorbing them on solid surfaces can also separate out specific particles and dissolved compounds. No chemical reactions of any significance occur in this type of treatment step.

Biological: In the presence of oxygen (*aerobic* conditions), bacteria and other small organisms can consume at least some of the organic content of the WW to form a biomass which is removed by physical means such as further sedimentation or flotation.

Chemical: Chemicals can be added which react with specific compounds in the WW to form new products which are either removed physically as solids or remain dissolved in the WW.

Each of these physical, biological, and chemical remediation techniques are incorporated into three key treatment steps—primary, secondary, and tertiary sewage treatment processes as summarised in Figure 6.1. In addition, an initial preliminary treatment step renders the raw wastewater (sewage *influent*) more acceptable for treatment with the subsequent treatment processes.

6.1.1 Preliminary treatment

This is a physical treatment process in which influent is passed through a series of coarse screens and filters to remove particles in the influent typically >2–5 cm in diameter (Figure 6.2).

Smaller mesh screens or a settling tank are then used to remove smaller particles such as grit. This is particularly important if the wastewater has any stormwater component which can contribute significant quantities of grit, particularly in the initial

The Life-Cycle of Pharmaceuticals in the Environment. http://dx.doi.org/10.1016/B978-1-907568-25-1.00006-2

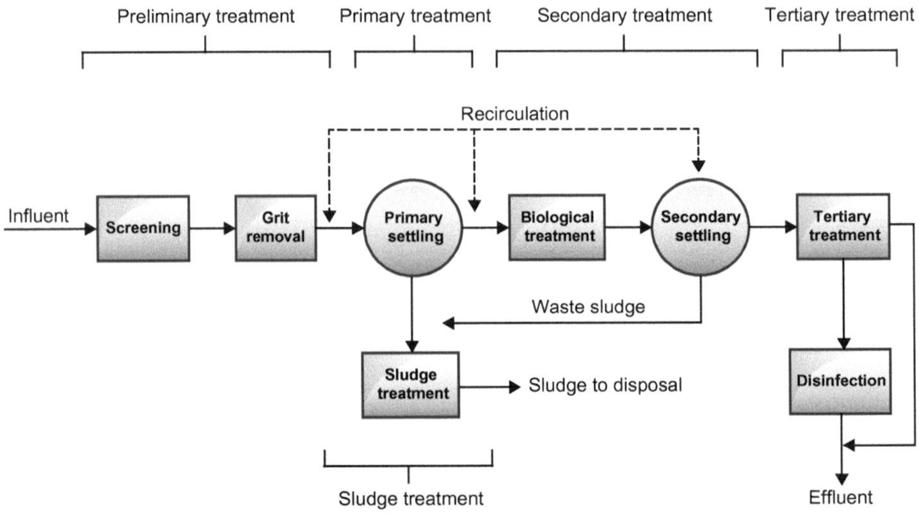

Figure 6.1 Basic steps in wastewater treatment.
Redrawn from Preston (1989).

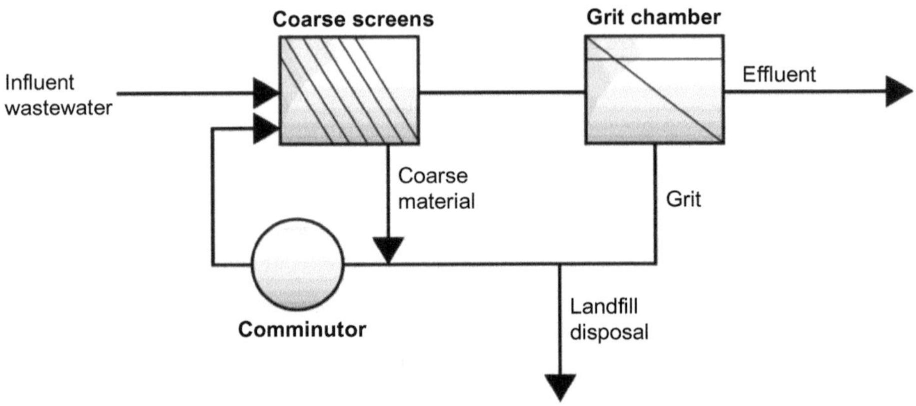

Figure 6.2 Pretreatment for removal of coarse material and grit.

stages of a rain or storm event or in winter when roads are gritted to combat ice. These solid materials are then separated from the remaining influent and typically macerated into smaller units by *comminuting* equipment. Macerated particulates are subsequently returned to the raw wastewater inlet for further treatment, or they are collected for disposal at a landfill site.

6.1.2 Primary treatment

This is a physical process in which the coarse-suspended material remaining after preliminary filtering and screening is removed by some means of physical separation, normally sedimentation and/or flotation. This takes place in a settling tank typically over a period of

3–5 h in quiescent conditions (i.e., no stirring) to remove 40–60% of the suspended solids (SSs) by gravity to form primary sewage sludge at the bottom of the tank (see Figure 6.3).

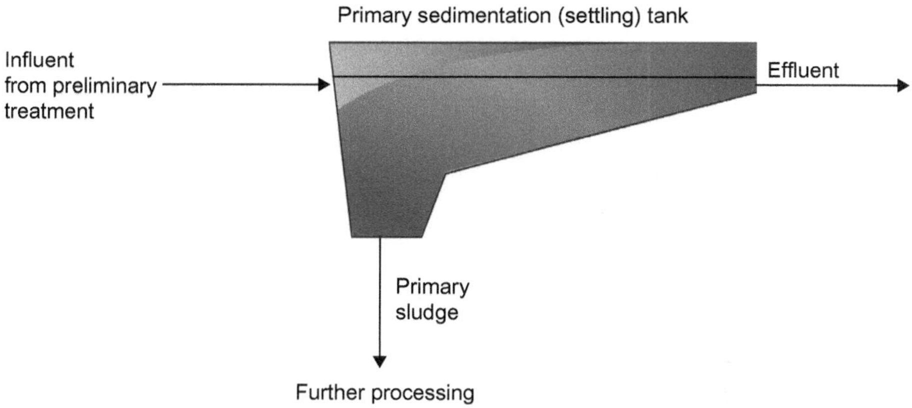

Figure 6.3 Primary treatment using a sedimentation tank.

Sedimentation sometimes is assisted by passing a mechanical skimmer across the surface and through the wastewater at various depths.

Treated primary effluent is either discharged directly into the aquatic or terrestrial environment or undergoes further treatment within the sewage treatment facility. The sewage sludge is removed from the bottom of the tank and undergoes further treatment or is incinerated or landfilled.

The effectiveness of primary treatment is evaluated by a decrease in the SS content and/or the biochemical oxygen demand (BOD) of the primary-treated effluent. BOD is an indirect measure of the overall organic carbon content of the wastewater obtained by monitoring the amount of oxygen consumed by microorganisms as they breakdown organic material in the wastewater. BOD is a function of temperature and duration of exposure and is typically measured at 20 °C over five days of incubation (i.e. BOD_5^{20}). An effluent with a high BOD value is more likely to have a negative impact on the environment compared to one with a low BOD. As such, many regulatory authorities specify that the quality of the effluent arising from sewage treatment facilities must not exceed maximum allowed levels for SS and BOD, which are typically both set at $100\,mg\,L^{-1}$.

6.1.3 Secondary treatment

Secondary treatment involves further removal of organic matter from primary-treated effluent through the use of microorganisms.

In sewage water, various components such as carbohydrates, fats and proteins, and even pharmaceuticals present in relatively low concentrations can be used by microorganisms as a food source. In addition, nitrogen and phosphorus in sewage organic matter are also converted by sewage microbes to ammonium (NH_4^+), nitrate (NO_3^-), and phosphate ions ($H_2PO_4^-$, HPO_4^{2-}), which are necessary for the growth and maintenance of microbial communities in the sewage treatment facility.

Organic matter in sewage is generally oxidized during secondary wastewater treatment (WWT) to carbon dioxide (CO_2), which is released into the atmosphere without any harmful effects, and other innocuous inorganic products which remain dissolved in the effluent. In addition, oxidation processes further reduce the BOD as the microbial communities in the sewage consume the available organic matter. Combined, both primary and secondary treatments typically reduce SS and BOD in the treated sewage water by 80–90%.

There are various types of secondary WWT processes including *suspended growth-activated sludge*, *oxidation ditches*, *fixed-film biological processes*, and *surface-aerated basins*. Most of these processes involve aerobic microorganisms but some involve anaerobic microorganisms.

6.1.3.1 Suspended growth-activated sludge

In suspended growth-activated sludge treatment, primary-treated effluent is mixed with a concentrated suspension of aerobic microorganisms in a tank and agitated or aerated by mechanical paddles or currents of air, respectively (see Figure 6.4).

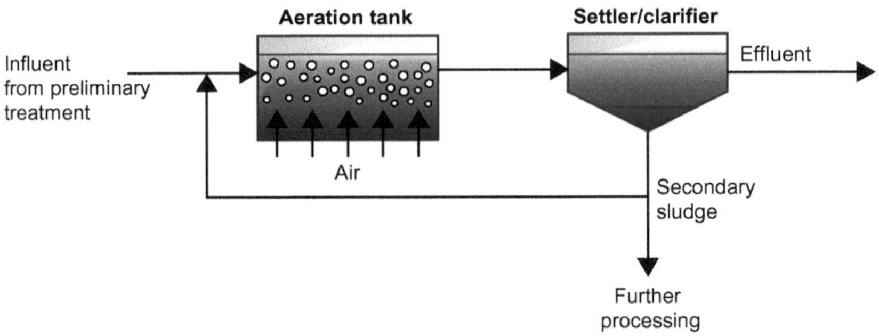

Figure 6.4 Conventional suspended growth-activated sludge process.

After several hours in the tank, the resulting effluent is transferred to a settling tank (clarifier), where any newly formed solid organic matter settles to the bottom and the clarified effluent is then discharged or subjected to further treatment. A proportion of the sludge still contains microorganisms and can be mixed with incoming primary effluent in the aeration tank, while the remaining sludge is also subjected to further treatment or discharged.

6.1.3.2 Oxidation ditch

An oxidation ditch is a variant of a suspended growth-activated sludge chamber and consists of one or more large ring- or oval-shaped channels open to the air (Figure 6.5).

In an oxidation ditch, rotating biological contactors (RBCs) agitate the influent sewage and enhance the rate of aerobic degradation by increasing the levels of dissolved oxygen in the sewage. In a similar manner to a suspended growth-activated sludge process, sewage treated in the ditch is transferred to a clarifier for separation of the sludge from the effluent.

Oxidation ditches are generally more efficient in decreasing the BOD of treated sewage compared to a conventional suspended growth-activated sludge process. However, the time it takes for processing the influent (i.e., the *hydraulic retention time*

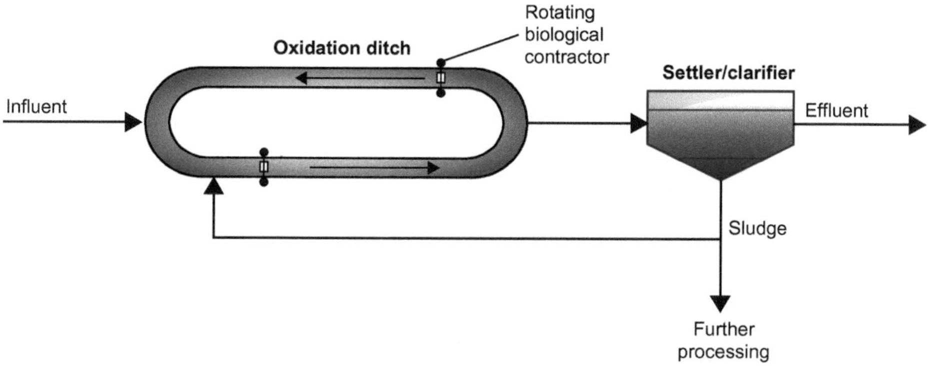

Figure 6.5 Oxidation ditch.

(HRT) or residence time) in an oxidation-ditch treatment process (24–48 h) is much longer than the 2–4 h for treatment in a conventional suspended growth-activated sludge process and the ditches occupy a greater land area (source: http://water.me.vccs.edu/concepts/oxidation.html, 10-30-12).

6.1.3.3 Attached growth processes

Sewage is treated in an *attached* growth or *fixed-film* treatment process by the action of a fixed slime (*biofilm*) of microorganisms on a bed of porous media. Under a combination of anaerobic and aerobic conditions, the biofilm reduces the BOD in the influent water by oxidizing organic matter as it comes into contact with the porous media.

There are two common variants to the attached growth process. In a *trickling* filter system, rotary sprayers distribute the primary-treated effluent over the surface of the trickling filter—a tank filled with media beds of porous stones, clinker, or plastic granules (Figure 6.6).

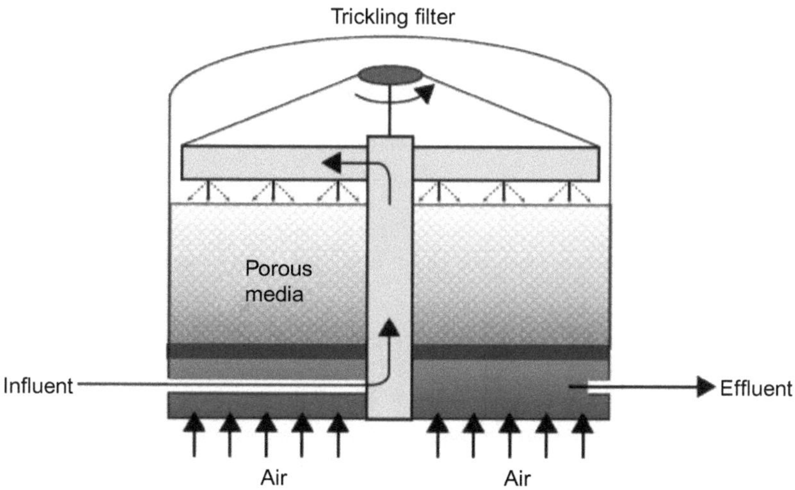

Figure 6.6 Trickling filter system.

The resulting solid biomass material is allowed to separate out over time by sedimentation and periodic removal from the media bed and/or by passing the effluent from the tank through a clarifier.

A *rotating biological concentrator* (RBC) treatment system (Figure 6.7) consists of groups of large plastic disks mounted close together on a rotating shaft which is positioned so that half of each disk is in contact with the influent wastewater and the other half exposed to air. The rotating disks aerate the sewage water and the biofilms on these disks facilitate the breakdown of organic matter in the influent.

Additionally, air can be pumped into the base of the media bed in a trickling filter or into the wastewater container of an RBC to enhance aerobic biodegradation processes.

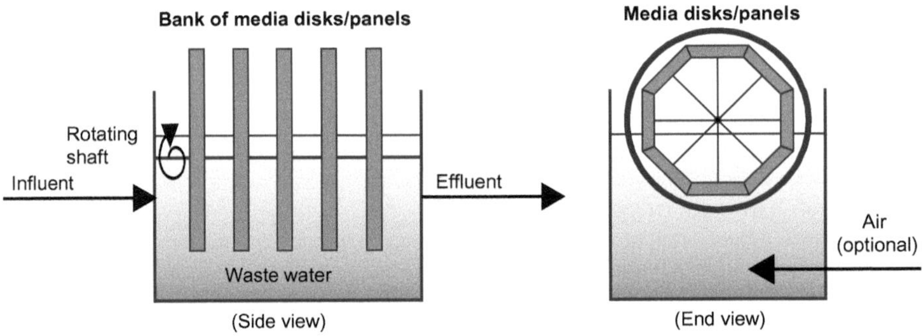

Figure 6.7 Rotating biological contactor (RBC).

A variation of the RBC system is the *moving bed biofilm reactor* (MBBR; *see* Figure 6.8). Here, thousands of engineered polyethene biofilm carriers are introduced into a treatment tank. Under constant motion typically achieved by moving air currents blowing into the tank, the high surface area of the carriers supports the formation of fixed biofilms of *heterotrophic* (i.e., organisms that require organic carbon for energy and growth) and *autotrophic* bacteria (i.e., organisms that can utilize CO_2 and sunlight for energy and growth) which degrade organic matter in the influent.

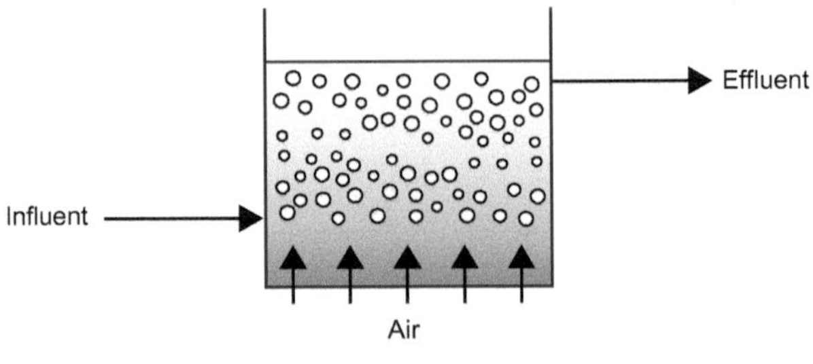

Figure 6.8 Moving bed biofilm reactor (MBBR).

6.1.3.4 Surface-aerated basin (oxidation/stabilization pond, lagoon)

Surface-aerated treatment can be considered to be a combined primary and secondary treatment process occurring in a relatively shallow-holding basin (an oxidation/stabilization pond or lagoon) (Figure 6.9).

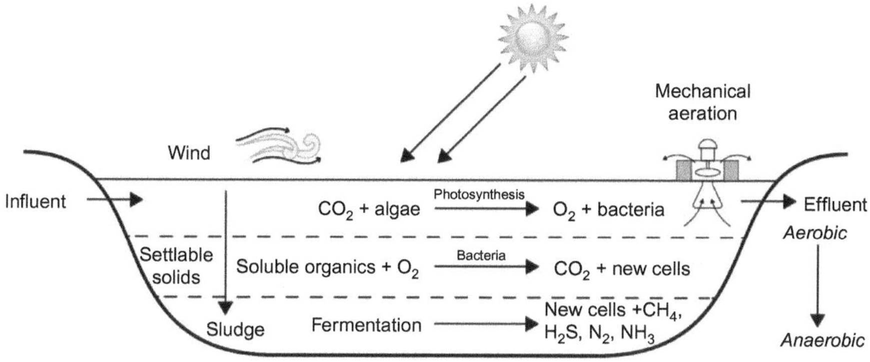

Figure 6.9 Reactions occurring in a surface-aerated basin (oxidation/stabilization pond, lagoon). Reprinted with permission from *Journal Water Pollution Control Federation*, Vol. 45, No. 2, Copyright February 1974, pp. 335. Water Environment Federation, Alexandria, Virginia, www.wef.org.).

Raw influent sewage is discharged into the pond and primary treatment occurs by sedimentation of sludge to the bottom of the pond. Secondary treatment of this sludge then occurs where it is decomposed via fermentation processes involving anaerobic bacteria to form various by-products, including the gases methane (CH_4) and hydrogen sulfide (H_2S) which rise up to the surface and escape into the atmosphere. For the remaining sewage fraction, secondary treatment also occurs in the surface layers of the pond where soluble organic components in the influent are digested by aerobic bacteria. These bacteria obtain their source of oxygen not only from the mechanical aeration of the stabilization pond achieved by wind currents above the pond, but also from the photosynthesis of algae in the sunlight-abundant surface layers or from artificially induced mechanical aeration using floating aerators.

Oxidation ponds are cheaper to construct and operate than the separate primary and secondary treatment processes carried out in mechanically-driven plants, but they do require large areas of land which makes them unsuitable for urban areas. The accumulated sludge typically only requires removal after a number of years. However, the degradation processes are relatively slow and strongly affected by seasonal temperatures with most operating in the range from +4 to 32 °C. Typical treatment times range from a few days in warm climates to up to months in cold climates, which means that oxidation ponds are generally restricted to warmer climate regions.

6.1.4 Tertiary treatment

In spite of the very significant decreases in the BOD and SS content that can be achieved by various combined primary and secondary sewage treatment processes that have been described above, there are occasions when the amounts of potentially harmful pollutants in the wastewater such as certain pharmaceuticals have not been reduced to any significant extent during primary and secondary treatment. This becomes particularly important if the effluent or the aquatic system into which the effluent is to be ultimately discharged, is to be used as a source of water for human consumption. In these situations, it becomes necessary to consider further treatment of the secondary-treated effluent using a variety of more advanced and often compound-specific *tertiary* treatment techniques typically involving both chemical and physical processes.

The contaminants removed by tertiary WWT fall into three main categories:

1. *Suspended solids*: These are in the form of finely divided material, the organic component of which is primarily responsible for the residual BOD of secondary-treated effluent.
2. *Dissolved inorganic materials*: Many of these can act as nutrients such as inorganic phosphorus (phosphates) and nitrogen-containing compounds such as nitrate and ammonium species. Algal blooms and their subsequent breakdown can occur if treated effluent containing these species is discharged into confined areas of freshwater such as small streams and lakes. Toxic metal species, such as cadmium, might also arise from industrial sources of wastewater and these contribute to this inorganic fraction.
3. *Dissolved organic compounds*: These can be numerous and in the context of this present text, include pharmaceuticals. A whole range of other types of organic compounds can also arise from industrial sources discharging into the wastewater system such as food-processing and agricultural-related industries. Many of these compounds are not removed to any significant extent (i.e., are *refractory)* from raw wastewater by primary settling or microbial action in secondary treatment.

A variety of physical and chemical processes are used to perform tertiary treatment of secondary-treated effluent including the following.

6.1.4.1 Filtration/membrane separation processes

Some solid particles and colloidal material arising from primary and secondary treatment are often too small (fine) in size to be removed by sedimentation or with a clarifier. Instead they can be physically removed either by retaining them on a physical barrier such as a membrane (*surface* filter) or on a bed of granular material such as sand (*depth* filter). The influent wastewater passes through the filter material under gravity or, more commonly, through mechanically applied pressure.

The size of the particles and colloidal material retained on the physical barrier provided by the filter or membrane material and hence the extent of the removal of waste material, depends on the pore size of the filter or membrane (Figure 6.10).

The oldest type of filter consists of sand which traps particles that are too large to pass through the pores between the packed grains of sand, or smaller particles which are adsorbed onto the grains of sand. The rate of flow of effluent through a sand filter is slow under a natural gravity feed but can be increased by additional mechanical pressure. In all types of sand filters, the top layer of sand containing the particulates must be periodically removed by reversing the direction of flow.

Membrane	Pore size	Species removed	Applied pressure
Sand filter	100–10 μm	Pollens, colloids, bacteria, algae	Gravity
Microfiltration (MF)	10–0.1 μm	Organic macro-molecules, colloids, bacteria	100–400 kPa
Ultrafiltration (UF)	0.1–0.001 μm	Organic macromolecules, viruses, organic compounds	200–700 kPa
Nanofiltration (NF)	0.01–0.0005 μm	Organic compounds	600–1,000 kPa
Reverse osmosis (RO)	>0.0001 μm	Dissolved salts	0.3–6.0 MPa

Figure 6.10 Types of some common filtration and membrane separation systems.

Membrane filtration treatment processes use flat sheets or hollow fibers of synthetic polymers, such as cellulose acetate, polysulfone, polyamide, or polycarbonate, to filter the influent sewage water. Under an applied pressure, the influent water is forced through from one side of the membrane and the purified filtrate (effluent) is collected from the other side. The membrane is periodically backwashed to prevent retentate material on the membrane from accumulating, and the collected retentates are added to the sludge waste stream for disposal.

Reverse osmosis is a form of membrane filtration in which the spontaneous diffusion of water molecules through a semipermeable membrane into a solution of higher solute concentration (*osmosis*) is reversed by the application of pressure to the more concentrated solution (*see* Figure 6.11). Purified water or "effluent" is collected from the less concentrated side of the semipermeable membrane. The efficiency of reverse osmosis (RO) is mainly governed by the osmotic pressure difference between the two solutions separated by the semipermeable membrane. Highly polluted water with high solute concentrations or suspended microbial loads will require a higher applied pressure before water purification occurs. This is in contrast to conventional membrane or sand filtration whereby water can be potentially purified irrespective of the magnitude of the applied pressure. However, while semipermeable membranes used in RO are potentially impermeable even to small ions and molecules, the range of materials that can be filtered by sand or other membrane filtration processes is limited by the sizes of the sand granules or the membrane pores, respectively (Figure 6.10).

Activated sludge treatment and membrane filtration can be combined together to form a *membrane bioreactor* (MBR). The membrane unit effectively undertakes the function of the settler/clarifier shown in Figure 6.5 and by being typically immersed in the aeration tank, it can avoid problems with poor settling of the sludge. Various types of membrane can be used for the MBR, including flat sheet (FS MBR) and hollow fiber (HF MBR).

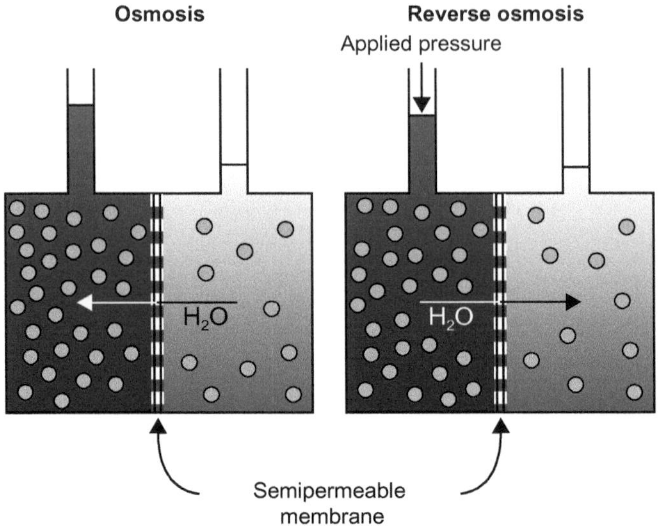

Figure 6.11 Principles of osmosis and reverse osmosis (RO).

6.1.4.2 Adsorption by activated carbon

Secondary-treated effluent can also be further treated by passing the wastewater over a bed or canister of activated carbon (AC) (Grassi et al., 2012). This form of carbon is made by heating organic matter such as wood, coal, or lignite with a supply of air sufficient enough to enable chemical degradation of the organic matter to form elemental carbon but not combustion to the extent of forming either CO or CO_2. The resulting carbon residue is then activated by partial oxidation with CO_2 or oxygen (O_2) in air to increase the surface area available for adsorption of dissolved organic compounds such as pharmaceuticals present in the wastewater (Manahan, 1990).

AC is used in a powdered form (PAC) which can be added into the liquid influent wastewater and then filtered off, or held in a granular form (GAC) in columns or fixed beds through which the influent flows. In the latter situation, when the carbon pores become saturated with adsorbed material, the GAC column must be removed and replaced. In either form, spent AC can be disposed of as a solid waste or the carbon can be regenerated by thermal or chemical means, which either desorb or chemically destroy any adsorbed organic compounds leaving about 90% of the AC for reuse (Martin and Wj, 1997).

AC is particularly effective for removing organic contaminants with poor water solubility but is expensive compared to other tertiary treatment processes such as membrane filtration.

6.1.4.3 Advanced oxidation processes

The tertiary treatment processes described above remove trace pharmaceuticals from the wastewater by physically removing them (filtration) or by transferring them from one phase to another (adsorption). However, there are some tertiary treatment techniques which can alter the chemical composition of the wastewater without requiring a change of phase. Some of these techniques oxidize pharmaceuticals to CO_2 (i.e. the process of *mineralization*) or other organic by-products in wastewater in a similar manner to the biodegradation of potential wastewater pollutants by secondary treatment. In place of the microbial colonies or biofilms, chemical oxidants such as chlorine (Cl_2), ozone (O_3), or hydrogen peroxide (H_2O_2) can be introduced into secondary-treated effluent for further degradation or removal of dissolved organic compounds.

Tertiary treatment is also often employed for the disinfection of secondary-treated effluent before it is discharged into the *receiving* environment. This need arises from the typically high pathogenic content of the raw wastewater, particularly that arising from human sources such as feces and urine, which is often not significantly reduced during any step in the WWTP and could adversely affect later use of the receiving environment as a drinking water source, recreational uses such as bathing and boating, irrigation, or aquaculture. Common disinfection processes involve treatment of the liquid effluent before ultimate discharge by chlorination or ozonation, or by UV irradiation (see Das et al. 2001; Leong et al. 2008), all of which can also lead to at least some degree of degradation of any pharmaceuticals present in the waste water

Oxidation and subsequent degradation of dissolved organic compounds can be induced by irradiation with ultraviolet (UV) light on its own, or in combination with additional reagents. For example, the UV irradiation of wastewater in the presence of H_2O_2 generates hydroxyl radicals (•OH) which are themselves chemical species that are short lived in the water but highly reactive toward many organic compounds (Glaze et al., 1987). The photochemical production of hydroxyl radicals can be further enhanced by the addition of Fe (II) to the H_2O_2 (the Fenton (Haber-Weiss) reaction) (Walling, 1975 and reviews by Neyens and Baeyens (2003) and Wadley and Waite (2004)):

$$Fe^{II} + H_2O_2 \rightarrow Fe^{III} + \bullet OH + OH^-$$

followed by

$$Fe^{III} + 2H_2O + h\nu \rightarrow Fe^{II} + \bullet OH + H_3O^+$$

In this photo-Fenton process, the iron species cycle between the Fe(II) and Fe(III) oxidation states in a *photocatalytic* reaction.

Other photocatalysts that have been used to produce hydroxyl radicals for the advanced oxidation of organic compounds include the semiconductor TiO_2 (Tong et al., 2012).

Ozone (O_3) can also oxidize an organic substrate (R) *directly*:

$$O_3 + R \rightarrow R_{ox}$$

but the reaction rate constants are generally slower than for reaction with •OH. Much faster oxidation rates are observed when the ozone reacts with the organic substrate *indirectly* by first forming the hydroxyl radical in a chain reaction initiated by the hydroxide ion present in water:

$$3O_3 + OH^- + H_3O^+ \rightarrow 2\bullet OH + 4O_2 + H_2O$$

The production of •OH from ozone is also enhanced in the presence of hydrogen peroxide:

$$2O_3 + H_2O_2 \rightarrow 2\bullet OH + 3O_2$$

or by *direct photolysis* of ozone with UV-C light (<254 nm):

$$O_3 + H_2O + h\nu \rightarrow H_2O_2 + O_2$$

with the resulting H_2O_2 going on to react with more O_3 as shown above.

For a full discussion of these ozone reactions and their application to water and wastewater treatment (see Gottschalk et al., 2008).

Other advanced redox methods that have been investigated for the degradation of pharmaceuticals in aqueous solutions include simultaneous oxidation with •OH radicals and reduction with solvated electrons (e^-_{aq}). Both these reactive species can be produced by high-energy *radiolysis* of water (Cooper et al., 2008):

$$H_2O \leadsto \bullet OH + e^-_{aq} + H_2O_2 + \bullet H + H^+ + H_2$$

6.1.5 Other sewage treatment methods

There is a wide number of other methods of WWT used throughout the world to achieve various levels of purification of wastewater (*see* EPA, 2004). Some of the more common include:

6.1.5.1 Land treatment

Untreated or partially treated wastewater is distributed directly onto land where many of the physical, chemical, and biological degradation processes discussed above can occur as the water passes across or through the soil. *Slow rate infiltration* involves natural filtering by the soil and microbial activity associated with plant action. Some of the water volume evaporates while the remainder is collected for surface discharge or enters the groundwater. *Rapid infiltration* is typically used to "polish" primary or secondary effluent for which large volumes are applied to very permeable soils and then collected in the underlying water table.

6.1.5.2 Wetlands

These are either natural or man-made water systems that can act as biofilters by removing or degrading pollutants in wastewater using microorganisms that are growing on the roots (*rhizosphere* region), stems, and leaves of vegetation. This breakdown can involve a wide variety of physical, chemical, and biological interactions between the various components of the natural vegetation and the wastewater and is equivalent to many of the conventional secondary or tertiary treatment steps outlined above.

The various types of wetland include (*see* Figure 6.12):

Surface-flow: influent flows *above* a wide variety of soil types in a marsh or swamp planted with various types of vegetation such as water hyacinth and cattails.

Subsurface-flow: influent flows either horizontally or vertically *through* a gravel (typically limestone or volcanic rock) or sand on which plants are rooted.

Overland-flow: periodic applications of influent are made to the upper surface of a sloping grass plot and then it flows down over the vegetated surface with some components soaking into the soil while the remainder are collected at the bottom as effluent to be discharged elsewhere.

Tidal-flow: the organic carbon in the pollutants is oxidized by the addition of nitrate through a series of flood and drain cycles from one side of the wetland to the other.

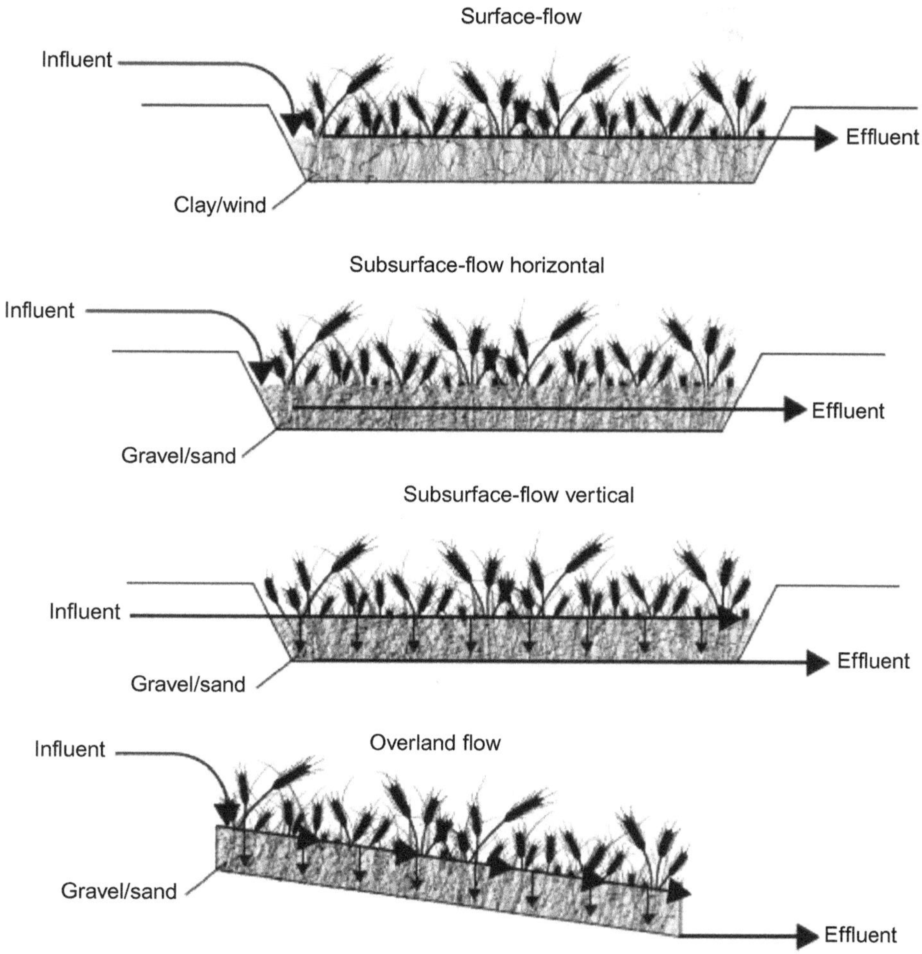

Figure 6.12 Some common types of natural and constructed wetlands.

A further distinction can made between wetlands operating under the above regimes that occur naturally and those that are artificial:

Constructed: These are marshes and swamp areas that are artificially constructed to duplicate and optimize the water purification processes that occur in natural treatment wetlands.

Engineered: A special form of a constructed wetland where the operating conditions are monitored, manipulated, and controlled more than in an ordinary constructed wetland.

The treatment performance of natural wetlands is much less predictable than more engineered treatment systems and so they are often used as a "polishing buffer" for the effluent arising from the use of a conventional treatment system before it enters the final receiving environment.

A comprehensive assessment of the design configuration of constructed wetlands for the removal of pharmaceuticals has been published by Hijosa-Valsero et al. (2010). Guidelines for suitable wetland planting at least in New Zealand conditions have been published by Tanner et al. (2006).

6.1.5.3 Onsite sewage treatment and disposal (septic tanks)

In many countries, individual homes that cannot be connected to any communal sewage collection and treatment system (typically in a more isolated rural environment), invariably use *septic tanks* which effectively involve mainly primary treatment on a very small, localized scale (for an overview *see* EPA, 1992; Toor et al., 2011a; EPA, 2011). Raw sewage discharged from a single household is piped into a settling tank sunk in nearby soil (Figure 6.13) where solids settle out and must be periodically removed as solid sludge. The effluent is then generally discharged into a subsurface land distribution system (drain field) where it can soak into the surrounding soil, or it is sprayed onto land. In either case, there is some potential for surface runoff into waterways with time.

Pharmaceutical residues present in the influent are removed in this type of onsite sewage treatment system by *sorption* onto solids which settle out as sludge in the septic tank

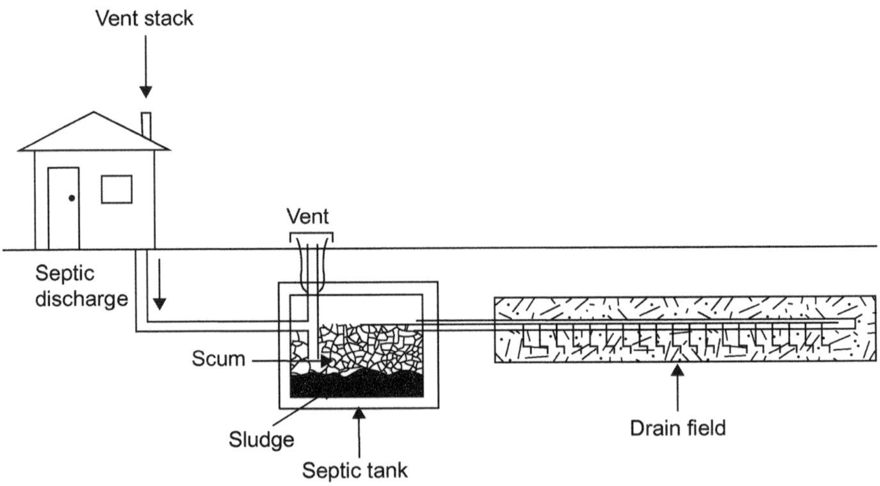

Figure 6.13 Conventional gravity onsite sewage treatment and disposal system (septic tank).

or onto the soil particles in the drain field, by *volatilization* of gaseous components which are released into the air, and by *biotransformation* in the septic tank before discharge of the tank effluent into the drain field (see Matamoros et al., 2009 and Toor et al., 2011b).

6.2 Removal and degradation of selected human pharmaceuticals

All of the above WWT process lead to a general decrease in the level of both inorganic and organic contaminants in any raw influent wastewater as typically monitored by significant decreases in the values for the BOD, chemical oxygen demand (COD), and level of SSs. Some typical BOD and SS values for primary and secondary WWT are given in Table 6.1 and demonstrate a significant reduction in at least these bulk parameters.

Table 6.1 Typical concentrations of bulk wastewater contaminants after primary and secondary biological treatment

	Concentration (mg L^{-1})			Overall % reduction
	Raw Influent	After Primary treatment	After Secondary biological treatment	
BOD	422	326	30	93
SS	321	183	46	86

As part of the reduction in organic content, many of the WWT processes also lead to pharmaceuticals present in the influent being removed or at least undergoing a variety of transformations including (Santos et al., 2010):

1. Mineralisation to form inorganic compounds such as CO_2 and water which obviously have little, if any, subsequent environmental impact;
2. Entrapment by suspended solids leading to sedimentation as sludge;
3. Chemical breakdown of the parent compound to form conjugate compounds whose concentration should still be minimized in the final effluent; and
4. Conversion to more water-soluble and persistent forms which are often still detectable in the final effluent.

There have been *many* reports of the removal and/or breakdown (degradation) of different pharmaceuticals present in wastewater during various levels of treatment. The extent of this removal/breakdown (degradation efficiency) varies with each pharmaceutical and also depends on many other factors including the type and extent of the WWT process(es), and the demographics of the community producing the wastewater containing the pharmaceuticals.

The overall aim of a treatment process is to degrade pharmaceuticals in any type of waste stream and at least convert them into a form which has a lower impact on the environment than the material initially present in the wastewater. Ideally this would involve conversion to harmless inorganic compounds such as CO_2, H_2O, and simple compounds as minor components which contain other elements such as N, P, S, and the halogens. Although this does occur in some WWT processes, it is hard to detect the formation of such inorganic products against the general inorganic and organic matrix background of any natural water and instead most reports indicate only that there is a decrease in the organic content as measured by the BOD or COD values. In relatively few cases, but particularly in laboratory studies, a variety of smaller, *intermediate* organic structures are detected as a result of the degradation process. Of course as discussed in Chapter 5, this situation then raises the question as to what are the ecotoxicological effects of these simpler breakdown products compared to the parent pharmaceutical? However, such information is rarely provided in any reports of pharmaceutical degradation in wastewater.

It is not possible in the present text to consider the details of *every* published report on the removal/degradation efficiencies for all the human pharmaceuticals that have been detected in different wastewaters and after various stages of treatment. In some cases, only the influent or the effluent concentration has been reported and so no degradation efficiency can be calculated. Also the details of WWT process(es) are often not specified other than the broad classification of some combination of primary, secondary, and/or tertiary treatment.

6.2.1 Degradation of some common pharmaceuticals in wastewater

Examples of the degradation/removal of a selection of common pharmaceuticals from wastewater using a variety of the main wastewater treatment processes are given in Table 6.2. Some comments on the behaviour of some of these pharmaceuticals are as follows.

6.2.1.1 Paracetamol (acetaminophen)

Almost 100% removal of the widely used pharmaceutical from wastewater can be achieved by a wide range of sewage treatment processes (see Table 6.2). Ternes (1998) noted that he could not detect any acetaminophen in surface waters in Germany which he attributed to "high removal efficiencies of municipal sewage treatment plants."

6.2.1.2 Diclofenac

As noted in earlier chapters, this is a classic drug that was one of the first to be detected in the influent and effluent wastewaters of a large number of German sewage treatment plants (Ternes, 1998) and shown to be only partially removed (69%) in a sewage treatment plant employing at least "preliminary and final clarification and an aerator tank." There have been many other studies of this drug using a range of treatment processes giving a large variation in the % removal values (see Table 6.2).

The failure of some sewage treatment plants to effectively remove diclofenac at all from the influent using at least secondary treatment processes is illustrated in Figure 6.14 for three sewage treatment plants in Switzerland (Tauxe-Wuersch et al., 2005).

The overall *increase* (i.e., a *negative* % removal value) in diclofenac concentration reported for a Swedish STP (Kristainstad) (Zorita et al., 2009) of −105% for

Table 6.2 **Removal/degradation efficiencies for treatment of some common pharmaceuticals in wastewater**

Pharmaceutical	Country	Concentration (ng L^{-1}) (mean)		% removal/ degradation	Type of treatment	Reference
		Influent	Effluent			
Paracetamol (acetaminophen)	South Korea	70	60	8.7	2×clarification+aeration	Han et al. (2006)
		11,500	<10–21	>99	MBR	Kim et al. (2007)
			<1	>99	RO, NF, RO-UV, NF-RO	
		59,944	<5.9	>99	ns	Choi et al. (2008)
	Spain	29,000–246,000	<32–4300	98	Secondary	Gómez et al. (2007)
		7,100–11,400	(9900)	>99	Conventional AS	Radjenovic et al. (2009)
				>99	Flat sheet MBR	
				>99	Hollow fiber MBR	
		1,571–37,458	<LOD	100	Secondary	Rosal et al. (2010)
	Taiwan	2,695	330	88	Anaerobic MBR	Dutta et al. (2014)
			7	98	Anaerobic fluidized BR	
	UK	5,529–69,750	<20	98	Tertiary	Roberts and Thomas (2006)
Aspirin	Australia	9.0–38.5	<LOD	>99	Tertiary incl. BAC MF, O_3, RO, Cl_2	Al-Rifai et al. (2007)
	Germany	180	31	81	Tertiary	Temes (1998)
	Japan	470–19,400	38–111	99	Secondary (AS)	Nakada et al. (2006)
Diclofenac	Australia	22	<LOD	>99	Tertiary incl. BAC MF, O_3, RO, Cl_2	Al-Rifai et al. (2007)
	Brazil	780	710	9	Activated sludge	Stumpf et al. (1999)
			190	75	Biological (trickling) filter	
	Canada	50–2,450	70–250	90	ns	Lee et al. (2005)
	Finland	350±100	–	9–60	Biological activated sludge	Lindqvist et al. (2005)
	Germany	100	30	69	Tertiary	Ternes (1998)

(Continued)

Table 6.2 Continued

Pharmaceutical	Country	Concentration ($ng\,L^{-1}$) (mean)		% removal/ degradation	Type of treatment	Reference
		Influent	Effluent			
		3,020	2,510	17	Various	Heberer (2002)
		1,000–1,600	(1,320)	22	Conventional AS	Radjenovic et al. (2009)
				66	Flat sheet MBR	
				63	Hollow fiber MBR	
	Greece	12–560	10–365	35	Secondary	Koutsouba et al. (2003)
	Luxembourg	2–43	3–78	ns	ns	Pailler et al. (2009)
	South Korea	2,590	1,970	24	2 × clarification + aeration	Han et al. (2006)
		10	22–25	−120[1]	MBR	Kim et al. (2007)
			<1	>99	RO, NF, RO-UV, NF-UV	
	Spain	200–3,600	2,220	39	Secondary	Gómez et al. (2007)
		<LOD-561	6–431	5	Secondary	Rosal et al. (2010)
	Sweden	160	120	25	Secondary(AS)	Bendz et al. (2005)
		230	100	56	Primary	Zorita et al. (2009)
			485	−110[1]	+Secondary (AS)	
			470	−105[1]	+Tertiary ($FeCl_3$ flocculation)	
			490	−113[1]	+Sand filter	
	Switzerland	470–1,920	310–930	5–52	3-stage biological	Buser et al. (1998)
	Taiwan	3–347	4–101	71	Secondary (AS)	Lin et al. (2009)
		58	40	31	Anaerobic MBR	Dutta et al. (2014)
			13	68	Anaerobic fluidised BR	
	UK	901–1,036	261–598	42	Tertiary	Roberts and Thomas (2006)
	US		47–150	80–100[2]	Ozonation	Wert et al. (2009)

Ibuprofen	Brazil	360	280	22	Activated sludge	Stumpf et al. (1999)
			90	75	Biological (trickling) filter	
	Canada	6,770	310	>99	ns	Lee et al. (2005)
	Germany	235	30	90	Tertiary	Ternes (1998)
	Finland	13,100	±400	78–100	Biological activated sludge	Lindqvist et al. (2005)
	Italy	ns	ns	0–100 (93) summer 25–72 (38) winter	Secondary	Castiglioni et al. (2006)
	Japan	381–1,130	1.4–177	84	Secondary (AS)	Nakada et al. (2006)
	Luxembourg	82–3,080	3–359	88	ns	Pailler et al. (2009)
	New Zealand	168	<8	>95	Land application on volcanic soil	Gielen et al. (2009)
			<8	>95	Sand	
		662	<8	>99	Volcanic soil	
			<8	>99	Sand	
	South Korea	300	70	78	2×clarification+aeration	Han et al. (2006)
		5,320	52–90	>99	MBR	Kim et al. (2007)
			<1	>99	RO,NF,RO-UV, NF-UV	
	Spain	14,600–31,300	(21,700)	>99	Conventional AS	Radjenovic et al. (2009)
				>99	Flat sheet MBR	
				>99	Hollow fiber MBR	
		<LOD-4,113	<LOD-653	95	Secondary	Rosal et al. (2010)
	Sweden	3,590	150	96	Secondary(AS)	Bendz et al. (2005)
		6,900	5,700	17	Primary	Zorita et al. (2009)
			88	99	+Secondary (AS)	
			61	>99	+Tertiary ($FeCl_3$ flocculation)	
			47	>99	+Sand filter	

(Continued)

Table 6.2 Continued

Pharmaceutical	Country	Concentration (ng L^{-1}) (mean)		% removal/ degradation	Type of treatment	Reference
		Influent	Effluent			
	Switzerland	170–4,500	1,200	73	Secondary(AS) $FeCl_3$	Tauxe-Wuersch et al. (2005)
	Taiwan	711–2,883	313–3,777	71–76	Secondary (AS)	Lin et al. (2009)
		2,500	1,155	54	Anaerobic MBR	Dutta et al. (2014)
			228	80	Anaerobic fluidised BR	
	UK	7,741–33,764	4,239	88	Tertiary	Roberts and Thomas (2006)
	US		5–85	80–100[2]	Ozonation	Wert et al. (2009)
	Sweden	3,650	250	93	Secondary(AS)	Bendz et al. (2005)
Atenolol	Canada	221–1090[a]	184–662	17	Secondary	Lee et al. (2007)
	Italy	ns	ns	36–76(55) summer 0–21(10) winter	Secondary	Castiglioni et al. (2006)
	Spain	84–280	(200)	61	Conventional AS	Radjenovic et al. (2009)
				77	Flat sheet MBR	
				69	Hollow fiber MBR	
		660–2,432	517–1,025	14	Secondary	Rosal et al. (2010)
	Taiwan	738–2,883	210–681	71–76	Secondary	Lin et al. (2009)
			260–980	80–100[2]	Ozonation	Wert et al. (2009)
Metoprolol	Canada	214–664[a]	177–402	17	Secondary	Lee et al. (2007)
	Finland	980–1,350	910–1,070	21		Vieno et al. (2006)
	Germany	380	110	83		Ternes (1998)
	Spain	26–63	39	25	Conventional AS	Radjenovic et al. (2009)

				44	Flat sheet MBR	
				30	Hollow fiber MBR	
		<LOD-52	<LOD-38	7	Secondary	Rosal et al. (2010)
	Taiwan	14–597	12–199	67	Secondary	Lin et al. (2009)
Amoxicillin	Australia	190	nd	100	Conventional AS	Watkinson et al. (2007)
			nd	100	Microfiltration/RO	
	India	173-ND	63-ND	~50	Aeration/biological	Mutiyar and Mittal (2013)
	Italy	ns	ns	100 (100) summer 49–100(75) winter	Secondary	Castiglioni et al. (2006)
Erythromycin	China	180	51	72	Activated sludge	Gao et al. (2012)
		260	120	54	Oxidation ditch	
	Hong Kong	810	850	<0	Primary treatment	Gulkowska et al. (2008)
		740	600	19	Secondary AS	
	South Korea	44	40	9	MBR	Kim et al. (2007)
			<1	>99	RO, NF, RO-UV, NF-UV	
	Spain	32–2,700	820	35	Conventional AS	Radjenovic et al. (2009)
				43	Flat sheet MBR	
				25	Hollow fiber MBR	
		<LOD-2310	<LOD-760	4	Secondary	Rosal et al. (2010)
	Switzerland	60–90	ns	−8 to 4	Primary treatment	Göbel et al. (2007)
				−22 to 7	+Conventional AS	
				−13 to 7	+Fixed bed reactor	
	Taiwan	226–1,537	361–811	22–65	Secondary (AS)	Lin et al. (2009)
		319	132	59	Anaerobic MBR	Dutta et al. (2014)
			44	86	Anaerobic fluidized BR	

(Continued)

Table 6.2 Continued

Pharmaceutical	Country	Concentration ($ng\,L^{-1}$) (mean)		% removal/ degradation	Type of treatment	Reference
		Influent	Effluent			
Sulfamethoxazole	Australia	360	270	25	Conventional AS	Watkinson et al. (2007)
			nd	100	Microfiltration/RO	
	China	490	220	55	Activated sludge	Gao et al. (2012)
		460	340	26	Oxidation ditch	
	Luxembourg	13–155	4–39	70	ns	Pailler et al. (2009)
	South Korea	194	58–70	64–70	MBR	Kim et al. (2007)
			<1.0	>99	RO, NF, RO-UV, NF-UV	
	Switzerland	230–570	ns	−21 to −5	Primary treatment	Göbel et al. (2007)
				−138 to 60	+Conventional AS	
				−61 to 29	+Fixed bed reactor	
	Taiwan	226–1,537	361–811	22–56	Secondary (AS)	Lin et al. (2009)
		312	201	40	Anaerobic MBR	Dutta et al. (2014)
			34	83	Anaerobic fluidised BR	
	Spain		250–1,300 (930)	74	Conventional AS	Radjenovic et al. (2009)
				81	Flat sheet MBR	
				78		
		162–530	104–370	17	Secondary	Rosal et al. (2010)
	Sweden	20	70	‡	Secondary(AS)	Bendz et al. (2005)
	Taiwan	179–1,760	47–964	29–88	Secondary (AS)	Lin et al. (2009)
	US		330–1,200	80–100[2]	Ozonation	Wert et al. (2009)

Trimethoprin	Australia	340	50	85	Conventional AS	Watkinson et al. (2007)
			5	~100	Microfiltration/RO	
	Hong Kong	120	170	<0	Primary treatment	Gulkowska et al. (2008)
		320	120	63	Secondary AS	
	Sweden	160	120	25	Secondary (AS)	Bendz et al. (2005)
	South Korea	21	28–31	−33 to 47[2]	MBR	Kim et al. (2007)
			<1	>99	RO, NF, RO-UV, NF-UV	
	Spain	150–430	204	40	Conventional AS	Radjenovic et al. (2009)
				67	Flat sheet MBR	
				47	Hollow fiber MBR	
		78–197	<LOD-148	5	Secondary	Rosal et al. (2010)
	Switzerland	210–440	ns	−13 to 31	Primary treatment	Göbel et al. (2007)
				−40 to 20	+Conventional AS	
				12–17	+Fixed bed reactor	
	Taiwan	259–949	203–415	0–56	Secondary (AS)	Lin et al. (2009) 312
		16	2	87	Anaerobic MBR	Dutta et al. (2014)
			<LOD	>99	Anaerobic fluidised BR	
Estrone (E1)	Brazil	566	242	58	Stabilisation ponds; AS + Cl_2	Marques et al. (2013)
	Canada	16	5	70	ns	Lee et al. (2005)
		49	17	65	Conventional AS; lagoon	Minim et al. (2009)
	Iran	7–19	1–5	75	AS	Logan et al. (2012)
	Italy	ns	ns	49–100 (75) summer 100 (100) winter	Secondary	Castiglioni et al. (2006)
	Japan	~10–12	~0.5–10	86	Secondary (AS)	Nakada et al. (2006)
	Luxembourg	1–9	1–14	‡	ns	Pailler et al. (2009)

(Continued)

Table 6.2 Continued

Pharmaceutical	Country	Concentration ($ng\,L^{-1}$) (mean)		% removal/ degradation	Type of treatment	Reference
		Influent	Effluent			
	Sweden	14	10	29	Primary	Zorita et al. (2009)
			70	−470[1]	+Secondary (AS)	
			43	−207[1]	+Tertiary ($FeCl_3$ flocculation)	
			3	78	+Sand filter	
	US		0.44–12	80–100[2]	Ozonation	Wert et al. (2009)
17β-estradiol (E2)	Brazil	143	48	66	Stabilization ponds; AS + Cl_2	Marques et al. (2013)
	Canada	22	2	290	ns	Lee et al. (2005)
		15.6	1.8	88	Conventional AS; lagoon	Minim et al. (2009)
	Iran	1–8	0.5–2.6	67	AS	Logan et al. (2012)
	Japan	20	7	90	Secondary (AS)	Nakada et al. (2006)
	Luxembourg	3–102	3–85	16	ns	Pailler et al. (2009)
	Sweden	3.2	9.2	−187[1]	Primary	Zorita et al. (2009)
			2.5	22	+Secondary (AS)	
			<LOD	>99	+Tertiary ($FeCl_3$ flocculation)	
			<LOD	>99	+Sand filter	
17α-ethinylestradiol (EE2)	Brazil	421	122	71	Stabilization ponds; AS + Cl_2	Marques et al. (2013)
	Iran	4.2–11.8	0.5–2.6	78	AS	Logan et al. (2012)
	Luxembourg	6–24	<6	75	ns	Pailler et al. (2009)

Oseltamivir	Germany	11.9 ± 4.2 (OSP)[3]	15.8 ± 4.0	None	Primary + Secondary	Prasse et al. (2010)
		42.7 ± 9.3 (OSC)[4]	17.3 ± 5.4	59		
	Japan	140–460 (OSC)		2–9	Primary	Ghosh et al. (2010)
				<20%	Conventional AS	
				20–37	Nutrient removal ST	
				<50 %	Primary + biological ST	
		38		90%	+Ozone (10 mins)	
	Norway	5–529 (OSP)		−0.8 to 8	Physical, biological, chemical	Leknes et al. (2012)
		28–1213 (OSC)		−14 to 0.6		
Fluoxetin	Canada	191 ± 35	122 ± 5	36	Conventional AS + tertiary	Metcalfe et al. (2010)
	Norway	1,300	1,200	8	ns	Vasskog et al. (2006)
		2,400	1,300	46	ns	Vasskog et al. (2006)
		1,100	600	45	Mechanical sieve	Vasskog et al. (2008)
	Spain	120–2,300 (573)		33	Conventional AS	Radjenovic et al. (2009)
				98	Flat sheet MBR	
				98	Hollow fiber MBR	
		<LOD-1,827	34–929	62	Secondary	Rosal et al. (2010)
	Sweden	11.0	<LOD	>99%	Primary	Zorita et al. (2009)
	US		30–82	80–100[2]	Ozonation	Wert et al. (2009)

ns = Not specified.
[1] Increase in concentration in effluent sample.
[2] Ozonation performed on the effluent.
[3] Oseltamivir phosphate.
[4] Oseltamivir carboxylate.

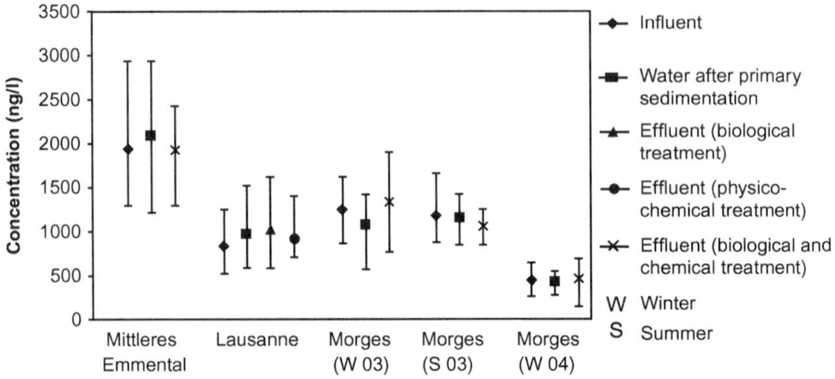

Figure 6.14 Ranges of diclofenac concentrations over 4–7 days for three sewage treatment plants in Switzerland and during winter and summer over two years.
From Tauxe-Wuersch et al. (2005a).

effluent at the conclusion of tertiary treatment disguises a removal value of +56% after primary treatment followed by progressive *increases* of 385%, −3%, and +4% after secondary activated sludge and sedimentation, $FeCl_3$-induced flocculation and sedimentation, and sand filtration, respectively (see Figure 6.15).

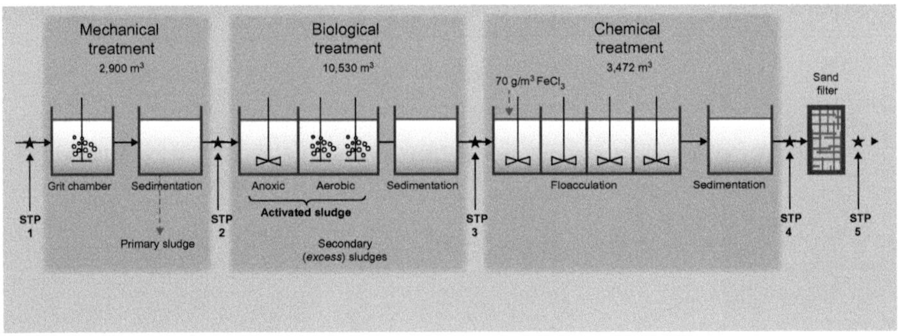

Figure 6.15 Schematic of Kristianstad sewage treatment plant.
From Zorita et al. (2009).

These variations were attributed to an initial adsorption to particles in the primary treatment step but no biodegradation in the secondary treatment step involving activated sludge. Indeed, the marked increase in the diclofenac concentration in the effluent sampled after activated sludge treatment was suggested may be due to "de-conjugation of glucuronidated or sulfonated diclofenac and/or its desorption from particles." A similar increase in diclofenac concentration had been reported earlier by (Kim et al., 2007) using a MBR. However, these authors also report nearly complete removal, if RO and nanofiltration both with and without UV irradiation was used.

Ozonation has been shown to degrade concentrations of diclofenac in the range 47–150 ng L^{-1} in different sources of wastewater effluent depending on the ratio of O_3 to the total organic carbon (TOC) content of the effluent (Wert et al., 2009).

6.2.1.3 *Ibuprofen*

Ibuprofen is another widely used NSAID which has been detected in many natural waters (see Chapter 4). It is almost completely removed using any form of WWT process as shown by the high % removal values listed for this pharmaceutical in Table 6.2.

Ozonation has been shown to almost completely degrade concentrations of ibuprofen in the range 5–85 ng L^{-1} in different sources of wastewater effluent depending on the ratio of O_3 to the TOC of the effluent (Wert et al., 2009).

Tertiary-treated effluent and a mixture with preliminary-treated effluent both containing detectable levels of ibuprofen, was applied to sandy and volcanic soils at a New Zealand site and >95% removal was observed in the soil leachate (Gielen et al., 2009).

The effect of dilution induced by a receiving stream/river on the concentrations of a number of these NSAID drugs present in effluent arising from a WWTP has been reported. For example, Stumpf et al. (1999) reported concentrations of diclofenac, ibuprofen, and naproxen in Pariba do Sul river water in Rio de Janiero which were much lower than in the effluent from a number of local treatment plants. Average decreases of 75% and 74% in the respective concentrations of ibuprofen and diclofenac downstream have been reported for effluent discharged from five WWTPs in England Wales (Ashton et al., 2004). The concentrations of these three pharmaceuticals were determined upstream, and at various distances downstream and at the point of discharge of effluent from three conventional secondary sewage treatment plants in Finland (Lindqvist et al., 2005). Allowing for the relative rates of river water and effluent flow, the results suggested that only dilution and not any further elimination of at least these three pharmaceuticals had occurred within the length of stream sampled.

6.2.1.4 *β-blockers*

Many common β-blockers such as atenolol and metoprolol are not completely removed by many conventional primary and secondary WWT processes as shown by the wide range of low values listed in Table 6.2. Some improvements of up to ~70% removal are achieved by using MBRs (Radjenovic et al., 2009).

Propranolol is extensively metabolized in the liver (Sweetman, 2011), and so the levels of the parent pharmaceutical that are detected in influent and effluent sewage are much less than might be predicted from the human usage figures available for this drug.

6.2.1.5 *Antibiotics*

As noted in earlier chapters, this class of pharmaceutical is among the most widely prescribed medication globally and is of special interest because of the potential impact of antibiotics on the development and the spread of antimicrobial resistance among human and other biological species. Following human consumption, antibiotics are excreted in urine and feces into sewage in both the unchanged parent form and as metabolites. As can be noted in the many entries listed in Table 6.2 for a small selection of common antibiotics, the resulting degradation efficiencies vary enormously among different countries and plant operational parameters but *in general* these values indicate, at best, incomplete removal during the various available sewage treatment processes.

In a number of these studies, the concentration of a specific antibiotic has been found to be higher in the effluent than the influent and this has been attributed to de-conjugation of a metabolite back to the parent antibiotic compound during waste water treatment. For example, an apparent *increase* in the range of +5% to +21% in the sulfamethoxazole content after just primary treatment in a Swiss STP has been attributed to the presence of metabolites of sulfamethoxazole such as N^4-acetylsulfamethoxazole in the influent which undergoes deconjugation to reform the parent compound during primary treatment (Göbel et al., 2005, 2007).

6.2.1.6 An antiviral drug (oseltamivir)

Oseltamivir phosphate (OSP, Tamiflu) is one of most commonly used antiviral drugs that has been developed in recent times for the prevention and treatment of pandemic influenza. After oral administration, OSP is converted into oseltamivir carboxylate (OSC) which undergoes extensive renal elimination via urine which then enters the sewage system as influent. Because millions of prescriptions of the drug could be taken over a relatively short period of time to counter any serious outbreak of influenza in a big city, the concentration of this OSC metabolite could reach the order of $\mu g\,L^{-1}$ in raw sewage and this level has the potential to have an adverse ecological effect on wildlife such as waterfowl living in the receiving waters (Fick et al., 2007; Singer et al., 2007). Accordingly there have been many studies reported on the efficiency of the removal of OSP from raw sewage using a variety of treatment processes, and many of them have indicated that conventional sewage treatment does not remove either OSP or OSC very efficiently.

For example, Fick et al. (2007) reported no degradation and/or adsorption of OSC using mechanical and chemical treatments or activated sludge treatment in batch experiments simulating these processes. Ghosh et al. (2010) measured OSC levels in influent sewage during two influenza pandemics in Japan and reported poor removal efficiencies (<50%) in the resulting effluent from primary and various combinations of secondary treatment processes. However, additional treatment of secondary-treated sewage involving 10 min exposure to O_3 improved this to >90% as confirmed by a reported rate constant for reaction of OC with O_3 of $(1.7 \pm 0.1) \times 10^5\,M^{-1}\,s^{-1}$ in a laboratory study by Mestankova et al. (2012). They also reported that treatment of a 1 mM solution of OSC in secondary effluent using a dose rate of 0.5 g O_3 g DOC^{-1} reduced the OSC concentration to below detection limits.

6.2.1.7 A psychotic drug (fluoxetine)

Almost complete (>90%) removal of fluoxetine (FXT) has been demonstrated after even just primary treatment in a study undertaken at the tertiary sewage treatment system of Kristianstad (Sweden) (*see* Figure 6.15) (Zorita et al., 2009). It was suggested that FXT was most likely removed by adsorption to the fat and primary sludge formed during this initial level of treatment, rather than any subsequent abiotic or biotic degradation as also suggested by Styrishave et al. (2011).

Ozonation has been shown in pilot studies to almost completely degrade concentrations of FXT in the range 30–82 $ng L^{-1}$ in different sources of wastewater effluent depending on the ratio of O_3 to the TOC content of the effluent (Wert et al., 2009).

6.2.2 Laboratory studies of pharmaceutical degradation in aqueous solution

As well as the many field measurements that have been made of the degradation of specific pharmaceuticals present in influent and during various types of wastewater treatment, there have also been many degradation studies undertaken in the laboratory on aqueous solutions of various pharmaceuticals.

The advantages of using a laboratory environment for such measurements are numerous including:

1. A specific initial composition for both the pharmaceutical(s) and other species present in the test solution can be carefully controlled and organized to minimize the analytical effort required to monitor their change with reaction time and hence determine the degradation kinetics.
2. The experimental degradation reaction conditions can be more readily measured and controlled such as monitoring the light levels used to induce photolytic degradation, and maintaining a constant pH and/or temperature.
3. The initial composition of the test solution can be simplified so that subsamples taken during the course of the degradation reaction can be more readily analyzed to identify decomposition products and hence possible reaction mechanisms using routine analytical techniques such as HPLC, LC-MS/MS (*see* earlier discussion in Chapter 3), and nuclear magnetic resonance (NMR).
4. The inherent environmental variability associated with any natural wastewater sample can be minimized using a test solution made up in the laboratory using pure reagents and solvents.
5. The influence of the background sample matrix (particularly the general natural organic matter (NOM); *see* Section 6.3) on the observed pharmaceutical degradation can be minimized, if not completely avoided.
6. Small quantities of reagents and small size reactors can be used which significantly lowers the cost of experimentation and minimizes any potential health and safety issues.

There are, however, significant disadvantages in trying to extrapolate such laboratory measurements to wastewater treatment processes in a plant including:

1. In order to simplify the analytical detection and quantification, the initial concentration of pharmaceutical used in a laboratory study is often 10^3 (or more) times higher than would be typically found in raw influent wastewater.
2. The test mixtures are often so "simple" in composition that the laboratory results will be very different from those conducted with "real" wastewater with its myriad of other compounds and background matrix involving NOM and inherent variability.
3. The reagents and equipment used in the laboratory to achieve significant degradation rates particularly for tertiary processes such as AOPs involving radiolysis or photocatalysis are likely to be too expensive to implement in a treatment plant.

In spite of these limitations, a large number of laboratory studies of pharmaceutical degradation have been reported using a wide range of processes and a selection of these results is presented in Table 6.3

For many of the laboratory studies reported in this table, fast degradation rates and hence short half lives have been measured for the specific reaction conditions, particularly those involving advanced oxidation/reduction processes.

Some specific examples of these laboratory studies include the following.

6.2.2.1 *Paracetamol*

Direct photolysis of aqueous solutions of paracetamol showed some degradation with UVC but not UVA radiation (Yang et al., 2007). However, in the presence of TiO_2, almost complete degradation occurred after 80 min of irradiation with UVC. The degradation rate constant decreased with an increase in the initial paracetamol concentration, but increased with increased oxygen concentration and light intensity. The active species responsible for the major degradation pathway in this TiO_2-assisted degradation were shown in a later study (Yang et al., 2008) to include e_{cb}, h_{vb}, •OH, •HO_2, O^{2-}, and H_2O_2.

The use of electro-Fenton and photoelectron-Fenton processes (de Luna et al., 2012) lead to almost complete paracetamol degradation although only an ~42% decrease in COD and ~20% decrease in the total organic content (TOC). The degradation was shown to involve initial •OH attack at the para position to give a range of toxic intermediate products such as hydroquinone, benzaldehyde, and benzoic acids which were then further degraded and ultimately mineralized to CO_2 in agreement with a previous study (Yang et al., 2007).

Significant degradation of paracetamol has also been observed for reaction with SO_4^{-} radicals generated from the uncatalysed and Co(II)-catalyzed decomposition of peroxymonopersulfate (OXONE) (Rivas et al., 2012).

6.2.2.2 *Diclofenac*

Generally, the low % removal of this commonly prescribed pharmaceutical in wastewaters (*see* entries for diclofenac in Table 6.2) has led to a large number of laboratory studies of its possible degradation. In terms of photolysis, diclofenac in aqueous solution exhibits a high-intensity UV absorption band at 275 nm ($\varepsilon \approx 10{,}000$) extending into the solar wavelength region >290 nm (*see* Figure 6.16) and hence there have been numerous reports of significant photodegradation of diclofenac in aqueous solution using natural sunlight (e.g., Buser et al., 1998; Packer et al., 2003; Andreozzi et al., 2003b; Vogna et al., 2004; Perez-Estrada et al., 2005b; Bartels and Von Tumpling, 2007) or laboratory UV/vis sources (e.g., Moore et al., 1990; Perez-Estrada et al., 2005a; Bartels and Von Tumpling, 2007; Felis et al., 2009; Eriksson et al., 2010; Rivas et al., 2010; Peuravuori, 2012).

Packer et al. (2003) reported that 100 μM solutions of diclofenac in both MQ and Mississippi River water react rapidly with a half-life of 39 min for direct sunlight irradiation. Surprisingly, addition of the •OH quencher isopropanol (IPA) accelerated the degradation in both media and this was attributed to the possible involvement of

Table 6.3 **Laboratory studies of some common pharmaceuticals in aqueous solution**

Pharmaceutical	Initial concentration	pH	Degradation process	Rate constant ($L^{-1}s^{-1}$)	% removal/degradation	Reference
Paracetamol (acetaminophen)	4.0 μM	5.3–3.3	Photolysis direct			Yang et al. (2008)
			at 365 nm	No reaction		
			at 254 nm	$(2.3 \pm 0.2) \times 10^{-3}$		
			TiO_2 (0.04 g L^{-1}) photolysis at 254 nm	$(4.9 \pm 0.2) \times 10^{-3}$		
				$(14.7 \pm 1.7) \times 10^{-3}$	~95 at 80 min	Yang et al. (2009)
	5 mM	3	Electro-Fenton	98%	ΔCOD-43%	de Luna et al. (2012)
			Photoelectron-Fenton	97%	ΔCOD-42%	
	50 μM	2	Potassium monopersulfate (0.12 M)	$(8.3 \pm 0.2) \times 10^{-3}$	~10 at 50 min	Rivas et al. (2012)
		7		0.17 ± 0.03	~80 at 50 min	
		9		00.42 ± 0.04	~100 at 50 min	Nfodzo and Choi (2011)
Salicylic acid	0.1–1 ppm		Direct photolysis	0.0042 min^{-1}		Gangwang et al. (2012)
			Indirect photolysis			
			$[NO_2^-]$ 24 μM	0.0068 min^{-1}		
			$[NO_3^-]$ 24 μM	0.0083 min^{-1}		
	1 mM		Solar/ZnO 320 min		COD 62.6–9.1 mg L^{-1} TOC 65.2–6.2 mg L^{-1}	Shinde et al. (2011b)
			Solar/TiO_2–$AlFe_2O_3$ 320 min		COD 62.6–7.4 mg L^{-1} TOC 65.2–7.2 mg L^{-1}	Shinde et al. (2011a)

(Continued)

Table 6.3 **Continued**

Pharmaceutical	Initial concentration	pH	Degradation process	Rate constant ($L^{-1}s^{-1}$)	% removal/ degradation	Reference
Diclofenac	Lake water 100 ngL^{-1} 1,000 ngL^{-1}		Sunlight >290 nm	~0.004 day^{-1} 0.80 h^{-1}		Buser et al. (1998)
	100 µM		Sunlight	MQ $(2.0 \pm 0.3) \times 10^2\ min^{-1}$ River water $(2.0 \pm 0.2) \times 10^2\ min^{-1}$		Packer et al. (2003)
	ns	7	UVB (254 nm)	3.18 $m^2 einstein^{-1}$ (26% depletion)		Canonica et al. (2008)
	4 µM	7.8	UV	$5.3 \times 10^{-3}\ cm^2 mJ^{-1}$		Baeza and Knappe (2011)
			UV/H_2O_2	($k_{\bullet OH}$) $9.3 \times 10^9\ M^{-1}s^{-1}$		
	~3,700 ngL^{-1}		UVA (315–800 nm) UVB (280–400 nm)	0.0734 day^{-1} 34.9 day^{-1}		Peuravuori (2012)
	280 ngL^{-1}		UV/TiO2	8.2 $m^3 kWh^{-1}$		Benotti et al. (2009) Achilleos et al. (2010)
Ibuprofen	75 µM		UV	MQ $(0.10 \pm 0.02) \times 10^2\ min^{-1}$ River water $(0.45 \pm 0.03) \times 10^2\ min^{-1}$		Packer et al. (2003)
	170 ngL^{-1}		UV/TiO_2	0.79 $m^3 kWh^{-1}$		Benotti et al. (2009)

Atenolol	0.2–1.0 mM	55.0	Pulse radiolysis			Song et al. (2008)
			•OH	$(7.05 \pm 0.27) \times 10^9 M^{-1} s^{-1}$		
			e^-_{aq}	$(0.591 \pm 0.021) \times 10^9 M^{-1} s^{-1}$		
	$1300 ng L^{-1}$		UV photolysis	$1.6 m^3 kWh^{-1}$		Benotti et al. (2009)
			UV/H_2O_2 (10 ppm)	5.1		
			UV/H_2O_2 (20 ppm)	7.2		
			UV/TiO_2	1.1		
	$100 mgC L^{-1}$		Electro-Fenton	$0.033 \times 10^{-5} min^{-1}$	22%	Isarain-Chavez et al. (2011)
			Solar electro-Fenton	$2.46 \times 10^{-5} min^{-1}$	88%	
	20 μM	7.0	UV/H_2O_2	$0.013 min^{-1}$		Liu (2013)
			UV/ peroxydisulfate	$0.048 min^{-1}$		
Metoprolol	0.2–1.0 mM	5.0	Pulse radiolysis			Song et al. (2008a, 2008b)
			•OH	$(8.39 \pm 0.0) \times 10^9 M^{-1} s^{-1}$		
			e^-_{aq}	$(0.173 \pm 0.003) \times 10^9 M^{-1} s^{-1}$		
	$100 mgC L^{-1}$		Electro-Fenton	$0.013 \times 10^{-5} min^{-1}$	9%	Isarain-Chavez et al. (2011)
			Solar electro-Fenton	$1.91 \times 10^{-5} min^{-1}$	88%	
Propranolol	0.2–1.0 mM	5.0	Pulse radiolysis			Song et al. (2008a)
			•OH	$(10.7 \pm 0.2) \times 10^9 M^{-1} s^{-1}$		
			e^-_{aq}	$(12.6 \pm 0.2) \times 10^9 M^{-1} s^{-1}$		
	$100 mgC L^{-1}$		Electro-Fenton	$0.09 \times 10^{-5} min^{-1}$	15%	Isarain-Chavez et al. (2011)
			Solar electro-Fenton	$2.37 \times 10^{-5} min^{-1}$	93%	

(Continued)

Table 6.3 Continued

Pharmaceutical	Initial concentration	pH	Degradation process	Rate constant ($L^{-1}s^{-1}$)	% removal/ degradation	Reference
Amoxicillin		2.5–4	Fenton 104 mg L^{-1}	100	None-48	Elmolla and Chaudhuri (2010)
		33.5–4	Photo-Fenton	100	33–58	
		3–11	UV/TiO_2	42–71	3.4–6.4	
		5	UV/TiO_2/H_2O_2	100	5.2–14.0	
		5–8	UV/ZnO	100	5.2–16.3	
	0.1–0.5 mM	pH 7	Pulse radiolysis	3.5–6.9 × 10^9 $M^{-1}s^{-1}$	95	Song et al. (2008)
	23–312 mg L^{-1}	pH 3–9	Photocatalysis (TiO_2)	0.006–0.014 min^{-1}		Basha et al. (2011)
			Adsorption	0.04–0.08 mg L^{-1} min^{-1}		
	260 ng L^{-1}		Constructed wetland (surface flow)	42% removal		Hijosa-Valsero et al. (2011)
Sulfamethoxazole	ns	7	UVB (254 nm)		81 m^2 $einstein^{-1}$ (7.4% depletion)	Canonica et al. (2008)
	270 ng L^{-1}		UV/TiO_2	1.7 m^3 kWh^{-1}		Benotti et al. (2009)
	4 μM	3.6	UV	9.8 × 10^{-3} cm^2 mJ^{-1}		Baeza and Knappe (2011)
		7.8		2.4 × 10^{-3} cm^2 mJ^{-1}		
		7.8	UV/H_2O_2	($k_{\bullet OH}$) 5.6 × 10^9 $M^{-1}s^{-1}$		

	260 $ng\,L^{-1}$		Constructed wetland (surface flow)	71% removal		Hijosa-Valsero et al. (2011)
	47 $ng\,L^{-1}$	6.8	Constructed wetland (vertical flow)	24% removal		Dan et al. (2013)
		7.2	(horizontal flow)	22% removal		
Trimethoprim	360 $ng\,L^{-1}$		UV/TiO2	1.6 $m^3\,kWh^{-1}$		Benotti et al. (2009)
	4 μM	3.6	UV	$1.4\times10^{-5}\,cm^2\,mJ^{-1}$		Baeza and Knappe (2011)
		7.8		$1.7\times10^{-5}\,cm^2\,mJ^{-1}$		
		7.8	UV/H_2O_2	($k_{\bullet OH}$) $5.70\times10^9\,M^{-1}\,s^{-1}$		
	100 $ng\,L^{-1}$		Constructed wetland (surface flow)	88% removal		Hijosa-Valsero et al. (2011)
	0.15 mM	7	UV/Direct photolysis	No measurable effect		Tong (2013)
		7	UV/H_2O_2	0.24 min^{-1}		
		3	Photo-Fenton	0.38 min^{-1}		
		5		0.23 min^{-1}		
		7		0.20 min^{-1}		
	112 $ng\,L^{-1}$	6.8	Constructed wetland (vertical flow)	31% removal		Dan et al. (2013)
		7.2	(horizontal flow)	95%		
Estrone (E1)	26 $ng\,L^{-1}$		UV/TiO_2	13 $m^3\,kWh^{-1}$		Benotti et al. (2009)
17β-estradiol (E2)			UV/TiO_2	0.0173 min^{-1}		Ohko et al. (2002)
	18 $ng\,L^{-1}$		UV/TiO_2	12 $m^3\,kWh^{-1}$		Benotti et al. (2009)

(Continued)

Table 6.3 **Continued**

Pharmaceutical	Initial concentration	pH	Degradation process	Rate constant ($L^{-1}s^{-1}$)	% removal/ degradation	Reference
17α-ethinylestradiol (EE2)	9.1 $mg L^{-1}$	6.5	O_3 (0.7 $\mu mol L^{-1}$)	0.019 min^{-1}		Zhang et al. (2006)
			(11.6 $\mu mol L^{-1}$)	0.176 min^{-1}		
	ns	7	UVB (254 nm)	7.1 $m^2 einstein^{-1}$	(0.7% depletion)	Canonica et al. (2008)
	140 $ng L^{-1}$	6.8	UV/TiO_2	9.8 $m^3 kWh^{-1}$		Benotti et al. (2009)
	0.61 $mg L^{-1}$		UVB (254 nm)	0.091 min^{-1}		Zhang et al. (2010)
	1.21 $mg L^{-1}$			0.0488 min^{-1}		
	1.95 $mg L^{-1}$			0.0371 min^{-1}		
	1.98 $mg L^{-1}$		H_2O_2 (20 $mg L^{-1}$)	0.0006 min^{-1}		
	2.0 $mg L^{-1}$		UVB/H_2O_2			
			(20 $mg L^{-1}$)	0.098 min^{-1}		
			(50 $mg L^{-1}$)	0.25 min^{-1}		
Oseltamivir	0.12 mM		Direct UV	0.21 ± 0.03 min^{-1}		Tong et al. (2011)
			UV + H_2O_2	1.56 ± 0.54 min^{-1}		
			UV + H_2O_2 + Fe(II)	1.75 ± 0.13 min^{-1}		
			UV + H_2O_2 + *t*-butyl alcohol	0.13 ± 0.505 min^{-1}		
Fluoxetine	0.3 $mg L^{-1}$	4–7	Natural sunlight in distilled water	0.0126 h^{-1}	55.2 h (*t*1/2)	Lam et al. (2005)
			Simulated sunlight in field water	0.03–0.13	5.5–22 h	
	5 $mg L^{-1}$	5–9	Simulated sunlight in aqueous buffer	0.002–0.007 day^{-1}	102–385 day	Kwon and Armbrust (2006)
			Synthetic humic in distilled water	0.034 day^{-1}	21 day	

			Natural lake water	$0.0062\,day^{-1}$	112–113 day	
			Adsorption from lake water to sediment		6.1 day	
	$54\,ng\,L^{-1}$		UV/TiO_2	$1.3\,m^3\,kWh^{-1}$		Benotti et al. (2009)
	0.11 mM	3	O_3 (dark)	$0.65\,min^{-1}$		Mendez-Arriaga et al. (2011)
	$34\,mg\,L^{-1}$	3	O_3/H_2O_2 (dark)	$0.39\,min^{-1}$		
		11	O_3 (dark)	$\gg 0.7\,min^{-1}$		
		11	UV/O_3	$\gg 0.6\,min^{-1}$		
		11	$UV/O_3/H_2O_2$	$0.56\,min^{-1}$		
		11	UV/TiO_2 $(0.01\,g\,L^{-1})/O_3$	$0.76\,min^{-1}$		
		11	UV/TiO_2 $(0.01\,g\,L^{-1})/O_3/H_2O_2$	$0.52\,min^{-1}$		
		11	UV/TiO_2 $(0.10\,g\,L^{-1})/O_3/H_2O_2$	$0.80\,min^{-1}$		

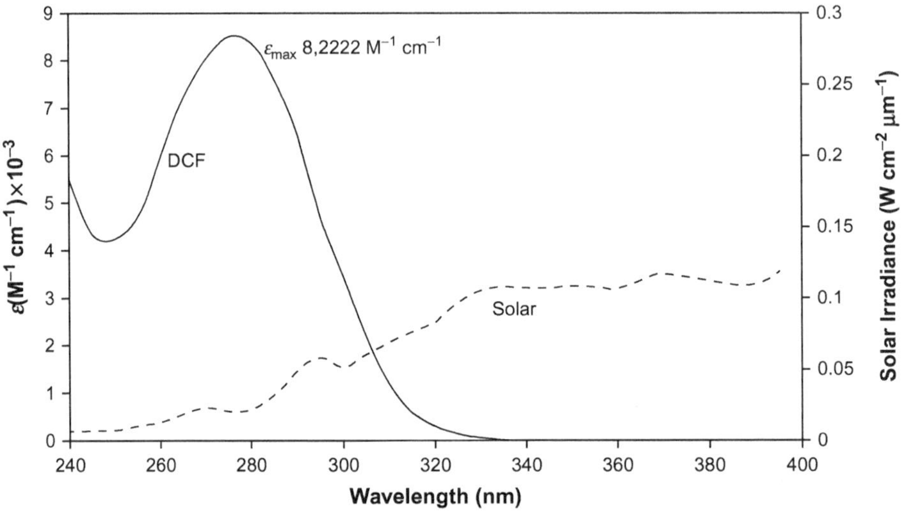

Figure 6.16 Absorption spectrum of a 0.1 mM aqueous solution of diclofenac (DCF) in phosphate buffer (pH 7.0) and corresponding solar spectral irradiance.
Data from Bobo et al. (2005).

a carbon-centered radical arising from the IPA leading to the formation of further reactive superoxide and hydroperoxyl radicals, acceleration of ring closure, and/or photo-initiated reduction between the IPA and diclofenac.

Many of these studies have progressively identified an increasing number of photoproducts including those associated with photocyclization of the diclofenac parent structure to form carbozole derivatives, and decarboxylation followed by oxidation and/or dechlorination (Peuravuori, 2012). The same study reported that neither UVA or UVB photolysis in the presence of natural chromophoric dissolved organic matter (CDOM) or nitrate ions (both photochemical sources of the •OH radical) led to any significant increase in the rate of degradation of diclofenac through indirect photolysis.

The roles of the triplet state of the deprotonated form of diclofenac and different radical and oxygenated intermediate species in the photodegradation pathway have been investigated theoretically using time-dependent density-functional theory (TD-DFT) (Musa and Ericksson, 2009).

6.2.2.3 Ibuprofen

Unlike diclofenac (*see* Figure 6.16), the UV absorption spectrum of ibuprofen exhibits little overlap with the solar spectrum and so ibuprofen exhibits minimal degradation in sunlight (Packer et al., 2003). Instead, experiments undertaken in the laboratory with a more intense UV lamp as the light source indicated some degradation for aqueous Milli-Q™ solutions of ibuprofen but this was enhanced for solutions in natural Mississippi River water which suggested radical involvement in this photolytic degradation. This was confirmed by a decrease in the degradation rate for both types of solutions in the presence of 1% of the •OH quencher, isopropanol (IPA).

6.2.2.4 β-Blockers

The relatively poor and variable removal efficiency of this class of pharmaceutical in waste water using conventional WWT processes (*see* Table 6.2) has led to a number of laboratory studies of alternative degradation methods with many involving tertiary processes such as advanced oxidation. For example, Song et al. (2008a) undertook pulse radiolysis measurements to determine the kinetic data listed in Table 6.3 for reaction of •OH and e^-_{aq} with the three main β-blocker pharmaceuticals, viz. atenolol, propranolol and metoprolol. All reactions with the β-blockers with these two ROS were very fast with the rate constants approaching the theoretical diffusion-controlled value of ~$10^{10} M^{-1} s^{-1}$ (Elliot et al., 1990). Trends in these rate constant values were rationalized in terms of the parent chemical structure of each pharmaceutical.

Mass spectral measurements of the stable •OH-induced products arising from ^{60}Co steady-state radiolysis identified specific degradation products for which reaction pathways could be proposed (e.g., see Figure 6.17 for •OH oxidation of atenolol).

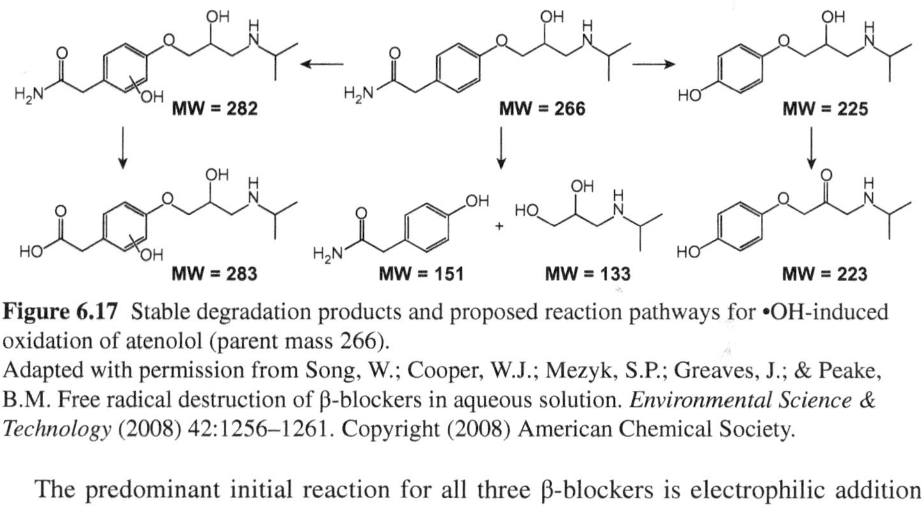

Figure 6.17 Stable degradation products and proposed reaction pathways for •OH-induced oxidation of atenolol (parent mass 266).
Adapted with permission from Song, W.; Cooper, W.J.; Mezyk, S.P.; Greaves, J.; & Peake, B.M. Free radical destruction of β-blockers in aqueous solution. *Environmental Science & Technology* (2008) 42:1256–1261. Copyright (2008) American Chemical Society.

The predominant initial reaction for all three β-blockers is electrophilic addition of •OH to an aromatic ring leading to the formation of a resonance-stabilized carbon-centered radical which reacts with oxygen to form a phenolic product and the hydroperoxyl radical (•HO_2) (as summarized by the general scheme shown in Figure 6.18).

In terms of using such oxidative and reductive processes to degrade β-blockers in real sewage, account must be taken of the competitive scavenging of •OH and e_{aq} by the

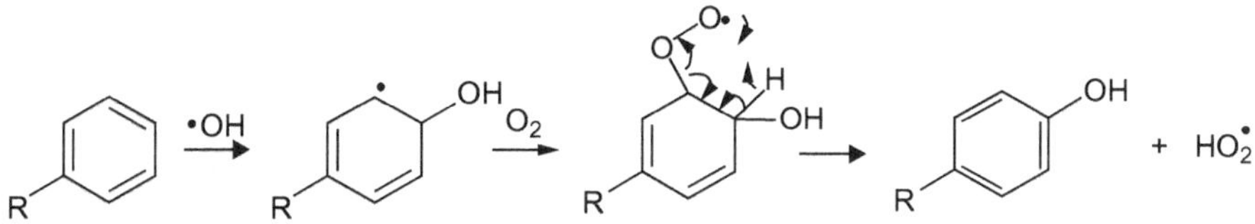

Figure 6.18 General scheme for •OH addition to an aromatic ring.
Adapted with permission from Song, W.; Cooper, W.J.; Mezyk, S.P.; Greaves, J.; & Peake, B.M. Free radical destruction of β-blockers in aqueous solution. *Environmental Science & Technology* (2008) 42:1256-1261. Copyright (2008) American Chemical Society.

much more abundant NOM and HCO_3^- species likely to be present in the wastewater. Simple kinetic modeling indicated that the fraction of •OH likely to be available to degrade atenolol, metoprolol, and propranolol is calculated to be only 0.19%, 0.23%, and 0.30%, respectively. Although these values are all low, they do provide a quantitative basis on which to assess the extent of this type of advanced oxidative/reductive treatment that would be required to improve on the low removal values reported for these pharmaceuticals using conventional treatment technologies (*see* entries in Table 6.2).

6.2.2.5 Oseltamivir

The poor removal efficiency reported for oseltamivir using conventional sewage treatment processes (Table 6.2) has led to a number of laboratory studies of its removal using a range of other treatment processes. For example, we have examined the degradation of aqueous solutions of the parent OSP induced by direct and indirect UV photolysis with the results shown in Figure 6.19 (Tong et al., 2011).

These degradation curves were analyzed in terms of first-order kinetic behavior (*see* Section 5.3.1) to give the apparent first-order degradation rate constants listed for OSP in Table 6.3.

The degradation rate increases with increased production of •OH radicals in the presence of H_2O_2 and then Fe(II) in the photo-Fenton system. The involvement of •OH radicals in the H_2O_2 systems was confirmed by the marked rate decrease upon addition of *t*-butyl alcohol which is a well-known quencher of this type of radical (Dixon and Norman, 1963; Andreozzi et al., 2003a; Kim et al., 2009).

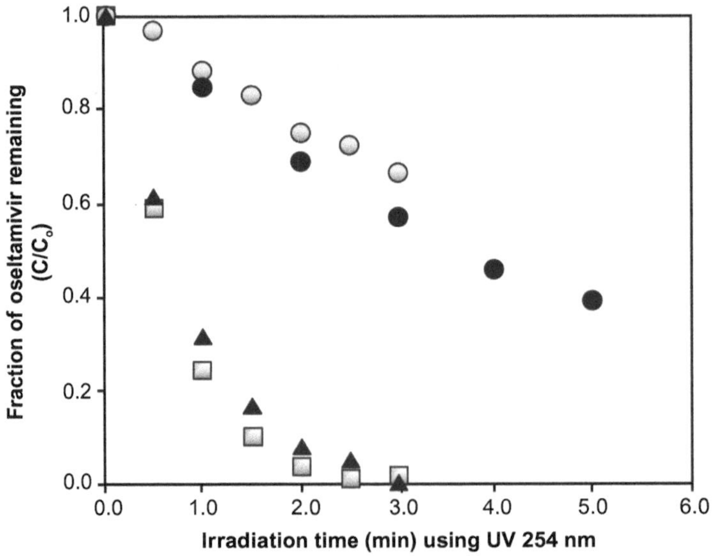

Figure 6.19 Time dependence of photolytic degradation at 254 nm of aqueous solutions of oseltamivir phosphate (OSP) (0.12 mML^{-1}) by UV alone (●), UV/H_2O_2 (▲), UV/H_2O_2/Fe(II) (□), and UV/H_2O_2 with 5% *t*-butyl alcohol (O).
(Modified from Tong et al., 2011)

NMR and mass-spectrometric measurements on samples of irradiated OSP material were used to identify some of the initial degradation products arising from direct photolysis of OSP, and these suggested the initial step in the decomposition pathway involves hydroxylation of the cyclohexadienyl ring (*see* Figure 6.20).

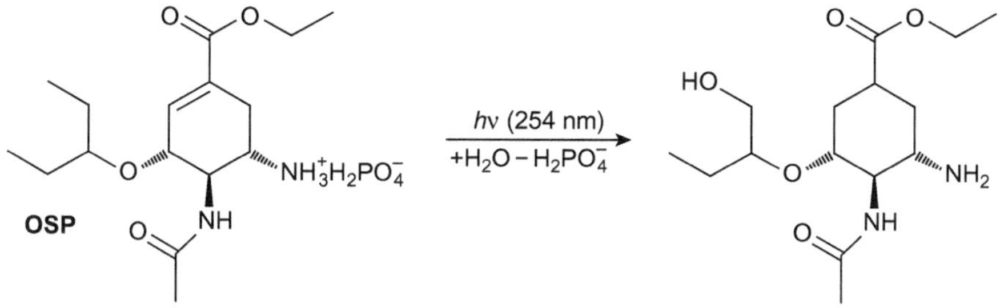

Figure 6.20 Initial step in the pathway for direct UV photolytic degradation of OSP in aqueous solution.
(Modified from Tong et al., 2011)

6.2.2.6 *Fluoxetine*

Direct sunlight-induced photochemical degradation of 10 μM fluoxetine (FXT) in pure water has been observed but it is as much as ten times slower than indirect photolysis involving the reaction of •OH, $^1O_2$, and $\bullet CO_3^-$ reactive species formed from irradiation of FTX in simulated field water containing varying amounts of NO_3^-, DOM, and HCO_3^- (Lam et al., 2005). LC-MS/MS techniques were used to identify degradation products (*see* Figure 6.21) including two O-dealkylated photoproducts (I and II) and a carboxylic acid photoproduct (III) from direct photolysis. A hydroxylated FXT product (IV) was identified from indirect photolysis.

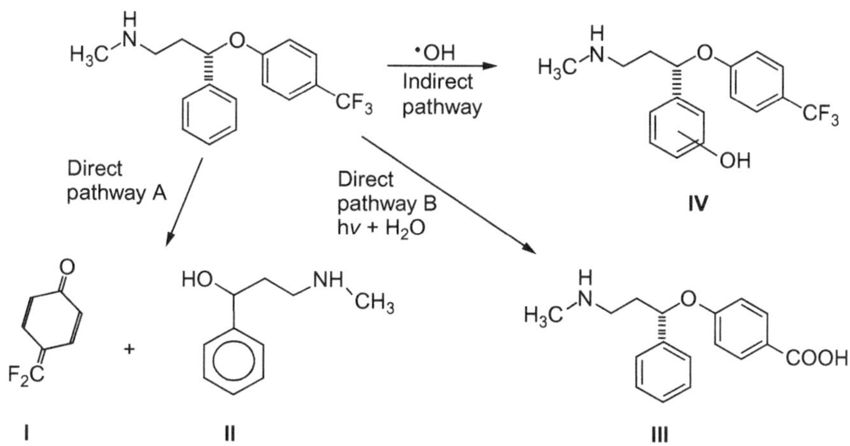

Figure 6.21 Some degradation products arising from direct and indirect photolysis of fluoxetine in aqueous solution.
Redrawn from Mendez-Arriaga et al. (2011).

Another simulated sunlight photolysis study of FXT showed much slower rates of degradation ($t_{½}$ in the range 102–385 days in aqueous buffer solution) (Kwon and Armbrust, 2006). The degradation rate increased somewhat in natural lake water and increased by as much as 13-fold when synthetic humic water was used ($t_{½}$ 112 and 21 days respectively). It was suggested that this may due to the presence of quenching agents for reactive oxygen species and attenuation of the light in the natural water.

A significant fraction (97–98%) of FXT present in lake water was removed by adsorption to sediment after 30 days ($t_{½}$ 6.1 days) which suggested that FXT will be completely removed from natural surface waters by this process rather than being photodegraded in the water using sunlight.

Advanced oxidation processes (AOPs) involving various combinations of O_3, UV, H_2O_2, and TiO_2 have been used to degrade 0.11 mM aqueous solutions of FXT (Mendez-Arriaga et al., 2011). The combination of UV/TiO_2/O_3 under alkaline (pH 11) conditions was found to give the optimal degradation rate. However, the formation of significant amounts of inorganic products suggested they would have a detrimental effect on the biodegradability of FXT.

Laboratory studies of sewage sludge containing FTX have shown significant degradation under aerobic conditions (Vasskog et al., 2009) but less under anaerobic composting conditions (Vasskog et al., 2009; Bergersen et al., 2012). This suggested that at least this psychiatric drug may bioaccumulate in sludge under anaerobic conditions as supported by field measurements (*see* Section 6.2.1.7) and thus further treatment of sludge containing this drug may be required to minimise any potential environmental impact.

6.3 Effect of the natural organic matrix on pharmaceutical removal and degradation in wastewater

Although many laboratory and pilot-scale studies of the removal/degradation of pharmaceuticals have been undertaken using "pure" aqueous solutions of the pharmaceuticals (typically buffered Milli-Q™ water), trace levels of these contaminants make up only a small part of a much larger matrix of natural organic matter (NOM) present in all forms of natural waters including wastewater. The chemical composition and concentration of the components of this NOM vary with the specific natural water and can have a marked influence on physicochemical and biological processes such as sedimentation, biological digestion, adsorption, filtration, and advanced oxidation involved in the various sewage and other natural water treatment processes. Such effects on the removal efficiency of the trace levels of pharmaceuticals present in the natural waters with the various treatment processes compared to pure aqueous solution have been the subject of many studies and, again, depend on the nature and concentration of the pharmaceutical and the type of natural water. For example, no change was reported in the removal efficiency arising from the photodegradation of diclofenac spiked in Mississippi River water compared to Milli-Q™, whereas there was a significant increase in removal for ibuprofen and a decrease for naproxen (Packer et al., 2003).

With the increasing interest in reusing the effluent water arising from various levels of wastewater treatment, there are also an increasing number of reports of the behavior of effluent organic matter (EfOM; *see* review by Longworth, 2001) and how any remaining pharmaceuticals can be removed from it. The level and chemical nature of a particular EfOM can influence the degradation of residual pharmaceuticals present in the effluent by acting both as a scavenger for •OH radicals produced in the many different types of AOPs (e.g., Westerhoff et al., 2007) and also as an additional source of •OH radicals produced by photochemical irradiation (e.g., West, 2002; Bucher et al., 2003). A study of the effluent samples collected from two different wastewater treatment plants in Colarado USA showed that the lower the apparent molecular weight of the humic acid component of the EfOM, the faster was the •OH photochemical formation rate and hence the more efficient is the potential photochemical removal of pharmaceuticals from the EfOM (Wang et al., 1999).

6.4 Summary and conclusions

This chapter has provided a broad overview of the wide range of treatment processes that are available to assist in the removal/degradation of the large suite of human pharmaceuticals that have been detected in wastewater. These processes can be broadly classified in terms of an increasing degree (and hence cost) of physicochemical and biological treatment: *primary* involving mainly solids removal by coagulation, flocculation, and sedimentation; *secondary* involving biological degradation in the presence of activated sludge, filters, and membranes; and *tertiary* involving physical methods that use finer filters, adsorption to AC, RO, and chemical methods such as advanced oxidation involving ozonation and photolysis.

The extent of removal of a given pharmaceutical varies with many factors including the chemical nature and influent concentration of the pharmaceutical, the presence of other pharmaceuticals and their metabolites, the background organic matrix, and the type of treatment process(es) used. Since one of the earliest reports (Ternes, 1998) of the detection of 32 drug residues from a wide range of different classes of human pharmaceuticals in the effluent from a selection of German municipal sewage treatment plants, there has been a multitude of reports on the removal efficiencies of an enormous range of pharmaceuticals using many different treatment processes either as full-scale operating plants, pilot plants or on even smaller scales in the laboratory. It is impossible to summarize all of these results given all these different factors but some very general trends have been summarized by Kravitz and Bell (2013).

Conventional WWT (i.e., primary sedimentation and secondary activated sludge) can remove/degrade up to 90% or more of the very commonly occurring pharmaceuticals in domestic wastewater influent including paracetamol (acetaminophenol), aspirin, salicylic acid, ibuprofen, estrone, 17β-estradiol, and fluoxetine. However, notable exceptions that are generally considered recalcitrant to conventional wastewater treatment include diclofenac, β-blockers such as metoprolol, and oseltamivir. To remove a significant proportion of these latter pharmaceuticals present in the effluent arising from conventional WWT, one can often use one or more of the many different tertiary processes such as nanofiltration, microfiltration, and RO, and AOPs such as ozonation and photochemical irradiation.

There are also an increasing number of reports of the use at least on a laboratory scale of an ever-increasing range of tertiary methods such as photocatalysis in the presence of Fe(II) and semiconductors such as TiO_2, along with many different experimental variations of these tertiary treatment methods, to remove a wide range of pharmaceuticals typically prepared as solutions in buffered Milli-Q™ water. How practical and economic many of these mainly laboratory/bench-scale methods and their results would be to implement in a real wastewater treatment plant is, however, often still unknown.

For a wastewater engineer faced with demands from governing bodies and the public to address the increasingly emotive issue of "emerging contaminants" in their wastewater treatment process(es), the wide variation in the removal/degradation results for different pharmaceuticals can potentially make for real difficulties in deciding "if" and "how" they should be removed from their wastewater. Perhaps one practical approach would be to first undertake an analysis of the average concentrations of the range of pharmaceuticals known to be commonly prescribed in their community, in both the raw influent and final effluent arising from their treatment plant. The effluent values could then be compared with the appropriate ecotoxicity data as discussed in Chapter 5 and if there was any concern, then *additional* WWT methods would need to be considered based on the vast and ever increasingly literature concerning different treatment methods, the main examples of which have been introduced in this chapter. However what might finally determine whether further treatment processes are implemented will be economic issues such as their cost and if the community can afford them.

References

Achilleos, A., Hapeshi, E., et al., 2010. Factors affecting diclofenac decomposition in water by UV-A/TiO_2 photocatalysis. Chem. Eng. J. 161 (1–2), 53–59.

Al-Rifai, J.H., Gabelish, C.L., et al., 2007. Occurrence of pharmaceutically active and non-steroidal estrogenic compounds in three different wastewater recycling schemes in Australia. Chemosphere 69, 803–815.

Andreozzi, R., Caprio, V., et al., 2003a. Paracetamol oxidation from aqueous solutions by means of ozonation and H_2O_2/UV system. Water Res. 37 (5), 993–1004.

Andreozzi, R., Marotta, R., et al., 2003b. Pharmaceuticals in STP effluents and their solar photodegradation in aquatic environment. Chemosphere 50 (10), 1319–1330.

Ashton, D., Hilton, M.J., et al., 2004. Investigating the environmental transport of human pharmaceuticals to streams in the United Kingdom. Sci. Total Environ. 333, 167–184.

Baeza, C., Knappe, D.R.U., 2011. Transformation kinetics of biochemically active compounds in low-pressure UV Photolysis and UV/H(2)O(2) advanced oxidation processes. Water Res. 45 (15), 4531–4543.

Bartels, P., Von Tumpling, W., 2007. Solar radiation influence on the decomposition process of diclofenac in surface waters. Sci. Total Environ. 374 (1), 143–155.

Basha, S., Barr, C., et al., 2011. On the adsorption/photodegradation of amoxicillin in aqueous solutions by an integrated photocatalytic adsorbent (IPCA): experimental studies and kinetics analysis. Photochem. Photobiol. Sci. 10 (6), 1014–1022.

Bendz, D., Paxeus, N.A., et al., 2005. Occurrence and fate of pharmaceutically active compounds in the environment, a case study: Hoje River in Sweden. J. Hazard. Mater. 122 (3), 195–204.

Benotti, M.J., Stanford, B.D., et al., 2009. Evaluation of a photocatalytic reactor membrane pilot system for the removal of pharmaceuticals and endocrine disrupting compounds from water. Water Res. 43 (6), 1513–1522.

Bergersen, O., Hanssen, K.O., et al., 2012. Anaerobic treatment of sewage sludge containing selective serotonin reuptake inhibitors. Bioresour. Technol. 117, 325–332.

Bobo, W.V., Miller, S.C., et al., 2005. The abuse liability of dextromethorphan among adolescents: a review. J. Child. Adoles. Subst. 14 (4), 55–75.

Bucher, H.C., Tschudi, P., et al., 2003. Effect of amoxicillin-clavulanate in clinically diagnosed acute rhinosinusitis: a placebo-controlled, double-blind, randomized trial in general practice. Arch. Intern. Med. 163 (15), 1793–1798.

Buser, H.R., Poiger, T., et al., 1998. Occurrence and fate of the pharmaceutical drug diclofenac in surface waters: rapid photodegradation in a lake. Environ. Sci. Technol. 32 (22), 3449–3456.

Canonica, S., Meunier, L., et al., 2008. Phototransformation of selected pharmaceuticals during UV treatment of drinking water. Water Res. 42 (1–2), 121–128.

Castiglioni, S., Bagnati, R., et al., 2006. Removal of pharmaceuticals in sewage treatment plants in Italy. Environ. Sci. Technol. 40 (1), 357–363.

Choi, K., Kim, Y., et al., 2008. Seasonal variations of several pharmaceutical residues in surface water and sewage treatment plants of the Han River, Korea. Sci. Total Environ. 405, 120–128.

Choina, J., Kosslick, H., et al., 2013. Photocatalytic decomposition of pharmaceutical ibuprofen pollutions in water over titania catalyst. Appl. Catal. B Environ. 129, 589–598.

Cooper, W.J., Mezyk, S.P., Peller, J.R., Cole, S.K., Song, W.H., Mincher, B.J., Peake, B., 2008. Studies in radiation chemistry: application to ozonation and other advanced oxidation processes. Ozone-Sci. Eng. 30, 58–64.

Dan, A., Yang, Y., Dai, Y., Chen, C., Wang, S., Tao, R., 2013. Removal and factors influencing removal of sulfonamides and trimethoprim from domestic sewage in constructed wetlands. Bioresour. Technol. 146, 363–370.

Das, T.K., 2001. Ultraviolet disinfection application to a wastewater treatment plant. Clean Techn. Environ. Policy 3, 69–80.

de Luna, M.D.G., Veciana, M.L., et al., 2012. Acetaminophen degradation by electro-Fenton and photoelectro-Fenton using a double cathode electrochemical cell. J. Hazard. Mater. 217, 200–207.

Dixon, W.T., Norman, R.O.C., 1963. Electron spin resonance studies of oxidation. 1. Alcohols. J. Chem. Soc. May, 3119–3124.

Duffy, S.J., 2005. Environmental Chemistry: A Global Perspective. Oxford University press, Oxford, UK.

Dutta, K., Lee, M.Y., et al., 2014. Removal of pharmaceuticals and organic matter from municipal wastewater using two-stage anaerobic fluidized membrane bioreactor. Bioresour. Technol. 165, 42–49.

Elliot, A.J., McCracken, D.R., Buxton, G.V., Wood, N.D., 1990. Estimation of rate constants for near-diffusion-controlled reactions in water at high temperatures. J. Chem. Soc. Farad. Trans. 86, 1539–1547.

Elmolla, E.S., Chaudhuri, M., 2010. Comparison of different advanced oxidation processes for treatment of antibiotic aqueous solution. Desalination 256 (1–3), 43–47.

EPA, 1992. Water treatment/disposal for small communities. EP/625/R-92/005. Available from, http://www2.epa.gov/nscep, accessed 9/17/15.
EPA, 2004. Primer for municipal waste water treatment systems. Available from, http://www.epa.gov/owm/primer.pdf.
EPA, 2011. A homeowner's guide to septic systems. EPA 832-B-02-005. Available from, http://www.epa.gov/owm/septic/pubs/homeowner_guide_long.pdf, accessed 9/17/15.
Eriksson, J., Svanfelt, J., et al., 2010. A photochemical study of diclofenac and its major transformation products. Photochem. Photobiol. 86 (3), 528–532.
Felis, E., Wiszniowski, J., et al., 2009. Advanced oxidation of diclofenac in various aquatic environments. Arch. Environ. Prot. 35 (2), 15–25.
Fick, J., Lindberg, R.H., et al., 2007. Antiviral oseltamivir is not removed or degraded in normal sewage water treatment: implications for development of resistance by influenza A Virus. PLoS One 2 (10), 5.
Gangwang, G., Liu, G., et al., 2012. Photodegradation of salicylic acid in aquatic environment: effect of different forms of nitrogen. Sci. Total Environ. 435, 573–577.
Gao, L.H., Shi, Y.L., et al., 2012. Occurrence of antibiotics in eight sewage treatment plants in Beijing, China. Chemosphere 86 (6), 665–671.
Ghosh, G.C., Nakada, N., et al., 2010. Occurrence and fate of oseltamivir carboxylate (Tamiflu) and amantadine in sewage treatment plants. Chemosphere 81 (1), 13–17.
Gielen, G.J.H.P., van den Heuvel, M.R., et al., 2009. Factors impacting on pharmaceutical leaching following sewage application to land. Chemosphere 74, 537–542.
Glaze, H., Kang, J., et al., 1987. The chemistry of water treatment processes involving ozone, hydrogen peroxide and ultraviolet radiation. Ozone Sci. Eng. 9, 335–352.
Göbel, A., Thmsen, A., McArdell, C.S., Joss, A., Giger, W., 2005. Occurrence and sorption behaviour of sulfonamides, macrolides, and trimethoprim in activated sludge treamtment. Cent. Eur. J. Chem. 39, 3981–3989.
Göbel, A., McArdell, C., Joss, A., Siegrist, H., Giger, W., 2007. Fate of sulphonamides, macrolkides, and trimethoprim in different waste water treatment technologies. Sci. Total Environ. 372, 361–371.
Gómez, M.J., Martínez Bueno, M.J., Lacorte, S., Fernández-Alba, A.R., Agüera, A., 2007. Pilot survey monitoring pharmaceuticals and related compounds in a sewage treatment plant located on the Mediterranean Coast. Chemosphere 66, 993–1002.
Gottschalk, C., Libra, J.A., et al., 2008. Ozonation of Water and Waste Water: A Practical Guide to Understanding Ozone and Its Applications. Wiley-VCH, Weinheim.
Grassi, N., Kaykioglu, G., et al., 2012. Removal of emerging contaminants from water and wastewater by adsorption process. In: Lofrano, G. (Ed.), Emerging Compounds Removal from Wastewater. Springer, New York.
Gulkowska, A., Leung, H.W., et al., 2008. Removal of antibiotics from wastewater by sewage treatment facilities in Hong Kong and Shenzhen, China. Water Res. 42 (1–2), 395–403.
Han, G.H., Hur, H.G., et al., 2006. Ecotoxicological risk of pharmaceuticals from wastewater treatment plants in Korea: occurrence and toxicity to Daphnia magna. Environ. Toxicol. Chem. 25 (1), 265–271.
Heberer, T., 2002. Tracking persistent pharmaceutical residues from municipal sewage to drinking water. J. Hydrol. 266 (3–4), 175–189.
Hendricks, D.W., Pote, D.W., 1974. Thermodynamic analysis of a primary oxidation pond. J. Wat. Pollution Control Federation 46 (2), 333–351.
Hijosa-Valsero, M., Matamoros, V., Sidrch-Cardona, R., Martin-Villacorta, J., Bacares, E., Bayona, J.M., 2010. Compreehensive assessment oof the design of constructed wetalns doe the removal of pharmaceuticals and presonal cae products from urban wastewater. Water Res. 44 (12), 3669–3678.

Hijosa-Valsero, M., Fink, G., Schlusener, M.P., Sidrach-Cardona, R., Martin-Villacorta, J., Ternes, T., Becares, E., 2011. Removal of antibiotics from urban wastewater by constructed wetland optimisation. Chemosphere 83, 713–719.

Isarian-Chavez, E., Rodriguez, R.M., Cabot, P.L., Centellas, F., Arias, C., Garrido, J.A., Brillas, E., 2011. Degradation of pharmaceutical beta-blockers by electrochemical advanced oxidation processes using a flow pant with a solar compound parabolic collector. Water Res. 45, 4119–4130.

Kim, S.D., Cho, J., et al., 2007. Occurrence and removal of pharmaceuticals and endocrine disruptors in South Korean surface, drinking, and waste waters. Water Res. 41 (5), 1013–1021.

Kim, I., Yamashita, N., et al., 2009. Photodegradation of pharmaceuticals and personal care products during UV and UV/H_2O_2 treatments. Chemosphere 77 (4), 518–525.

Koutsouba, V., Heberer, T., et al., 2003. Determination of polar pharmaceuticals in sewage water of Greece by gas chromatography–mass spectrometry. Chemosphere 51, 69–75.

Kravitz, R.L., Bell, R.A., 2013. Media, messages, and medication: strategies to reconcile what patients hear, what they want, and what they need from medications. BMC Med. Inform. Decis. Mak. 13, 8.

Kwon, J.W., Armbrust, K.L., 2006. Laboratory persistence and fate of fluoxetine in aquatic environments. Environ. Toxicol. Chem. 25 (10), 2561–2568.

Lam, M.W., Young, C.J., et al., 2005. Aqueous photochemical reaction kinetics and transformations of fluoxetine. Environ. Sci. Technol. 39 (2), 513–522.

Lee, H.B., Peart, T.E., et al., 2005. Determination of endocrine-disrupting phenols, acidic pharmaceuticals, and personal care products in sewage by solid-phase extraction and gas chromatography–mass spectrometry. J. Chromatogr. A 1094, 122–129.

Lee, H.B., Sarafin, K., et al., 2007. Determination of beta-blockers and beta(2)-agonists in sewage by solid-phase extraction and liquid chromatography-tandem mass spectrometry. J. Chromatogr. A 1148 (2), 158–167.

Leknes, H., Sturtzel, I.E., et al., 2012. Environmental release of oseltamivir from a Norwegian sewage treatment plant during the 2009 influenza A (H1N1) pandemic. Sci. Total Environ. 414, 632–638.

Leong, L., Jeff Kuo, J., et al., 2008. Disinfection of Wastewater Effluent—Comparison of Alternative Technologies. Water Intelligence Online.

Lin, A.Y.C., Yu, T.H., et al., 2009. Removal of pharmaceuticals in secondary wastewater treatment processes in Taiwan. J. Hazard. Mater. 167, 1163–1169.

Lindqvist, N., Tuhkanen, T., et al., 2005. Occurrence of acidic pharmaceuticals in raw and treated sewages and in receiving waters. Water Res. 39 (11), 2219–2228.

Liu, S., Ying, G.-G., Zhao, J.-L., Zhou, L.-J., Yang, Chen, Z.-F., Lai, H.-J., 2012. Occurrence and fate of androgens, estrogens, glucocorticoids and progestagens in two different types of municipal wastewater treatment plants. J. Environ. Monit. 14, 482–491.

Logan, B.K., Yeakel, J.K., et al., 2012. Dextromethorphan abuse leading to assault, suicide, or homicide. J. Forensic Sci. 57 (5), 1388–1394.

Longworth, D.L., 2001. Microbial drug resistance and the roles of the new antibiotics. Cleve. Clin. J. Med. 68 (6), 496.

Manahan, S.E., 1990. Hazardous Waste Chemistry, Toxicology and Treatment. Lewis, Michigan, USA.

Marques, J.G., Calado, G., et al., 2013. Tonic water: a rare cause of exanthema. Allergol. Immunopathol. 40 (1), 60–61.

Martin, R.J., Wj, N., 1997. The repeated exhaustion and chemical regeneration of activated carbon. Water Res. 21 (8), 961–965.

Matamoros, V., Arias, C., Brix, H., Bayona, J.M., 2009. Preliminary screening of small-scale domestic wastewater treatment systems for removal of pharmaceutical and personal care products. Water Res. 43: 55–62.

Mendez-Arriaga, F., Otsu, T., et al., 2011. Photooxidation of the antidepressant drug Fluoxetine (Prozac (R)) in aqueous media by hybrid catalytic/ozonation processes. Water Res. 45 (9), 2782–2794.
Mestankova, H., Schirmer, K., et al., 2012. Removal of the antiviral agent oseltamivir and its biological activity by oxidative processes. Environ. Pollut. 161, 30–35.
Metcalfe, C.D., Chu, S.G., et al., 2010. Antidepressants and their metabolites in municipal wastewater, and downstream exposure in an urban watershed. Environ. Toxicol. Chem. 29 (1), 79–89.
MfE, 2003. Sustainable Wastewater Management: A Handbook for Smaller Communities. Ministry for the Environment, Wellington, New Zealand.
Minim, V.P.R., Ribeiro, M.M., et al., 2009. Tonic water: acceptance and analysis time-intensity of the bitter taste. Ciênc. Tecnol. Aliment. 29 (3), 567–570.
Moore, B.E., Thomson, S.R., et al., 1990. Photochemical studies of the anti-inflammatory drug diclofenac. Photochem. Photobiol. 52, 685–690.
Musa, K.A.K., Ericksson, L.A., 2009. Photodegradation mechanism of the common non-steroid ani-inflammatory drug diclofenac and its carbazole photoproduct. Phys. Chem. Chem. Phys. 11, 4601–4610.
Mutiyar, P.K., Mittal, A.K., 2013. Occurrences and fate of an antibiotic amoxicillin in extended aeration-based sewage treatment plant in Delhi, India: a case study of emerging pollutant. Desalin. Water Treat. 51 (31–33), 6158–6164.
Nakada, N., Tanishima, T., et al., 2006. Pharmaceutical chemicals and endocrine disrupters in municipal wastewater in Tokyo and their removal during activated sludge treatment. Water Res. 40, 3297–3303.
Neyens, E., Baeyens, J., 2003. A review of classic Fenton's peroxidation as an advanced oxidation technique. J. Hazard. Mater. 98 (1–3), 33–50.
Nfodzo, P., Choi, H., 2011. Sulfate radicals destroy pharmaceuticals and personal care products. Environ. Eng. Sci. 28 (8), 605–609.
Ohko, Y., Iuchi, K.I., et al., 2002. 17 beta-estradiol degradation by TiO_2 photocatalysis as means of reducing estrogenic activity. Environ. Sci. Technol. 36 (19), 4175–4181.
Packer, J.L., Werner, J.J., et al., 2003. Photochemical fate of pharmaceuticals in the environment: naproxen, diclofenac, clofibric acid and ibuprofen. Aquat. Sci. 65, 342–351.
Pailler,J.Y.,Pfsiter,K.L.,etal.,2009.Solidphaseextractioncoupledtoliquidchromatography-tandem-mass spectrometry analysis of sulfonamides, tetracyclines, analgesics and hormones in surface water and wastewater in Luxembourg. Sci. Total Environ. 407, 4736–4743.
Perez-Estrada, L.A., Malato, S., et al., 2005a. Photo-fenton degradation of diclofenac: identification of main intermediates and degradation pathway. Environ. Sci. Technol. 39 (21), 8300–8306.
Perez-Estrada, L.A., Maldonado, M.I., et al., 2005b. Decomposition of diclofenac by solar driven photocatalysis at pilot plant scale. Catal. Today 101 (3–4), 219–226.
Peuravuori, J., 2012. Aquatic photochemistry of diclofenac in the presence of natural dissolved organic chromophoric material and nitrate. Int. J. Environ. Anal. Chem. 92 (13), 1470–1492.
Prasse, C., Schlusener, M.P., et al., 2010. Antiviral drugs in wastewater and surface waters: a new pharmaceutical class of environmental relevance? Environ. Sci. Technol. 44 (5), 1728–1735.
Preston, M., 1989 Marine Pollution. In: Riley, J. P. (Ed.), Chemical Oceanography. Academic Press, London, UK. Vol. 9, pp. 54–196.
Radjenovic, J., Petrovic, M., et al., 2009. Fate and distribution of pharmaceuticals in wastewater and sewage sludge of the conventional activated sludge (CAS) and advanced membrane bioreactor (MBR) treatment. Water Res. 43 (3), 831–841.
Rivas, J., Gimeno, O., et al., 2010. Influence of oxygen and free radicals promoters on the UV-254nm photolysis of diclofenac. Chem. Eng. J. 163 (1–2), 35–40.

Rivas, F.J., Gimeno, O., et al., 2012. Aqueous pharmaceuticals compounds removal by potassium monopersulfate. Uncatalysed and catalyzed semicontinuous experiments. Chem. Eng. J. 192, 326–333.
Roberts, P.H., Thomas, K.V., 2006. The occurrence of selected pharmaceuticals in wastewater effluent and surface waters of the lower Tyne catchment. Sci. Total Environ. 356, 143–153.
Rosal, R., Rodriguez, A., Perdigon-Melon, J.A., Petre, A., Garcia-Calvo, E., Gomez, M.J., Aguera, A., Fernandez-Alaba, A.R., 2010. Occurrence of emerging contaminants in urban wastewater and their removal through biological treatment followed by ozonation. Water Res. 44, 578–588.
Santos, L.H.M.L.M., Araújo, A.N., et al., 2010. Ecotoxicological aspects related to the presence of pharmaceuticals in the aquatic environment. J. Hazard. Mater. 175, 45–95.
Shinde, S.S., Bhosale, C.H., et al., 2011a. Photocatalytic oxidation of salicylic acid and 4-chlorophenol in aqueous solutions mediated by modified $AlFe_2O_3$ catalyst under sunlight. J. Mol. Catal. A: Chem. 347 (1–2), 65–72.
Shinde, S.S., Shinde, P.S., et al., 2011b. Zinc oxide mediated heterogeneous photocatalytic degradation of organic species under solar radiation. J. Photochem. Photobiol. B Biol. 104 (3), 425–433.
Singer, A.C., Nunn, M.A., et al., 2007. Potential risks associated with the proposed widespread use of Tamiflu. Environ. Health Perspect. 115 (1), 102–106.
Song, W., Cooper, W.J., et al., 2008a. Free radical destruction of β-blockers in aqueous solution. Environ. Sci. Technol. 42, 1256–1261.
Song, W.H., Chen, W.S., et al., 2008b. Free-radical destruction of beta-lactam antibiotics in aqueous solution. J. Phys. Chem. A 112 (32), 7411–7417.
Stumpf, M., Ternes, T.A., et al., 1999. Polar drug residues in sewage and natural waters in the state of Rio de Janeiro, Brazil. Sci. Total Environ. 225 (1–2), 135–141.
Styrishave, B., Hallling-Sorenson, B., Ingerslev, F., 2011. Enviornemntal risk assessment of three selective sertonin reuptake inhibitors in the aquatic environment. A case study including a cocktail scenario. Environ. Toxicol. Chem. 30, 254–261.
Sweetman, S.C., 2011. Martindale: The Complete Drug Reference. Pharmaceutical Press, London, UK.
Tanner, C.C., Champion, P.D., et al., 2006. New Zealand Constructed Wetland Planting Guidelines. National Institute of Water and Atmospheric Research report published in association with the New Zealand Water & Wastes Association (available from http://www.nzwwa.org.nz/Constructed_Wetland_Planting_Guidelines.pdf), accessed 9/18/15.
Tauxe-Wuersch, A., De Alencastro, L.F., et al., 2005. Occurrence of several acidic drugs in sewage treatment plants in Switzerland and risk assessment. Water Res. 39 (9), 1761–1772.
Ternes, T.A., 1998. Occurrence of drugs in German sewage treatment plants and rivers. Water Res. 32, 3245–3260.
Tong, A.Y.C., 2013. Pharmaceuticals in sewage: sources, concentrations and degradation. PhD, Otago.
Tong, A.Y.C., Braund, R., et al., 2011. UV-induced photodegradation of oseltamivir (Tamiflu) in water. Environ. Chem. 8 (2), 182–189 (http://www.publish.csiro.au/paper/EN10095.htm).
Tong, A.Y.C., Braund, R., et al., 2012. TiO_2-assisted photodegradation of pharmaceuticals—a review. Cent. Eur. J. Chem. 10 (4), 989–1027.
Toor, G.S., Lusk, M., Obreza, T., 2011a. Onsite sewage treatment and disposal systems: An Overview. Document SL347. IFAS Extension, University of Florida. Available from, http://edis.ifas.ufl.edu/ss549, accessed 9/17/15.
Toor, G.S., Lusk, M., Obreza, T., 2011b. Onsite sewage treatment and disposal systems: Toxic organic chemicals. Document SL352. IFAS Extension, University of Florida. Available from, http://edis.ifas.ufl.edu/ss552, accessed 9/17/15.

Vasskog, T., Berger, U., et al., 2006. Selective serotonin reuptake inhibitors in sewage influents and effluents from Tromso, Norway. J. Chromatogr. A 1115 (1–2), 187–195.

Vasskog, T., Anderssen, T., et al., 2008. Occurrence of selective serotonin reuptake inhibitors in sewage and receiving waters at Spitsbergen and in Norway. J. Chromatogr. A 1185 (2), 194–205.

Vasskog, T., Bergersen, O., et al., 2009. Depletion of selective serotonin reuptake inhibitors during sewage sludge composting. Waste Manag. 29 (11), 2808–2815.

Vieno, N.M., Tuhkanen, T., Kronberg, L., 2006. Analysis of neutral and basic pharmaceuticals in sewage treatment plants and in recipient rivers using solid phase extraction and liquid chromatography-ion trap tandem mass spectrometry. J. Chromatogr. A 1116, 193–203.

Vogna, D., Marotta, R., et al., 2004. Advanced oxidation of the pharmaceutical drug diclofenac with UV/H_2O_2 and ozone. Water Res. 38 (2), 414–422.

Wadley, S., Waite, T.D., 2004. Fenton processes. In: Parsons, S.A. (Ed.), Advanced Oxidation Processes for Waste Wastewater Treatment. IWA Publishing, London, UK, pp. 111–136, Chapter 5.

Walling, C., 1975. Fentons reagent revisited. Acc. Chem. Res. 8 (4), 125–131.

Wang, E.E.L., Einarson, T.R., et al., 1999. Antibiotic prescribing for Canadian preschool children: evidence of overprescribing for viral respiratory infections. Clin. Infect. Dis. 29 (1), 155–160.

Watkinson, A.J., Murby, E.J., et al., 2007. Removal of antibiotics in conventional and advanced wastewater treatment: implications for environmental discharge and wastewater recycling. Water Res. 41 (18), 4164–4176.

Wert, E.C., Rosario-Ortiz, F.L., et al., 2009. Effect of ozone exposure on the oxidation of trace organic contaminants in wastewater. Water Res. 43 (4), 1005–1014.

West, J.V., 2002. Acute upper airway infections. Br. Med. Bull. 61, 215–230.

Westerhoff, P., Mezyk, S.P., et al., 2007. Electron pulse radiolysis determination of hydroxyl radical rate constants with Suwannee river fulvic acid and other dissolved organic matter isolates. Environ. Sci. Technol. 41 (13), 4640–4646.

Yang, L.M., Yu, L.Y.E., et al., 2007. Photocatalytic oxidation of paracetamol in water. In: Wilson, J. (Ed.), Proceedings of the Second IASTED International Conference on Water Resources Management. Acta Press Anaheim, Anaheim, pp. 1–6.

Yang, L.M., Yu, L.E., et al., 2008. Degradation of paracetamol in aqueous solutions by TiO_2 photocatalysis. Water Res. 42 (13), 3480–3488.

Yang, L.M., Yu, L.E., et al., 2009. Photocatalytic oxidation of paracetamol: dominant reactants, intermediates, and reaction mechanisms. Environ. Sci. Technol. 43 (2), 460–465.

Zhang, X., Chen, P., Wu, F., Deng, N., Liu, J., Fang, T., 2006. Degradation of 17α-ethinylestradiol in aqueous solution by ozonation. J. Hazard. Mater. B133, 291–298.

Zhang, Z., Feng, Y., Lui, Y., Sun, Q., Gao, P., Ren, N., 2010. Kinetic degrdadation model and estrogenicity changes of EE_2 (17α-ethinylestradiol) in aqueous solution by UV and UV/H_2O_2 technology. J. Hazard. Mater. 181, 1127–1133.

Zorita, S., Martensson, L., et al., 2009. Occurrence and removal of pharmaceuticals in a municipal sewage treatment system in the south of Sweden. Sci. Total Environ. 407 (8), 2760–2770.

Regulatory practices to control the discharge of pharmaceuticals into the environment

7

7.1 Overview

The detection of pharmaceuticals in the environment is the result of a chain of processes leading to their eventual discharge. The increasing use of a wide range of pharmaceuticals in recent years has led to the appearance of active pharmaceutical ingredients (APIs) and their metabolites in sewage water mainly due to human excretory processes. In addition, prescription medications are often inappropriately prescribed or supplied in excess (Chapter 2), often leading to their improper disposal (Chapter 3). Subsequently, while their real or predicted environmental concentrations in rivers, lakes, estuarine and coastal environments, ground water, and drinking water (Chapter 4) are generally low, there is often the potential for unacceptable ecotoxic effects arising from pharmaceuticals and their metabolites either on their own or in mixtures (Chapter 5). Furthermore, many of these compounds are not significantly removed or degraded by conventional and sometimes even more advanced sewage treatment processes (Chapter 6).

There is currently no single practice that completely prevents the environmental discharge of pharmaceuticals whether it be how they are used or how they are treated in waste water. In addition, because environmental pollution resulting from emerging contaminants is a multifactorial problem, it will require the collaboration and co-operation of various bodies to adequately deal with the emerging problem and offer a range of solutions to ameliorate it. Therefore, any potential solution will need to be a concerted effort from health professionals (who prescribe and dispense medications), householders and patients (who use, store, and dispose of unwanted medications), wastewater industry scientists (who can develop new strategies to minimize the discharge of emerging contaminants into the environment), and government bodies and regulatory authorities who can turn the potential solution resulting from any concerted effort into a reality by implementing official regulatory guidelines and legislation that will continue to encourage environmentally practices practices from all parties involved.

As explained previously, there is still a paucity of information in many parts of the world regarding the best possible options for the disposal of unused medications, and confusion still exists as to the most environmentally-friendly disposal method. Confusion in dealing with pharmaceutical waste is a major contributor to the problem. As will be apparent in this chapter, many acts of legislation around the world have not considered or catered for pharmaceuticals as environmental pollutants and specific guidelines addressing their presence in the waste stream are still few and far between. In this chapter,

The Life-Cycle of Pharmaceuticals in the Environment. http://dx.doi.org/10.1016/B978-1-907568-25-1.00007-4

we will review some of the regulations and guidelines presently existing in various parts of the world to control in various ways the environmental discharge of wastes, whether they apply to pharmaceutical contaminants, and whether they can be used to combat the problem of pollution resulting from the environmental discharge of pharmaceuticals.

7.2 The United States of America and Canada

The establishment of the Federal Water Pollution Act in 1972 has helped the formal protection of waterways in the United States from pollution (available http://www2.epa.gov/laws-regulations/summary-clean-water-act, accessed 09.09.14). This Act was amended in 1977 and has been known since then as the Clean Water Act 1977. The primary objective of the Act is to "restore and maintain the chemical, physical and biological integrity of US waters." The Act is enforced by the National Pollutant Discharge Elimination System (NDPES), which controls the quality of wastes that are released into waterways from point sources such as sewage effluent but also wastes from pipes, tunnels, containers, or various other vessels. A permit system coordinated by the US Environmental Protection Agency (EPA) is regulated by the NDPES. Permits are required under the NDPES for any discharge of waste into waterways, but are not required if wastes from a point source are discharged into the general sanitation (i.e., sewage treatment) system. A permit to discharge effluent from point sources is issued provided the "best removal technology available" is used to treat the waste. This practical requirement is based on cost-effectiveness and the permit is granted if a sewage treatment facility endeavors to use the best possible but still affordable technology for wastewater. However, the Clean Water Act *does* require that most sewage treatment facilities should have at least "secondary treatment processes" (such as activated sludge processes, Chapter 6) to adequately meet this "best removal technology" requirement.

Despite these legislative requirements, most US sewage treatment facilities probably do not use the "best technology available" as sewage facilities can always claim a lack of funding for investing in that best technology available. Furthermore, the exact meaning of the term "best technology available" is debatable for any given sewage treatment facility. Because of this, the amendments made to the Clean Water Act in 1987 provided the use of state revolving funds (SRFs) in place of grants as a means to allow the upgrading of substandard centralized sewage treatment facilities. This has led to the improvement of many centralized sewage treatment facilities in the United States with a number of them utilizing more effective treatments than secondary treatment (i.e., they often include at least some form of tertiary treatment). However, if the water quality of the treated effluent is unsatisfactory even with the best affordable treatment technology, the NDPES only grants a limited permit which requires additional measures to be put in place by the sewage treatment facility to ultimately achieve an acceptable water quality. The NDPES and the Clean Water Act monitors the maximum levels and outputs of only 65 toxic pollutants and does not include the monitoring of pharmaceutical compounds.

One of the major issues not addressed by the NDPES/Clean Water Act is the inability of the Act to control the entry of pharmaceuticals into the environment before they become contaminants. Furthermore, the legislation was only designed to monitor conventional contaminants, such as pesticides or heavy metals—substances in discharge levels more likely to induce observable acute toxicity at typical environmental levels. As a result, it has become difficult for the Clean Water Act to be used to monitor the discharge of pharmaceuticals and establish guidelines for their discharge primarily because they are unlikely to induce acute toxicity and much uncertainty still exists regarding the maximum levels that would induce observable toxicity in aquatic organisms (see Chapter 5). In addition, the Clean Water Act makes no allowance for mixture toxicity, and thus risks associated with such pharmacological activity cannot be regulated or controlled by the Act.

A similar piece of legislation for drinking water quality is the Safe Drinking Water Act 1974, which regulates the maximum concentration limits (MCLs) for monitored contaminants in drinking water. An alphabetical listing of monitored contaminants and their MCLs is available on the EPA Web site (available http://water.epa.gov/drink/contaminants/index.cfm#List, accessed 02.11.13). However, this list contains only pesticides, heavy metals, and other inorganic contaminants and no pharmaceuticals. Again, this lack of regulation of pharmaceuticals may be due to the paucity of data concerning their presence in waterways. However, the EPA has also established contaminant candidate lists (CCLs) (available http://water.epa.gov/scitech/drinkingwater/dws/ccl/basicinformation.cfm, accessed 02.11.13) which are lists of contaminants not currently regulated by government authorities but at the same time are recognized as being currently detectable in at least some waterways where they could lead to potentially adverse human health effects.

The primary goal of the publication of CCLs is to initiate direct research and allow decisions to be made about the monitored contaminants. The research is directed primarily to determining whether there is enough information known about the contaminants to warrant regulation under the regulatory determination criteria set forth in the Safe Drinking Water Act section 1412(b)(1). A determination to formally regulate a chemical depends on three criteria: (1) evidence to show that the contaminant has adverse health risks related to exposure in drinking water; (2) evidence to show that the contaminant is prevalent in drinking water; and (3) whether regulation of the contaminant would present a "meaningful opportunity" for reducing risks to health. Only then is a decision made as to whether to regulate the contaminant or not. When a formal recommendation is made to regulate the contaminant, the setting of an MCL is based upon the evidence collected about the environmental nature of the contaminant. The latest rendition of this list (CCL3) includes the pharmaceuticals erythromycin, estradiol, estriol, estrone, 17α-ethinylestradiol, nitroglycerin (glyceryl trinitrate), and quinolone.

Methods that can be used to determine whether a contaminant should be listed in a CCL are constantly being restructured and improved. In 2001, The National Academy of Sciences (NAS) suggested that all potential future CCL contaminants should be screened in a preliminary process, whereby:

> *"a broadly defined 'universe' of potential drinking water contaminants is identified, assessed, and reduced to a preliminary CCL (PCCL) using simple screening criteria that indicate public health risk and the likelihood of occurrence in drinking water.*

All of the contaminants on the PCCL would then be assessed in more detail using a classification approach and tools along with expert judgment to evaluate the likelihood that specific contaminants could occur in drinking water at levels and at frequencies that pose a public health risk" (available http://water.epa.gov/scitech/drinkingwater/dws/ccl/basicinformation.cfm, accessed 09.09.14).

This step-wise and tiered approach for identifying and addressing emerging drinking water contaminants has also been endorsed by the National Drinking Water Advisory Council (NDWAC) which has provided a possible framework for the implementation of the PCCL approach. However, other routes are available to nominate contaminants for consideration in the CCLs. For example, public nominations for contaminants considered environmentally problematic were also included in the CCL3. This process has also been endorsed by the NAS and NDWAC, who consider it important to include such recommendations in the early stages of a PCCL process.

Despite the potential regulatory determinations placed on certain pharmaceuticals under the SDWA, one of the major challenges facing the EPA is how to identify which pharmaceuticals have the potential to cause risk to human health and which can be excluded from monitoring because they are considered safe. Also, there is no current process to establish MCLs for pharmaceuticals both in effluent discharge and in drinking water based on therapeutic dose (Conerly and Ohanian, 2010). The drafting of legislation to regulate the environmental presence of pharmaceuticals has also been hampered by classification. Should toxicity and acceptable levels of pharmaceuticals be defined singularly or by class of medication grouped by pharmacology? As explained in Chapter 5, it is often important to also consider synergistic toxicological effects. Furthermore, the Food and Drug Administration (FDA) excludes the use of environmental testing if the pharmaceutical is expected to be released into the environment from point sources at concentrations of <1 μg/L (Conerly and Ohanian, 2010). As most pharmaceuticals are detected at such environmental concentrations, it is unlikely that there would be a coordinated effort by US regulatory authorities (Office of the Water, OW, and the FDA) to harmonize the interpretation and implications arising from exposure and toxicological effect studies to set appropriate regulatory guidelines.

Processes leading to meaningful conclusions from risk assessments of pharmaceuticals are also currently incomplete. The EPA established the Endocrine Disruptor Screening Program in 1998 to identify screening methods and toxicity-testing strategies that can be used to determine whether chemicals are endocrine disruptors; however, there still needs to be a consensus on interpreting the data arising from the testing strategies (Snyder et al., 2003). This lack of consensus is also apparent in the various strategies developed by the European Organization for Economic Cooperation and Development to identify endocrine-disrupting compounds (EDCs). The predominant issue arising from such a lack of consensus is the inability of various authorities to standardize the definition of an EDC. Some naturally occurring and man-made chemicals are widely considered to be endocrine disruptors, including pesticides, industrial chemicals, combustion byproducts, phytoestrogens, and hormones excreted by animals and humans in addition to certain pharmaceuticals. There are many other chemicals for which there is limited or incomplete evidence of potential endocrine disruption activity or for which the evidence is controversial. An even greater number of chemicals have not yet been tested for potential endocrine activity using any of the available ecotoxicological assessment methods (see Chapter 5). Therefore, any list of chemicals defined as EDCs is likely to be incomplete.

In the United States, the discharge and landfilling of solid wastes into the environment is regulated by the Resource Conservation and Recovery Act (RCRA) (available http://www2.epa.gov/laws-regulations/summary-resource-conservation-and-recovery-act, accessed 9-9-14). This Act regulates the disposal of hazardous substances from the pharmaceutical and healthcare industries, but currently does not include provisions for the regulation of household wastes. The management of medication wastes has only been recently addressed by the Safe and Secure Drug Disposal Act 2010; however, it is only pertinent to the responsible community disposal of controlled drugs (Chapter 3).

The RCRA classifies hazardous waste according to the following criteria: (1) ignitability—wastes that can readily catch fire and maintain combustion; (2) corrosive character—wastes that are acidic or alkaline (basic); (3) reactivity—wastes that readily explode or undergo violent reactions or react by releasing toxic gases or fumes; and (4) toxicity—wastes likely to leach dangerous concentrations of toxic chemicals into groundwater. While unused pharmaceutical dosage forms do not fall under the first three criteria for hazardous waste, they may potentially be toxic to aquatic life. The EPA also categorizes wastes according to the following categories: (1) F-list (wastes from nonspecific sources)—hazardous wastes from certain common industrial and manufacturing processes; (2) K-list (wastes from specific industries); and (3) P- and U-lists of chemicals—known hazardous wastes. Chemicals are included on the U- and P-lists if they are acutely toxic. P-list chemicals are regulated at lower environmental levels than U-list chemicals (available http://www.epa.gov/oig/reports/2012/20120525-12-P-0508.pdf, accessed 2-11-13). In terms of these classifications, pharmaceuticals are likely to be included on the P- and U-lists as they can induce various adverse effects on human health and they can potentially exhibit acute toxicity at the prescribed dose.

The situation regarding solid wastes is similar to that of pharmaceutical residues in aquatic environments. In 1980, while the EPA identified the possibility that 31 chemicals used as pharmaceuticals qualified under the definition of hazardous waste, it has not since reviewed whether these pharmaceuticals should be included in one of the above listings nor decided on whether the environmental discharge of these pharmaceuticals under the RCRA should be monitored. Subsequently, as many more drugs enter the market, there is a lack of EPA guidelines pertaining to the disposal of these new drugs. In addition, the P- and U-lists have not been updated since 1980. This means many of the newer drugs, which may be as toxic or even more toxic than some of the older ones already listed on the P- and U-lists, are not included on these lists and thus it has been impossible to use the listings objectively to determine the environmental risk associated with the environmental discharge of many currently used pharmaceuticals in solid wastes. Pharmaceuticals currently in the P- and U-categories of hazardous waste under the RCRA in the Code of Federal Regulations (CFR) 40 provision 261.33 are listed in Tables 7.1 and 7.2.

Table 7.1 Pharmaceuticals listed as P-type hazardous substances in the RCRA Code of Federal Regulations

Pharmaceutical	Therapeutic class/drug type
Epinephrine (adrenaline)	Endocrine hormone—sympathomimetic
Nitroglycerine (glyceryl trinitrate)	Anti-anginal and vasodilator
Physostigmine	Cholinesterase inhibitor
Strychnine (and salts)	Stimulant and laxative
Nicotine	Stimulant
Warfarin (and salts)	Anticoagulant

(Available http://www.epa.gov/osw/laws-regs/regs-haz.htm, accessed 09.09.14)

Table 7.2 Pharmaceuticals listed as U-type hazardous substances in the RCRA Code of Federal Regulations

Pharmaceutical	Therapeutic class/drug type
Chlorambucil	Alkylating agent/antineoplastic
Daunomycin	Antineoplastic
Lasiocarpine	Anticholinergic
Melphalan	Alkylating agent/antineoplastic
Methylthiouracil	Anti-thyroid agent
Mitomycin C	Antineoplastic
Phenacetin	Analgesic
Reserpine	Antipsychotic/antihypertensive
Streptozotocin	Antineoplastic (pancreatic cancer)
Uracil mustard	Antineoplastic
Warfarin (and salts)	Anticoagulant

(Available http://www.epa.gov/osw/laws-regs/regs-haz.htm, accessed 09.09.14)

It is apparent from the entries in both these tables that they include substances which are acutely toxic when used at therapeutic doses—many of these include alkylating and antineoplastic agents used for chemotherapy. However, the current RCRA regulations do not deal with many other medications commonly used in modern clinical therapy. Another important point is the number of medications listed in the RCRA that are no longer commonly used in clinical therapy. For example, paracetamol (acetaminophen) is *not* listed in any of the P- or U-lists but a structurally similar analgesic, phenacetin, *is* listed. Phenacetin was one of the first non-opioid analgesics to be marketed and, like paracetamol, was an over-the-counter analgesic and antipyretic (fever reducer) used alone and in combination with other analgesics such as aspirin (acetylsalicylic acid). However, phenacetin is now no longer commonly used as an analgesic as it was observed to cause renal cancer and kidney failure in some patients. Subsequently, products containing phenacetin were withdrawn from the US market in 1983. Similarly, lasiocarpine is now less commonly used than the related pilocarpine (which is not included in either the P- or U-lists) and the neurotoxin, strychnine, which was commonly used as a stimulant and laxative in times past, is no longer prescribed.

While there is no formal legislation regulating the MCLs for individual pharmaceuticals in US drinking water, *drinking water equivalent levels* (DWELs) based on the maximum presumed safe dosage for various pharmaceuticals upon regular ingestion of drinking water have been calculated (Bruce et al., 2010). Many of these calculated concentrations are in the μg/L range and are much above the maximum water concentrations measured in many drinking water sources in the United States and other countries (*see* Table 7.3).

In summary, current US legislation has no provisions to deal with medicines (both used or improperly disposed of) that are flushed into the sewage system and/or present in drinking water. While the EPA is working on formalizing the maximum acceptable limits for some pharmaceuticals through the CCL system, not all pharmaceuticals are currently addressed in the present legislation. In addition, there is a need to harmonize the evaluation and interpretation of risk and toxicological data before the US EPA can objectively set regulatory guidelines. This has been made difficult, however, because there is still a lack of knowledge concerning the chronic adverse effects of pharmaceuticals at low environmental concentrations (*see* Chapter 5). While legislation has been put in place to classify a limited set of pharmaceuticals as hazardous waste, it has not been updated by the RCRA to include the many new and hazardous pharmaceuticals that have come on the market since the 1980s. There seems little incentive to do this given that there is currently no evidence to suggest that the trace amounts of pharmaceuticals measured in wastewater and drinking water or even landfill leachate are a real or potential acute human health concern.

Table 7.3 Drinking water equivalent levels (DWELs) and maximum concentrations for some pharmaceuticals detected in drinking water

Pharmaceutical	DWEL (μg/L)	Maximum drinking water concentration (μg/L)[a]
Atenolol	70	<0.00005 (Italy)
Carbamazepine	12	0.25 (US)
Diazepam	35	0.0235 (Italy)
Fluoxetine	34	<0.014 (US)
Norfluoxetine	34	N/A
Gemfibrozil	14	<0.015 (US)
Meprobamate	260	N/A
Phenytoin	2.0	N/A
Risperidone	0.49	N/A
Sulfamethoxazole	18,000	<0.023 (US)
Atorvastatin	4.9	N/A
Diclofenac	2300	0.93 (US)
Enalapril	8.1	N/A
Naproxen	20,000	<0.005 (France)
Simvastatin	4.9	N/A
Trimethoprim	6,700	<0.010 (US)

[a] No specification in terms of water type.
From Bruce et al. (2010)

In Canada, the Canadian Environmental Protection Act 1999 (CEPA) (available http://laws-lois.justice.gc.ca/eng/acts/c-15.31/, accessed 09.09.14) has provisions to cover the environmental discharge of chemical substances. The Wastewater Systems Effluent Regulations 1985 (available http://laws-lois.justice.gc.ca/eng/regulations/sor-2012-139/index.html, accessed 9-9-14) makes allowances for the minimal standard for the monitoring of acute toxicities and water quality parameters (such as biological oxygen demand (BOD)) in effluents. Although a number of toxic substances are listed in the Act, it currently does not include any pharmaceuticals. A substance is defined to be environmentally toxic in Canada in Section 64 of the Act if: (1) it has or may have an immediate or long-term harmful effect on the environment or its biological diversity; (2) it constitutes or may constitute a danger to the environment on which life depends; or (3) it constitutes or may constitute a danger in Canada to human life or health. To determine whether a particular substance meets each of these three criteria, a substance must be registered on the domestic substances list (DSL) which controls the importation and use of new chemicals and organisms in Canada (Breton and Boxall, 2003). A new substance is screened using a preliminary environmental risk assessment before it is allowed to enter the Canadian marketplace. This is monitored by the new substances notification regulations and is designed to keep acutely toxic substances out of Canada. This initial environmental risk assessment is based on the potential environmental concentration (PEC) and known acute or chronic ecotoxicological data (see Chapter 5).

For evaluation purposes, the substance on the DSL in question is placed on the priority substances list (PSL) by the Canadian Minister of Health and the Minister of Environment. Substances on the PSL are evaluated based on an extensive environmentally assessment within five years. The extensive environmental assessment, termed *screening level risk assessment*, correlates predicted environmental effects with any known information on the environmental behavior of the contaminant. Once a substance is considered toxic, recommendations will be issued concerning the release of the substance into the environment. These recommendations may include regulations which impose restrictions on an activity related to the use of the substance, set limits on environmentally acceptable concentrations or establish appropriate control measures which can effectively deal with the toxic substance should an accidental spill arise.

The lack of regulations concerning the use and acceptable environmental levels of pharmaceuticals in Canada is not surprising, given that the initial risk assessments are based on exposure related to PECs. Because most pharmaceuticals are present in the environment at subtherapeutic concentrations both in sewage and surface water (*see* Miao et al., 2004; Metcalfe et al., 2010; and Chapter 4 for some typical Canadian levels), there is probably no reason to suggest that they are toxic and need immediate evaluation in terms of the existing Canadian regulations.

7.3 European Union (EU)

Member countries of the European Union (EU) comply with the water framework directive (WFD) (Directive 2000/60/EC—OJL 327, 22nd December 2000, pp. 1–73) (available http://eur-lex.europa.eu/legal-content/EN/TXT/?uri=CELEX:32000L0060, accessed 10.09.14) enforced by the European Commission (EC). This framework, which aims for "good" status of surface waterways in member countries, is designed "to protect surface waters, transitional waters, coastal waters and groundwater in order to prevent and reduce pollution, promote sustainable water use, protect the aquatic environment, improve the status of aquatic ecosystems and mitigate the effects of floods and droughts" (Fuerhacker, 2009). The WFD assesses the physical-chemical and biological qualities of waterways and grades their quality as "good" or "poor." Adherence to "good" quality status requires the waterway to comply with *environmental quality standard* (EQS)s, the maximum environmental concentration limits set by the EU pertaining to particularly persistent, bioaccumulative, and toxic contaminants. Environmental contaminants, which "good" quality status is dependent upon, are based upon a PSL (Table 7.4). Every substance on the PSL has an EQS. However, in addition to the 33 PSL, an EC directive adopted in 2006 also requires the EQS monitoring of eight other pollutants which are considered to pose a particular risk to aquatic life or human health (Table 7.5).

There are also EQSs for some families or classes of pollutants in surface water (Fuerhacker, 2009). If a water body is compromised, member states have a duty to ensure water quality is improved by using the relevant best available removal techniques in sewage treatment facilities or other best environmental practices to minimize the discharge of the contaminant in waterways acceptable to their EQS. The EU hopes that this harmonized effort to deal with contaminants under the WFD will allow the waterways of all member countries to achieve "good status" by 2015. In addition, the EU will use the WFD as an instrument to prevent the direct environmental discharge of contaminants by 2025 (Fürhacker, 2008).

Currently, there are no pharmaceuticals on the priority list and thus no EQS minimal requirements on pharmaceuticals that have to be met by the authorities responsible for the management of waterways to achieve "good" quality status. However, a draft proposal placed before the European Parliament (EP) on 6 September 2012 by the EC suggested 28 amendments to the WFD. As well as reducing the monitoring of ubiquitously persistent, bioaccumulative, and toxic substances (PBT, Chapter 5), the EC recommended in the document COM(2011) 876, the monitoring of new priority substances in addition to the existing priority substances with EQSs. Fifteen substances were considered for the priority list, including three pharmaceutical compounds (17α-ethinylestradiol—EE2, 17β-estradiol—E2, and diclofenac) (Johnson et al., 2013). The anticipated annual EQS concentrations for the three pharmaceuticals are listed in Table 7.6 and it was considered that these would take effect in 2018. In addition, the EC proposal recommended the establishment of a contaminant "watch list" for other pharmaceuticals which were considered potentially detrimental to the environment.

Among the pharmaceuticals proposed in this "watch list" were carbamazepine and ibuprofen, as well as the personal care product triclosan. Environmental lobbyists and the European water industry thought that the EP would approve the regulation of these three pharmaceuticals and set appropriate EQS concentrations for them. However, this has been opposed by various European pharmaceutical industries who claim that there is still not enough evidence to support their inclusion in the WFD priority list. In addition, the setting of an EQS for a particular environmental contaminant usually

Table 7.4 European Union (EU) priority substances

Chemical substance	CAS number	Chemical substance	CAS number
Alachlor	15972-60-8	Hexachlorocyclohexane	608-73-1
Anthracene	120-12-7	Isoproturon	34123-59-6
Atrazine	1912-24-9	Lead and compounds	7439-92-1
Benzene	71-43-2	Mercury and compounds	7439-97-6
Brominated diphenyl ethers	32534-81-9	Naphthalene	91-20-3
Cadmium and compounds	7440-43-9	Nickel and compounds	7440-02-0
Chloroalkanes	85535-84-8	Nonylphenols	25154-52-3
Chlorfenvinphos	470-90-6	Octylphenols	1806-26-4
Chlorpyrifos	2921-88-2	Pentachlorobenzene	608-93-5
1,2-Dichloroethane	107-06-2	Pentachlorophenol	87-86-5
Dichloromethane	75-09-2	Polyaromatic hydrocarbons	Not applicable
Di(2-ethylhexyl)phthalate	117-81-7	Simazine	122-34-9
Diuron	330-54-1	Tributyltin compounds	36643-28-4
Endosulfan	115-29-7	Trichlorobenzenes	12002-48-1
Fluoranthene	206-44-0	Trichloromethane	67-66-3
Hexachlorobenzene	118-74-1	Trifluralin	1582-09-8
Hexachlorobutadiene	87-68-3		

(Available http://ec.europa.eu/environment/water/water-framework/priority_substances.htm, accessed 10.09.14)

Table 7.5 Additional chemical substances not on the EU priority list but considered by the EU to pose significant risks to wildlife

Chemical substance	CAS Number	Chemical substance	CAS number
Tetrachloromethane	56-23-5	Endrin	72-20-8
DDT	50-29-3	Isodrin	465-73-6
Aldrin	309-00-2	Tetrachloroethylene	127-18-4
Dieldrin	60-57-1	Trichloroethylene	79-01-6

(Available http://ec.europa.eu/environment/water/water-framework/priority_substances.htm, accessed 10.09.14)

Table 7.6 European Commission (EC) proposed environmental quality standard (EQS) concentrations for 17α-ethinylestradiol, 17β-estradiol, and diclofenac in EU surface water

Pharmaceutical	Proposed EQS concentration (annual average) (ng/L)
17α-Ethinylestradiol	0.035
17β-Estradiol	0.4
Diclofenac	100

(Data sourced from http://ec.europa.eu/smart-regulation/impact/ia_carried_out/docs/ia_2012/com_2011_0876_en.pdf, accessed 09.19.15)

means that the monitored contaminants are stigmatized and subsequently phased out of production. Such a move would probably not be beneficial to healthcare given the benefits of these three drugs in pharmacotherapy (Johnson et al., 2013). After much consultation, debate, and compromise, the EP finally rejected the classification of these three pharmaceuticals as priority substances. However, the proposal for a new "watch list" of potentially detrimental substances came into force on 28 November 2012 and the three pharmaceuticals were added to this list. This outcome means that EQSs will only be set if there is more evidence that they are significantly detrimental to aquatic and human life (Johnson et al., 2013).

While the WFD is the main European legislation relating to water quality, there are also other directives pertinent to water quality and waste discharge in the EU (Fürhacker, 2008). These include:

The Drinking Water Directive (DWD) (Directive 98/93 EC)
The Bathing Water Directive (Directive 2006/7/EC)
The Groundwater Directive (2006/118/EC)
The Urban Wastewater Treatment Directive (Directive 91/271/EEC)

The Drinking Water Directive has provisions to ensure clean water in EU countries and the monitoring of 48 microbial and chemical parameters. The Bathing Water Directive has provisions to ensure the minimization of microbial contamination in waterways while the Groundwater Directive monitors the input of pollutants into ground water. The major purpose of the Urban Wastewater Treatment Directive is much more specific—it ensures that sewage treatment facilities are capable of treating water to a standard that prevents oxygen depletion in waterways arising from the discharge of poorly treated effluent. The EU also has a Dangerous Substances Directive (76/464/EEC) which concerns the monitoring of pollution associated with highly toxic industrial chemicals and heavy metals.

Within the EU, individual member countries also have acts and regulations pertaining to water quality specific to their country. In the Dutch Drinking Water Act, all anthropogenic chemicals and organic contaminants are allowable at levels $\leq 1\ \mu g/L$. However, there are no maximum contaminant limits pertaining to specific pharmaceuticals (available http://www.rivm.nl/en/Documents_and_publications/Scientific/Reports/2013/januari/The_quality_of_drinking_water_in_the_Netherlands_in_2011, accessed 10.09.14).

Even though there are no official guidelines for the maximum concentrations of pharmaceuticals that can be discharged into European waterways, provisional *guideline value* (GLV)s have been developed for maximum concentrations in surface, ground, and drinking water based on available toxicological data (Schriks et al., 2010). These GLVs are defined in terms of the World Health Organization (WHO) drinking water guideline, namely "the concentration of a constituent pharmaceutical that does not exceed tolerable risk to the health of the consumer over a lifetime" (Schriks et al., 2010). The relevant toxicological parameters used in the calculations are based on the tolerance daily intake, *acceptable daily intake* (ADI), or reference dose value for a given pharmaceutical as recommended by an international organization (such as WHO). This calculated GLV is then compared with the concentration data available from the Dutch Association of River Water Companies (RIWA) and the German Association of River Water Companies (ARW). A *benchmark quotient* (BQ) is then calculated for a given pharmaceutical using the equation (Schirmer et al., 2011):

$$\text{BQ} = \text{Measured concentration in water} / \text{Guideline value}$$

A $\text{BQ} \geq 1$ suggests that the pharmaceutical may be of potential health concern if it is consumed over a period of a lifetime. If a BQ value was determined to be ≥ 0.1 in drinking water, then the pharmaceutical would warrant further investigation to determine whether there is a significant risk to human health. A calculation of $\text{BQ} \leq 0.2$ in ground water/surface water and ≤ 0.1 in drinking water suggests that there are no concerns to human health.

From the Dutch and German occurrence data, the GLV and BQ values for aspirin (acetylsalicylic acid), carbamazepine, clofibric acid, metoprolol, phenazone, and sulfamethoxazole were evaluated for ground water/surface water and drinking water (Table 7.7).

The only compound listed in Table 7.7 which may warrant any public health concern is carbamazepine in surface water (BQ=0.2). Even then, there is no evidence to suggest that its presence is a concern to human health. The presence of all other pharmaceuticals in surface, ground, and drinking water were considered to be well within safe levels.

Table 7.7 Guideline (GLV) and benchmark quotient (BQ) values for pharmaceuticals in surface, ground, and drinking water estimated from Dutch and German occurrence data

		GLV	BQ	BQ
Pharmaceutical	**Therapeutic class**	**Drinking water (μg/L)**	**Ground water/ surface water**	**Drinking water**
Aspirin	NSAID	25	0.003	0.005
Carbamazepine	Antiepileptic	1	0.2	0.03
Clofibric acid	Lipid regulator	30	0.003	0.005
Metoprolol	Antihypertensive	50	0.004	0.04
Phenazone	Analgesic	125	0.0009	0.0002
Sulfamethoxazole	Antibacterial agent	440	0.0003	0.00007

(From Schriks et al., 2010)

However, the GLV and BQ values were based very much on variable toxicological evidence while the occurrence data were sourced from only two countries. As such, the applicability of the guidelines to environmental situations in other countries is questionable (Schirmer et al., 2011). Therefore, it is important for the water industry in the near future to set evidence-based guidelines that "clearly define and rigorously adhere to commonly agreeable toxicological data sets" because without these, a clear risk assessment to human health will always be inconclusive or incomplete (Schirmer et al., 2011; Scrimshaw, 2011).

In the Netherlands, Mons et al. (2013) reported on some provisional drinking water limits (DWLs) based on the ADI value. The maximum observed concentrations in drinking water for a range of pharmaceuticals were used to estimate the accumulative amount of drug one would ingest in an average lifetime (70 years, I_{70}) and are very similar to the provisional GLVs reported by Schriks et al. (2010) (Table 7.8).

These data indicate that the amount of drug accumulated by the body over a lifetime is a small fraction of the therapeutic dose recommended for any of these pharmaceuticals. It is, therefore, not expected that a lifetime consumption of one of these pharmaceuticals through consumption of drinking water will result in *any* therapeutic effects from this mode of ingestion. Despite these data, the uncertainty associated with toxicological and occurrence parameters does not give consumers the confidence that there will be no adverse effects arising from the consumption of a "cocktail" mixture of pharmaceuticals (Chapter 5). Mons et al. (2013) have suggested that a baseline value of the threshold of toxicological risk for particular groups of emerging contaminants (including pharmaceuticals) be used as guidelines for the acceptable level of these contaminants in drinking water. Such an approach has been successfully used by the food industry for food contaminants (such as acrylamide and benzene) and because the *sum* of particular groups of organic contaminants can be used in the calculation of a guideline (unlike the CCL approach in the United States, *see* above), this can be undertaken using incomplete toxicological data. Target values of 0.1 and 1.0 μg/L for nongenotoxic and nonsteroid compounds (which include most pharmaceuticals) were set for the individual and collected concentrations, respectively, of substances in drinking water.

Table 7.8 Safe drinking water limit (DWL), accumulative lifetime dose (I_{70}), therapeutic dose (TD), and the proportion of therapeutic dose (I_{70}/TD (%)) accumulated over a lifetime for a range of pharmaceuticals used in the Netherlands

Pharmaceutical	DWL (μg/L)	I_{70} (mg)	TD (mg/day)	I_{70}/TD (%)
Aspirin	25	6.2	20	30
Bezafibrate	35	1.0	67	1.5
Carbamazepine	50	4.6	100	5.0
Clofibric acid	30	6.9	1200	0.6
Diclofenac	7.5	0.9	15	6.0
Fenofibrate	50	1.1	100	1.1
Fluoxetine	10	0.5	20	2.5
Ibuprofen	150	1.4	300	0.5
Lincomycin	30	1.1	1200	0.1
Metoprolol	50	1.3	100	1.3
Paracetamol	150	1.7	1200	0.15
Phenazone	125	1.5	250	0.6
Sulfamethoxazole	75	2.0	2000	0.1

Adapted from Mons et al. (2013).

7.4 China and other parts of Asia

Various acts and regulations exist in China for the protection of waterways at the national and local level, including:

EQSs for surface water (GB3838-2002)
Integrated wastewater discharge standard (GB8978-1996)
Discharge standard of pollutants for municipal wastewater treatment plants (GB18918-2002)
Various acts and regulations on discharge standards for industries (pulp, chemical, etc.)

The environmental water quality for surface and ground waters is separated into Classes I–V (EQS for surface water (GB3838-2002)). Class I water is the quality standard attributed to pristine water sources from protected reserves. Class II water is relatively pristine water from fish spawning areas and water reservoirs; however, the regulatory control on emissions is less stringent than that for Class I. Class III water describes areas of water suitable for direct human contact (i.e., recreational water for swimming and bathing). Class IV water is industrial and recreational water not fit for bathing or swimming and Class V is the quality describing agricultural water only suitable for irrigation and is substantially less pristine than Class IV water. Polluted water is considered below Class V quality.

Many of China's rivers are polluted and 45% of the total river water is considered unfit for human consumption (available http://chinawaterrisk.org/wp-content/uploads/2011/06/Chinas-Water-Crisis-Part-2.pdf, accessed 10-9-14). Some major pollutants derived from the discharge of sewage and industrial waste include organic matter, heavy metals, and petroleum. An assessment of the quality of some of China's major waterways is shown in Table 7.9.

Table 7.9 Surface water quality of some major Chinese waterways

River	Proportion of river length (%)					Worse than Class V
	Class I	Class II	Class III	Class IV	Class V	
Songhua	0.0	0.0	4.0	67.0	21.0	8.0
Liaohe	4.5	2.3	4.5	22.7	4.5	61.7
Haihe	5.0	19.0	4.0	10.0	9.0	53.0
Huaihe	0.0	11.0	17.0	18.0	6.0	48.0
Yellow	0.0	24.0	5.0	47.0	12.0	12.0
Yangtze	4.0	67.0	4.0	11.0	10.0	4.0
Pearl	29.0	36.0	7.0	22.0	2.0	4.0
Total	8.5	21.7	6.7	18.3	7.1	37.7

Adapted from Ouyang and Wang (2000) (available http://www.biotechnology.uni-koeln.de/inco2-dev/common/contribs/03_ouyan.pdf, accessed 12.11.13).

For all these classes of water bodies, the quality standards (available http://kjs.mep.gov.cn/hjbhbz/bzwb/shjbh/shjzlbz/200206/t20020601_66497.html, accessed 11-4-14) list a fundamental set of requirements which must be adhered to including that they should not contain:

(a) organic sediments from anthropogenic sources;
(b) floating materials originating from organic pollutants such as oil or slag;
(c) substances which may affect the natural odor or color of the water;
(d) substances that are harmful, toxic, or have negative physical effects on human beings, animals, and plants;
(e) substances that promote the growth of unwanted aquatic species which decreases water quality.

While these are broad categories for consideration and pharmaceuticals are not specifically mentioned as substances which may be harmful, toxic, or have negative effects on humans and aquatic species, their presence in Chinese-classified waterways certainly poses concern under category (d) above.

In China, human health risk assessments for new pharmaceuticals are conducted under the requirements of the Pharmaceutical Administration Law, but no environmental risk assessments are currently necessary in order to market a new chemical compound in the country (Liu and Wong, 2013). In addition, while there are environmental management regulations for hazardous chemicals, including the recently introduced Environmental Management Method for New Chemicals (2010) and Environmental Management and Registration Method for Hazardous Chemicals (2013), there is still a lack of environmental monitoring of pharmaceuticals under any of these regulations (Liu and Wong, 2013).

China has one of the largest economies in the world with a major contribution from industry involved in the manufacture of pharmaceuticals. Limits on the resources used for industrial processes are therefore regulated. The Integrated Water Discharge Standard (GB8978-1996) (available http://www.shinemem.com/uploads/soft/26%20Integrated%20Wastewater%20Discharge%20Standard%20GB8978-1996.pdf,

Table 7.10 Maximum environmental water discharge volumes in China arising from the production of various pharmaceuticals (from the Integrated Water Discharge Standard (GB8978-1996)

Pharmaceutical	Maximum water discharge (m^3/t)
Penicillin	4,700
Streptomycin	1,450
Riomistin	1,300
Neotetrine (tetracycline)	1,900
Lincomycin	9,200
Aureomycin	3,000
Gentamicin	20,400
Chloromycetin (chloramphenicol)	2,700
Sulfamethoxazole	2,000
Methamizole	180
Phenacetin	750
Furoxone	2,400
Caffeine	1,200

accessed 11-11-13) lists maximum volumes of water that can be discharged by various industries, including the water utilized for the production (per ton) of particular pharmaceuticals (Table 7.10).

The discharge of wastewater arising from drug manufacture is managed by the Ministry of Environmental Protection (Liu and Wong, 2013). While no particular guidelines pertain to the maximum allowable concentrations that can be discharged into the environment from sewage treatment facilities, there is an agreed guideline on the chemical oxygen demand (COD) quality of the water discharged from pharmaceutical and other industries in Classes I, II, III, IV, or V environments. The maximum COD limit is set at <15, <15, 15, 20, and 25 mg/L for pharmaceutical water discharge into Classes I, II, III, IV, and V environments, respectively.

While there appears to be few guidelines which govern the maximum discharge limits for pharmaceuticals in surface and drinking water in any Asian country, attempts have been made by some research groups to quantify safe limits and the risk posed by the environmental presence of specific pharmaceuticals. Toxicological data have been used to develop a DWEL to estimate of the safe amount of a pharmaceutical that could be ingested by various age groups (Leung et al., 2013) and some of these values are listed in Table 7.11.

All DWEL values are well above their average measured levels in drinking water with the exception of dimetridazole which suggests that this particular pharmaceutical has the potential for significant environmental risk at least to wildlife.

A further assessment of potential adverse health effects arising from ingestion of a particular pharmaceutical as presented by Leung et al. (2013) is provided by the classification of risk quotients (RQs) for the various DWEL "life stages" into three risk groups (Table 7.12).

Table 7.11 Drinking water equivalent level (DWEL) ranges for various age groups (from Leung et al., 2013)

Pharmaceutical	Therapeutic class	DWEL range (μg/L, various ages)
Trimethoprim	Antibacterial agent	20–140
Metoprolol	Antihypertensive	68–450
Clofibric acid	Lipid regulator	49–320
Caffeine	Stimulant	73–484
Carbamazepine	Antiepileptic	1.7–11
Diclofenac	NSAID/analgesic	327–2,160
Salicylic acid	NSAID	127–839
Clarithromycin	Antibacterial agent	0.98–6.5
Roxithromycin	Antibacterial agent	2.0–12.9
Azithromycin	Antibacterial agent	8.2–55.0
Tylosin	Antibacterial agent	4.1–27.4
Sulfathiazole	Antibacterial agent	244–1,613
Sulfamethazine	Antibacterial agent	7.8–51.1
Sulfamethoxazole	Antibacterial agent	634–4,194
Thiamphenicol	Antibacterial agent	4.4–29.0
Dimetridazole	Antibacterial agent	0.03–0.2
Metronidazole	Antibacterial agent	2.9–19.0

Table 7.12 Pharmaceutical risk groups defined by risk quotients (RQs) evaluated for Chinese drinking water sources (Leung et al., 2013)

Risk quotient (RQ)[a]	Group no.	Definition
At least one life-stage ≥ 0.01	1	Potentially risky to health
0.01 > all life stage RQs ≥ 0.0001	2	Little evidence for risk to health
Most life stage RQs < 0.0001	3	No evidence to suggest risk to health

[a] For a definition of risk quotient, *see* Chapter 5. The definition of a life-stage in Leung et al. (2013) is whether a person is an infant, child, or adult.

Most pharmaceuticals were classified in group 3 or group 2 which suggested little risk to health; however, dimetridazole, thiamphenicol, sulfamethazine, and clarithromycin were classified in group 1 and hence considered potentially detrimental to human health. However, whether these potential risks are likely to be a concern or not is unknown.

In contrast to China, Japan has an extensive regulatory system which monitors the discharge of wastewater from households and industries. There are several main pieces of legislation maintaining the quality of Japanese waterways and also regulations limiting the discharge of contaminants into the environment:

Basic Environment Law (and subsequent amendments and revisions to incorporate water standards) (available http://www.env.go.jp/en/laws/policy/basic/, accessed 11-9-14);

The Water Pollution Control Law (1970) (available http://www.wepa-db.net/policies/law/japan/wpctop.htm, accessed 11-9-14);

EQSs for Human Health and the Living Environment (available http://www.emecs.or.jp/01cd-rom/section_3_e/3top_a_ro_b_e.html, accessed 11-9-14).

The Water Pollution Control Law regulates effluent discharge from households, factories, and other industrial processes and sets maximum discharge limits whereas the EQSs set contaminant guidelines for monitoring the quality of waterways and lakes. In addition, the "Nation-wide Effluent Water Quality Standards for the Conservation of the Living Environment" has predetermined limits on a number of heavy metal and organic contaminants while in local prefectures a separate set of limits, which are usually more stringent than the national standards, are enforced. For example, this legislation sets a maximum daily COD and BOD concentration that can be discharged from wastewater as 160 mg/L and the loading of suspended solids (SS) as 200 mg/L (available http://www.wepa-db.net/policies/law/japan/standards7.htm, accessed 11-9-14). However, in the Shiga Prefecture, the COD and BOD must be $\leq$120 mg/L and loading of SS $\leq$150 mg/L (available http://www.wepa-db.net/policies/law/japan/wpc1.htm, accessed 11-9-14). No pharmaceuticals have been specifically monitored for either of these parameters in effluent discharge or in EQSs.

Effluent discharge is also regulated in other South East Asian countries, such as Singapore. However, like China and Japan, only the discharge of certain priority contaminants (pesticides, organophosphates, and heavy metals) is monitored and these seemingly do not include any pharmaceuticals (Singapore Effluent Discharge Standards; available http://www.water-treatment.com.cn/resources/discharge-standards/singapore.htm, accessed 12-11-13). The discharge standards for trade effluents include data for 30 organic chemicals for which the environmental discharge is completely prohibited, but this does not include any pharmaceuticals. However, there is a clause in the Effluent Discharge Standards for the monitoring of substances that could potentially encompass pharmaceuticals:

> *"... [a substance] that either by itself or in combination or by reaction with other waste or refuse may give rise to any gas, fume, odor or substance which is or is likely to be a hazard to human life, a public nuisance, injurious or otherwise objectionable, or which prevents or is likely to prevent entry into the public sewer by workmen maintaining or repairing it."*

Pharmaceuticals are considered unlikely to be present at concentrations which are objectionable or cause immediate human risk in Singapore; however, the ubiquitous lack of toxicological data means such conclusions cannot be drawn with any confidence. For example, a recent study of pharmaceutical concentrations in raw wastewater, surface water, and ground water in a Singaporean catchment area recorded typical environmental concentrations of the order of ng/L and raw sewage concentrations of μg/L (Toor et al., 2011).

7.5 Australia

Australia does not have any specific federal regulations pertaining to the maximum concentration of any particular pharmaceutical that has been discharged into the environment. However, many states in Australia have their own set of guidelines for the monitoring of effluent from point sources (such as the *Tasmanian Emission Limit Guidelines for Sewage Treatment Plants that Discharge Pollutants into Fresh and Marine Waters* published by the Tasmanian Department of Primary Industries, Water and Environment, available http://epa.tas.gov.au/documents/emission_limit_guidelines_june_2001.pdf, accessed 1-9-14). The National Water Quality Management Strategy—Australian guidelines for sewerage systems—Effluent Management (1997) (available http://www.environment.gov.au/system/files/resources/e52e452b-a821-4abe-9987-23988790e353/files/sewerage-systems-effluent-man-paper11.pdf, accessed 14-11-13) does have a section on toxicants and issues related to their environmental discharge in sewage effluents. Toxicants are defined generically as substances which can influence human health, "either as acute or chronic effects." Chronic effects are only considered by the guidelines to be of a concern when they:

(a) persist in the aquatic environment;

(b) are bioconcentrated several thousand-fold;

(c) exert a toxic effect after prolonged exposure to low concentrations of the toxic constituent.

While no maximum environmental discharge limits have been set for pharmaceuticals, they can be defined as toxicants as some pharmaceuticals can potentially "exert a toxic effect [at] low concentrations." The guidelines only recommend controls for toxicants when acute effects are encountered. Trigger values for toxicants based on No Observed Effect Concentrations (*see* Chapter 5) are presented in the Australian Water Quality Guidelines for Fresh and Marine Waters (available http://www.environment.gov.au/system/files/resources/53cda9ea-7ec2-49d4-af29-d1dde09e96ef/files/nwqms-guidelines-4-vol1.pdf, accessed 14-11-13); however, the list of trigger values does not include pharmaceuticals.

The Australian Guidelines for Drinking Water Recycling—Augmentation of drinking water supplies (2008) published by the Australian Environment Protection and Heritage Council (available http://www.stanford.edu/group/narratives/classes/08-09/CEE215/Projects/greendorm/water/GraywaterCD/AussieWatRecycl08/200805_WQ_GL_Final_AGWR_ADWS.pdf, accessed 7-11-13) lists acceptable GLVs for pharmaceuticals in drinking water guidelines (DWGs) (see Tables 7.13 and 7.14).

7.6 New Zealand

The Resource Management Act (RMA, 1991) and its subsequent revisions control all human activities associated with the New Zealand environment (available http://www.legislation.govt.nz/act/public/1991/0069/latest/DLM230265.html, accessed 29–14). Particularly pertinent to the environmental discharge of contaminants is Section 15, which states:

Table 7.13 Acceptable daily intakes (ADIs) and recommended Australian safe drinking water guidelines (DWGs) for pharmaceuticals

Pharmaceutical	ADI (µg/kg/day)	DWG (µg/L)
Amoxicillin	0.43	1.5
Anhydro-erythromycin A	5	17.5
Azithromycin	11	3.9
Cephalexin	10	35
Chlorotetracycline	30	105
Doxycycline	3	10.5
Enrofloxacin	6.2	22
Erythromycin	5	17.5
Lincomycin	1,000	3,500
Monensin	10	35
Penicillin G	0.43	1.5
Penicillin V	0.43	1.5
Sulfadimethoxine	10	35
Sulfamethazine	10	35
Sulfamethiazole	10	35
Sulfamethoxazole	10	35
Oxytetracycline	30	105
Tetracycline	30	105
Trimethoprim	20	70
Tylosin	300	1,050
Aspirin	8.3	29
Diclofenac	0.5	1.8
Dipyrone	150	525
Ketoprofen	1	3.5
Tolfenamic acid	5	17.5
Carazolol	0.1	0.35
17α-Estradiol	–	0.175[a]
17β-Estradiol	0.05	0.175
Progesterone	30	105
Testosterone	2	7.0
Clenbuterol	4.2	15
Paracetamol	50	175

[a] No ADI information, assumed similarity to 17β-estradiol.
Adapted from Australian Drinking Water Recycling Guidelines (2008).

Table 7.14 **Lowest daily therapeutic dose (LDTD) and recommended Australian drinking water guidelines (DWG) for human pharmaceuticals**

Pharmaceutical	LDTD (mg/day)	DWG (μg/L)
Cefaclor	500	250
Chloramphenicol	3,500	175
Ciprofloxacin	500	250
Clarithromycin	500	250
Clindamycin	600	300
Demeclocycline	600	300
Nalidixic acid	2,000	1000
Norfloxacin	800	400
Roxithromycin	300	150
Fenoprofen	900	450
Ibuprofen	800	400
Indomethacin	50	25
Naproxen	440	220
Betaxolol	20	10
Bisoprolol	1.25	0.63
Metoprolol	50	25
Nadolol	40	20
Propranolol	80	40
Timolol	20	10
17α-Ethinylestradiol	0.03	0.0015
Estriol	1	0.05
Estrone	0.6	0.03
Mestranol	0.05	0.0025
Norethindrone	5	0.25
Alprazolam	0.5	0.25
Antipyrine	2,000	1000
Atorvastatin	10	5
Bezafibrate	600	300
Carbamazepine	200	100
Cimetidine	400	200
Clofibric acid	1,500	750
Codeine	100	50
Cyclophosphamide	70	3.5
Dehydro nifedipine	40	20
Diazepam	5	2.5
Diltiazem	120	60
Enaloprilat	2.5	1.25
Fluoxetine	20	10
Gemfibrozil	1,200	600
Isophosphamide	70	3.5
Metformin	500	250
Salbutamol	6	3
Sulfasalazine	1,000	500
Temazepam	10	5
Terbutaline	9	4.5

Adapted from Australian Drinking Water Recycling Guidelines (2008).

(1) No person may discharge any—

(a) *contaminant or water into water; or*
(b) *contaminant onto or into land in circumstances which may result in that contaminant (or any other contaminant emanating as a result of natural processes from that contaminant) entering water; or*
(c) *contaminant from any industrial or trade premises into air; or*
(d) *contaminant from any industrial or trade premises onto or into land—*
unless the discharge is expressly allowed by a national environmental standard (NES) or other regulations, a rule in a regional plan as well as a rule in a proposed regional plan for the same region (if there is one), or a resource consent.

The RMA defines a contaminant to be (Section 2):

"...any substance (including gases, odorous compounds, liquids, solids, and microorganisms) or energy (excluding noise) or heat, that either by itself or in combination with the same, similar, or other substances, energy, or heat—

(a) *when discharged into water, changes or is likely to change the physical, chemical, or biological condition of water; or*
(b) *when discharged onto or into land or into air, changes or is likely to change the physical, chemical, or biological condition of the land or air onto or into which it is discharged."*

Therefore, emerging contaminants of significance, such as pharmaceuticals, must be addressed under the RMA if they are discharged into the environment at concentrations that will induce physical, chemical, or biological change.

To date, the chemical parameters specified for the monitoring of treated effluent in resource consents issued for waste water treatment in New Zealand under the RMA Act have mainly concerned the maximum allowed organic content as specified by the biochemical oxygen demand (BOD), the maximum loading of suspeneded solids (SS), and, in more recent times, maximum concentrations of dissolved nutrients such as phosphate and nitrogen species. However, should the consenting authorities become aware of maximum levels of one or more human pharmaceuticals or their metabolites to avoid unacceptable ecotoxicological effects in the receiving environment arising from treated effluent discharge or even discharge to land, then extra conditions could be added to the appropriate regional plans and resource consent(s) to minimize this environmental risk.

The RMA is administered by the regional councils covering the whole of New Zealand. These councils issue resource consents for the discharge of treated water to ensure compliance with any NESs, regional policies and rules, Section 15 of the RMA (in cases where there is no regional rule), and Section 107 of the RMA which is particularly relevant for real or potential contaminants such as pharmaceuticals:

"(1) Except as provided in subsection (2), a consent authority shall not grant a discharge permit or a coastal permit to do something that would otherwise contravene Section 15 [or Section 15A] allowing—

(a) *The discharge of a contaminant or water into water; or*
(b) *A discharge of a contaminant onto or into land in circumstances which may result in that contaminant (or any other contaminant emanating as a result of natural*

processes from that contaminant) entering water; or if, after reasonable mixing, the contaminant or water discharged (either by itself or in combination with the same, similar or other contaminants or water), is likely to give rise to all or any of the following effects in the receiving waters;

(c) *The rendering of fresh water unsuitable for consumption by farm animals;*
(d) *Any significant adverse effects on aquatic life…"*

One aspect of this section of the RMA pertaining to the environmental discharge of any type of water is particularly important:

How does a consent authority define a "reasonable mixing zone"?

This issue has been considered in some detail by Selvarajah (2011) for the Otago province particularly involving discharge of waste water into streams and rivers:

> *"It has been perceived by most RMA practitioners that a mixing zone shall always be provided as a 'noncompliance zone.' In the past, there have been attempts made by technocrats and bureaucrats to define an acceptable mixing zone. Many consultants still require or recommend the regional councils to grant long mixing zones (which are typically several hundred metres)."*

The Otago Regional Council's Regional Plan: Water (Water Plan) (available http://www.orc.govt.nz/Publications-and-Reports/Regional-Policies-and-Plans/Regional-Plan-Water/, accessed 24.9.14) incorporates a comprehensive policy on mixing zones. The Water Plan Policy 7.7.6 states:

> *"…where a mixing zone is required for the discharge of contaminants to water, [it must] take account of:*
>
> ***(a)*** *The sensitivity of the receiving environment;*
> ***(b)*** *The natural and human use values identified in Schedule 1 [of the RMA];*
> ***(c)*** *The natural character of the water body;*
> ***(d)*** *The amenity values supported by the water body;*
> ***(e)*** *The physical process acting on the area of discharge; and*
> ***(f)*** *The particular discharge, including contaminant type, concentration, and volume…"*

Notwithstanding the Water Plan policy on the specification of a reasonable mixing zone, the legal advice obtained on this matter by the Otago Regional Council emphasizes that a consent authority could set higher discharge requirements than provided in Section 107 of the RMA. In other words, if the Council chooses not to provide for a noncompliance zone in a resource consent that it is issuing, such a practice will not breach Section 107 of the RMA. Furthermore, if a Water Plan policy requires a waterway to be managed for a particular use (e.g., contact recreation which may result in some form of human contact with water such as swimming, fishing, or boating), it is assumed that the whole of the waterway is accessible to the community for contact recreation rather than only some parts.

As New Zealand regional councils have the authority to regulate the details of mixing zones in the resource consents under their jurisdiction, it means that the level of a given pharmaceutical discharged into any New Zealand water body (river, lake, estuarine, coastal area) could potentially be controlled by specifying in the resource consent a specific maximum contaminant level for any discharged effluent at any stage of the consent activity.

If a mixing zone is permitted, then obviously the concentrations of any pharmaceutical will be greater in that zone than outside it. The issue then becomes how to assess whether there are any adverse effects of the discharge on aquatic organisms likely to occur within this permitted mixing zone? There have been very few reported studies on the presence or persistence of pharmaceuticals in surface water, estuarine environments, rivers, lakes, or coastal areas in New Zealand water bodies; thus, the knowledge and understanding of any detrimental effects of even traces of pharmaceuticals in these mixing zones at least within New Zealand is limited. Also, it is not currently possible to determine if there are "any significant adverse effects on aquatic life" arising from the postulated discharge of pharmaceuticals from sewage effluent, given that there are no reports of direct ecotoxicological measurements determined for any New Zealand aquatic organism.

The lack of conclusive evidence that pharmaceuticals pose a significant environmental risk in New Zealand may also explain why there is a lack of direction in any regional plans or resource consent conditions concerning an acceptable water quality pertaining to the environmental discharge of pharmaceuticals. As such, any official methodologies to monitor pharmaceuticals in New Zealand water (which can specify the nature and frequency of sampling) and test for their environmental effects on aquatic life have yet to be established. In setting such monitoring regimes under the RMA, one has to also ask how practical such monitoring might be. Given the sensitivity and sophisticated methodologies required for detection (liquid chromatography-tandem mass spectrometry (LC–MS/MS), Chapter 4), it may well be that the consenting authorities such as a regional council could be convinced by applicants that it is currently too expensive and hence not practical to undertake such analyses as part of the monitoring conditions specified in a regional plan or resource consent.

7.7 Challenges and conclusions

The detection and quantification of selected pharmaceuticals in the environment has led to widespread debate pertaining to whether these compounds should be environmentally monitored and regulated. It can be argued that the regulation of maximum environmental discharge limits for pharmaceutical compounds is pointless, impractical, and not cost-effective. Current evidence around the world suggests that the ingestion of drinking water over a lifetime would contribute to only a small fraction of a total therapeutic dose for all pharmaceuticals that have been detected and quantified to date in drinking water. In addition, ADI values for many individual pharmaceuticals, derived from human and animal toxicological data, are up to a thousand-fold higher than the predicted concentrations in sewage, surface, lake, estuarine, or other surface water environments and/or drinking water. Because pharmaceuticals are detected in low concentrations, sophisticated instrumentation (Chapter 4) is usually necessary to detect and quantify them and this makes their regular monitoring a costly exercise in many situations. As previously discussed, even modern analytical instrumentation would probably struggle to detect the presence of 17α-ethinylestradiol at 0.035 ng/L even though this is the EQS concentration proposed by the EU (Johnson et al., 2013). Furthermore, it is still unclear if these low concentrations are significantly detrimental to aquatic life. Because of incomplete evidence on their detrimental effects, it has been difficult to set any guidelines for their environmental discharge, let alone regulate them anywhere around the world.

On the other hand, the use of pharmaceuticals in developed and developing countries is on the rise and new therapeutic agents are registered on the market every year. We do not know whether there will be another ecological disaster associated with the use of a new but environmentally hazardous pharmaceutical, such as that experienced with vultures in Pakistan (Chapter 1). Another question concerns mixture and additive pharmacological effects arising from the subtherapeutic concentrations of multiple pharmaceuticals in a water body (Chapter 5). This idea would make any discharge regulations for *single* pharmaceuticals a moot point. Also, given the analytical challenges, it is not possible even for a multiresidue method to detect, let alone quantify, *all* the pharmaceuticals in a given body of sewage and hence reliably assess its absolute risk to the human or aquatic organism. As explained in Chapter 4, liquid-chromatographic procedures currently employed for the detection of pharmaceuticals in sewage or surface water can only detect the substances that the instrumentation is "trained" to look for (i.e., from LC–MS/MS optimization procedures and the use of analytical standards to confirm positive detection). As well as the expense, method development, and validation for the instrumental detection and quantitation of new pharmaceuticals take considerable amounts of and so research and development into analytical capabilities for the detection of new pharmaceuticals in environmental matrices are always one step behind the development and marketing of these substances for therapeutic use.

The challenge for legislators around the globe is therefore to become adequately aware of the levels of human pharmaceuticals that have been shown to be of real (or potential) concern for the health of the environment and to be able to regulate their use through enactment of appropriate legislation in different countries.

References

Breton, R., Boxall, A., 2003. Pharmaceuticals and personal care products in the environment: regulatory drivers and research needs. QSAR Comb. Sci. 22 (3), 399–409.

Bruce, G.M., Pleus, R.C., Snyder, S.A., 2010. Toxicological relevance of pharmaceuticals in drinking water. Environ. Sci. Technol. 44 (14), 5619–5626.

Conerly, O., Ohanian, E.V., 2010. Pharmaceuticals in the environment: regulatory challenges ahead. Hum. Ecol. Risk Assess. Int. J. 16 (6), 1234–1241.

Fuerhacker, M., 2009. EU water framework directive and Stockholm convention. Environ. Sci. Pollut. Res. 16 (1), 92–97.

Fürhacker, M., 2008. The water framework directive—can we reach the target? Water Sci. Technol. 57 (1), 9–17.

Johnson, A.C., Dumont, E., Williams, R.J., Oldenkamp, R., Cisowska, I., Sumpter, J.P., 2013. Do concentrations of ethinylestradiol, estradiol, and diclofenac in European rivers exceed proposed EU environmental quality standards? Environ. Sci. Technol. 47 (21), 12297–12304.

Leung, H.W., Jin, L., Wei, S., Tsui, M.M.P., Zhou, B.S., Jiao, L.P., Cheung, P.C., Chun, Y.K., Murphy, M.B., Lam, P.K.S., 2013. Pharmaceuticals in tap water: human health risk assessment and proposed monitoring framework in China. Environ. Health Perspect. 121 (7), 839–846.

Liu, J.-L., Wong, M.-H., 2013. Pharmaceuticals and personal care products (PPCPs): a review on environmental contamination in China. Environ. Int. 59, 208–224.

Metcalfe, C.D., Chu, S.G., Judt, C., Li, H.X., Oakes, K.D., Servos, M.R., Andrews, D.M., 2010. Antidepressants and their metabolites in municipal wastewater, and downstream exposure in an urban watershed. Environ. Toxicol. Chem. 29 (1), 79–89.

Miao, X.S., Bishay, F., Chen, M., Metcalfe, C.D., 2004. Occurrence of antimicrobials in the final effluents of wastewater treatment plants in Canada. Environ. Sci. Technol. 38 (13), 3533–3541.

Mons, M.N., Heringa, M.B., van Genderen, J., Puijker, L.M., Brand, W., van Leeuwen, C.J., Stoks, P., van der Hoek, J.P., van der Kooij, D., 2013. Use of the threshold of toxicological concern (TTC) approach for deriving target values for drinking water contaminants. Water Res. 47 (4), 1666–1678.

Ouyang, Z., Wang, R., 2000. Water environmental problems and ecological options in China. In: Ecosystem Service and Sustainable Watershed Management in North China International Conference. Available from: http://www.biotechnology.uni-koeln.de/inco2-dev/common/contribs/03_ouyan.pdf (accessed 12.11.13).

Schirmer, M., Martienssen, M., Schirmer, K., 2011. Comment on Schriks, M., Heringa, M.B., van der Kooi, M.M.E., de Voogt, P., van Wezel, A.P., 2010. Toxicological relevance of emerging contaminants for drinking water quality. Water Res. 44, 461–476; 45 (3), 1512–1514.

Schriks, M., Heringa, M.B., van der Kooi, M.M.E., de Voogt, P., van Wezel, A.P., 2010. Toxicological relevance of emerging contaminants for drinking water quality. Water Res. 44 (2), 461–476.

Scrimshaw, M.D., 2011. Comment on "Toxicological relevance of emerging contaminants for drinking water quality" by M. Schriks, M.B. Heringa, M.M.E. van der Kooi, P. de Voogt and A.P. van Wezel [Water Res. 44 (2010) 461–476]. Water Res. 45 (9), 3012–3013.

Selvarajah, S., 2011. Key wastewater discharge consents granted in the Otago Region in the past 10 years. Unpublished Report Presented to the Regulatory Committee of the Otago Regional Council on 31 August 2011.

Snyder, S.A., Westerhoff, P., Yoon, Y., Sedlak, D.L., 2003. Pharmaceuticals, personal care products, and endocrine disruptors in water: implications for the water industry. Environ. Eng. Sci. 20 (5), 449–469.

Toor, G.S., Lusk, M., Obreza, T., 2011. Onsite Sewage Treatment and Disposal Systems: An Overview. University of Florida.

Green chemistry, green pharmacy, and life-cycle assessments

8

8.1 Green chemistry and green pharmacy

The comment that "the prevention of environmental problems associated with pharmaceuticals is better than the cure" has often been made (e.g., Kummerer, 2009; Daughton and Ruhoy, 2009). Possible preventive measures that can avoid environmental contamination resulting from pharmaceutical use and disposal include the use of "green chemistry" and related "green pharmacy" concepts in the synthesis and use of pharmaceuticals, respectively. (http://www2.epa.gov/greenchemistry; accessed 30.7.15).

8.1.1 Introduction

Green chemistry can be broadly defined (Manahan, 2009) as: "the sustainable practice of chemical science and technology within the framework of good practice of industrial ecology in a manner that is safe and non-polluting, that consumes minimum amounts of materials and energy while producing little or no waste material, that minimizes the use and handling of hazardous substances and does not release such substances to the environment." The concept of green chemistry grew out of the US EPA and its response to the "Pollution Prevention Act" 1990 which led to the American Chemical Society (ACS) symposia on "Environmentally Benign Chemical Synthesis" in 1993 and 1994 where the term "green chemistry" was widely used for the first time.

The 12 principles of green chemistry were published in the first handbook on green chemistry (Anastas and Warner, 1998) and include:

- It is better to prevent waste than to treat or clean up waste after it is formed.
- Wherever practicable, synthetic methodologies should be designed to use and generate substances that possess little or no toxicity to human health and the environment.
- Chemical products should be designed to preserve efficacy of function while reducing toxicity.
- Energy requirements should be recognized for their environmental and economic impacts and should be minimized.
- A raw material of feedstock should be renewable rather than depleting wherever technically and economically practicable.
- Catalytic reagents (as selective as possible) are superior to stoichiometric reagents.
- Chemical products should be designed so that at the end of their function, they do not persist in the environment, and break down into innocuous degradation products.

Clearly many of these principles are applicable to aspects of the environmental life-cycle of pharmaceuticals, particularly with regard to their initial molecular design and synthesis to minimize their potential environmental impact during both manufacture and use.

The Life-Cycle of Pharmaceuticals in the Environment. http://dx.doi.org/10.1016/B978-1-907568-25-1.00008-6

There appears to be no comparable widespread definition of *green pharmacy* but one could use the above broad definition of green chemistry as the basis for one: "the use of sustainable practices in the design, manufacture, prescription, and disposal of a pharmaceutical to maximize its medical efficacy, reduce its healthcare cost and minimize its impact on the environment." Clearly the sustainable design and manufacture of a pharmaceutical would be covered by the application of green chemistry principles but not its subsequent prescription, efficacy, healthcare cost, and disposal into the environment in a sustainable manner which could be considered as important differences between the two terms.

One could also devise some principles of green pharmacy analogous to those for green chemistry that might include:

- It is better to prevent pharmaceutical waste than to treat it after it has been formed.
- Wherever practicable, the design and manufacture of a pharmaceutical should be undertaken in a sustainable manner and minimize its toxicity to human health and the environment.
- Naturally occurring materials and their extracts should be used in preference to the chemical synthesis of a pharmaceutical.
- Prescription of a pharmaceutical should ensure adequate medical efficacy but minimize creation of unused material.
- Disposal of an unused pharmaceutical should be in a sustainable manner that minimizes its impact on the environment.
- A pharmaceutical should be designed so that at the end of its function, it does not persist in the environment, and can be readily broken down into innocuous degradation products in waste treatment processes.

The origins of the term "green pharmacy" are also not as obvious as those of green chemistry but one of the first uses of the term appears in 2003 by Daughton (2003a) to refer to the contents of two papers he published (Daughton 2003b,c) involving the extensive discussion of a "cradle-to-cradle" stewardship programme for the management of pharmaceuticals and personal care products (PPCPs) with the objective of "…seizing opportunities to lessen environmental impact while at the same time improving conditions for people and industry—managing all materials as perpetual resources rather than wastes." Daughton concluded that "…surprisingly little discussion has been devoted to actions that could be taken in the near-term and longer-term to eliminate the introduction of pharmaceuticals and personal care products to the environment. After all, pollution prevention (or source reduction, minimization, or elimination) embodies many advantages compared with waste treatment or environmental remediation" (2003a).

Some of the terminology associated with green chemistry and green pharmacy can be illustrated using the diagram shown in Figure 8.1 which depicts the general life-cycle of a product such as a pharmaceutical. In particular, the term "cradle" refers to the initial step involving resource extraction, "grave" refers to the final disposal step back into the environment and all intermediate steps are referred to as "gates." This terminology gives rise to expressions such as "cradle-to-gate" for the manufacture of a product from raw materials to the factory gate (i.e., before transporting to the consumer), "gate to gate" when only a small number of intervening steps and their interconnection are considered, and "cradle-to-grave" which covers all steps from resource extraction,

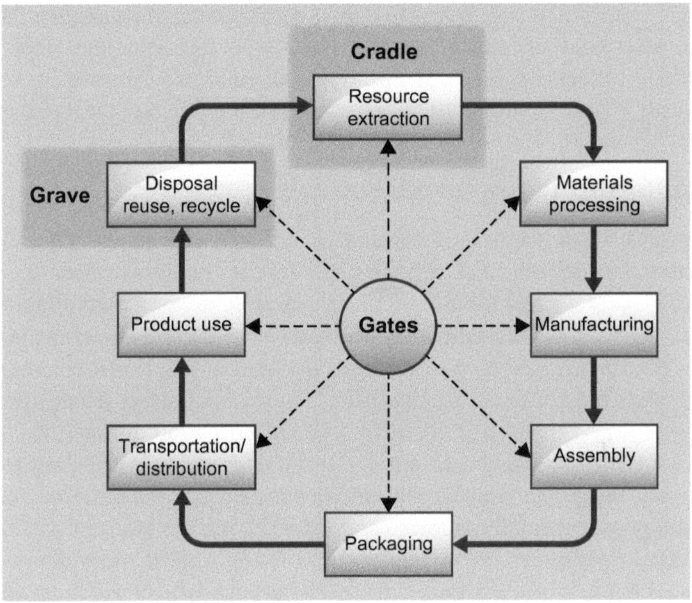

Figure 8.1 General life-cyle for a product.
(Modified from https://commons.wikimedia.org/wiki/File:Life_Cycle_Assessment_Overview.jpg, accessed 9/19/15)

manufacture, transportation, and use through to disposal. The joining up of the disposal gate with the resource extraction gate gives rise to the term "cradle-to cradle" which highlights the ultimate circular nature of this cycle. In the present text we have used a "gate-to-grave" approach by not considering any aspect of the manufacture, packaging, or distribution of pharmaceuticals, but rather just aspects of how they are used (prescribed) and disposed of along with associated environmental consequences.

The EU project on pharmaceutical products in environmental waters (Knowledge and Need Assessment on Pharmaceutical Products in Environmental Waters (KNAPPE)) undertaken over the period from 2007 to 2008 helped clarify some priority objectives in order to achieve these "green" aims. In summary, some of the key challenges that were identified as facing the pharmaceutical industry were: (1) the use of more renewable resources for the synthesis of chemicals and the use of cleaner synthetic routes for the production of active pharmaceutical ingredient (API); (2) the energy needed to synthesize them; and (3) whether they could be rapidly degraded in the environment or in the sewage treatment facility (i.e., their biodegradability) but at the same time retain their pharmacological efficacy.

Green pharmacy in principle can be practiced in the clinic by a prescribing doctor in the same way as an environmentally aware person does shopping. Just as an environmentally conscious shopper may want to look for produce which is locally produced (and thus is fresh and has not accumulated carbon miles resulting from transportation), an environmentally conscious prescriber may want to prescribe medications which have low toxicity to the patient and high efficacy, but at the same time, can be efficiently removed by conventional and advanced sewage treatment technologies (Chapter 6). To implement such green pharmacy principles would require the incorporation of environmental awareness courses in medical and pharmacy programmes, which has been tried to some degree with pharmacy students in the United States (Chapter 3).

8.1.2 Application to pharmaceuticals

The development of pharmaceuticals which are not only efficacious but also minimises their environmental impact poses a huge challenge to medicinal chemists even at the initial design stage, let alone their subsequent manufacture and marketing. In order for an active pharmacological ingredient (API) to achieve a therapeutic effect in the body, it must be absorbed and reach receptors on cells in target organ(s) before being cleared by the human body's metabolizing mechanisms. Thus, many drugs are designed to avoid metabolism in the body. This usually means the incorporation of functional groups into their chemical structure which are not easily biodegradable. Boethling et al. (2007) describes some "rules of thumb" in the effect of chemical structure on degradability potential. Halogens, especially chlorine and fluorine, are generally resistant to aerobic biodegradation and a molecule will be even more resistant if more than one halogen atom is present in the molecule. Other functional groups which also increase resistance to biodegradability include polycyclic structures and azo, nitroso, tertiary amine, nitro arylamino, imidazole, and aliphatic ether groups.

The incorporation of these groups into the chemical structure of a new pharmaceutical under development is generally done to enhance the therapeutic response of the API although they may also lessen the ultimate biodegradability of the drug. For example, the addition of fluorine atoms generally increases the potency of an API and hence leads to a reduction in the dose required to achieve the desired level of efficacy using the particular medicine (see Section 3.1 of Khetan and Collins, 2007). Subsequently, this minimizes the potential side effects from the medicine (such as the reduction in CYP450 interaction potential of the antifungal drug fluconazole, a fluorinated compound, compared to its older cousin ketoconazole when taken orally).

The presence of fluorine on the molecule also often means it is not easily degraded or removed from aquatic environments. For example, as noted in Chapter 5, the fluorinated selective serotonin reuptake inhibitor (SSRI) antidepressant fluoxetine has been shown to be resistant to both hydrolysis and photolysis over 30 days in aqueous solutions except in the presence of humic substances, and the biodegradability of the drug was considered poor (Kwon and Armbrust, 2006). The major mechanism for its removal from waste water was determined to be adsorption onto sediments. It remains photostable even under UVA irradiation except when irradiated in the presence of metal ions (Maślanka et al., 2013) and has been previously predicted to accumulate in the environment (Sebastine and Wakeman, 2003). Indeed, fluoxetine was observed to be persistent in sewage sludge-amended soil cultures for over 200 days (Redshaw et al., 2008) and was photolyzed more efficiently in the presence of humic substances (Lam et al., 2005).

Similarly, chlorinated drugs such as sertraline (another SSRI antidepressant) have also been shown to be fairly resistant to degradation even in the presence of oxidation catalysts (such as Iron-TAML) (Shen et al., 2011). Nevertheless, as noted in Chapter 6, advanced oxidation techniques have had some success in removing these compounds from wastewater. However, whether these techniques remain cost-effective and actually "environmentally friendly" remains to be seen.

Functional groups which can be biodegradable include esters and amides. A typical example of an ester-containing drug is pethidine (meperidine). This μ-opioid agonist is short-acting as the cleavage of the ester functional group renders the drug inactive. Thus, the use of these particular functional groups in drug design could allow for an environmentally friendlier alternative than a drug which does not possess these functional groups and is therefore less readily degraded (Eissen and Backhaus, 2011). However, it must be noted that the presence of biodegradable functional groups usually also means that they are sites of instability for storage and dosing purposes so that their pharmacological efficacy may be in question.

Despite the apparent opposing desires generally associated with biodegradability and efficacy, some medications currently on the market are purported to be both highly biodegradable and efficacious. For example, valproic acid is an effective antiepileptic drug suitable for treating various types of seizure episodes, and yet it is also extremely biodegradable (Lubick, 2008). Therefore, various factors can affect whether a pharmaceutical is biodegradable or not depending on "the circumstances" for which the pharmaceutical is used and the ambient conditions or stresses the API is placed in.

Even if a parent API is degradable, do we know enough about its transformation products to determine whether they have the potential to persist and induce aquatic toxicity in the environment? For example, while valsartan (an antihypertensive of the angiotensin II receptor blocker class) can be readily biodegraded, it is converted into a persistent transformation product through N-dealkylation, subsequent amide hydrolysis, and hydrolysis/oxidation to valsartan acid (Berkner and Thierbach, 2013). Thus, any pharmaceutical cannot be considered "green" without complete toxicological assessment of the transformation products, and there is currently a shortage of data on the toxicity of transformation products in the aquatic environment.

Furthermore, as explained in Chapter 1, pharmaceuticals are usually excreted from the body as metabolites as well as the unchanged parent drug. Many of these metabolites are bioactive, and indeed may transform back into the parent pharmaceutical given the various environmental processes acting on the molecule as it is excreted out of the body and transported into the sewage. This situation is illustrated by the conjugates of active pharmaceuticals. These compounds are Phase II metabolites of unchanged parent pharmaceuticals or their Phase I hydroxylation/oxidation products, produced by glucuronidation or sulfation. Most of these glucuronides or sulfates are inactive, although there are a select few glucuronides which possess bioactivity (such as morphine-6-glucuronide). However, excrement from feces in household sewage contains *Escherichia coli* bacteria which contain β-D-glucuronidase. This enzyme can cleave glucuronic acid from the conjugate and yield the free pharmaceutical or Phase I metabolite. The conversion of conjugates back to their free pharmaceutical forms increases the level of pharmaceutical that is discharged from the sewage treatment facility if the sewage treatment facility is not capable of removing the free pharmaceutical.

Currently, there is still little information on glucuronide or sulfate back-transformation in sewage treatment facilities for most pharmaceuticals which can be metabolized to Phase II conjugates. This in turn has been used as the reason for the appearance of higher concentrations of the parent pharmaceutical in sewage effluent compared to influent in some reported studies of conventional (primary or secondary) or even tertiary sewage treatment (Bendz et al., 2005; Miao et al., 2005; Leclercq et al., 2009).

There some direct evidence of glucuronide back-transformation for selected pharmaceuticals. The conversion of paracetamol (acetaminophen) glucuronide to the free pharmaceutical was observed in a batch reactor employing secondary activated sludge treatment (Möhle et al., 1999). Direct evidence of the conversion of sex steroids to their free pharmaceutical forms has also been observed in wastewater (Gomes et al., 2009; Kumar et al., 2012) and in soils (Zheng et al., 2013). In addition, the back-transformation of diclofenac β-O acyl glucuronide to free diclofenac was observed in a microbial-activated sludge closed-bottle test with OECD 301C specifications (Lee et al., 2012).

Phase II acetylated metabolites have been observed to back-transform to their parent pharmaceutical during sewage treatment. For example, Clara et al. (2005) and Göbel et al. (2005) observed the back-transformation of a major sulfamethoxazole Phase II metabolite, *N*-4-acetylsulfamethoxazole, back to the free pharmaceutical.

The transformation of metabolites other than sulfate, glucuronide, or acetylated conjugates back to their free pharmaceutical can also occur outside of sewage treatment facilities. The back-transformation of 4-nitroso-sulfamethoxazole to sulfamethoxazole was observed during direct photolysis in lake water (Bonvin et al., 2012). It was postulated that this could potentially be a source of sulfamethoxazole in the environment and thus is a reason for its persistence. Clearly, while much remains to be understood about this topic, these results collated from the small number of studies available suggest that in order for a pharmaceutical to be "environmentally friendly" or "green", there needs to be a better understanding of the conditions that mediate the back-transformation of metabolites to their free pharmaceutical before an overall environmental risk and hazard assessment for them can be undertaken.

Another aspect that can be addressed to determine whether a pharmaceutical is "green" is whether conventional sewage treatment processes—or the most common processes currently in place in sewage facilities—can remove them. While much information concerning removal efficiencies of a number of common pharmaceuticals has been discussed in Chapter 6, the removal efficiencies vary highly between different facilities and no substantive conclusion regarding the "environmental friendliness" of the pharmaceutical can be made. While some differences can be attributed to the type of sewage treatment that is available at a particular plant (i.e., some only have secondary activated sludge; others have tertiary treatment such as ozonation, chlorination, and/or UV irradiation), the removal efficiencies for the same processes at different plants may vary (Chapter 6). In conventional sewage treatment, such as activated sludge processes, the type of bacteria and ambient conditions (aerobic vs. anaerobic degradation) can affect the removal of a pharmaceutical or a metabolite. For example, the removal of sulfamethoxazole with the activated sludge with and without bacteria *Rhodococcus equi* alone and in combination was investigated by Larcher and Yargeau (2011) and only a maximum of 5% removal after 24h treatment was obtained by either treatment. This poor removal is consistent with other studies of the degradation of sulfamethoxazole in aerobic and anaerobic sludge conditions where in most cases only a partial removal of the pharmaceutical was observed (Xu et al., 2007; Chang et al., 2008). Given that biodegradation may be the predominant mechanism for removal by conventional sewage treatment processes as sulfamethoxazole adsorption to biosolids and sludge is limited (Müller et al., 2013), it is dubious whether sulfamethoxazole is an environmentally friendly or green pharmaceutical if only conventional sewage treatment is applied.

Finally, a number of the synthetic pathways used to manufacture pharmaceuticals often involve many individual reaction steps each one of which have their own efficiency and involve consumption of starting materials and production of by-products other than the target product. All of these processes consume energy and can impact on the environment. To minimize these effects and maximize the overall product yield, there are increasing efforts by synthetic chemists to develop alternative steps in a synthesis which incorporate at least some of the principles of green chemistry.

Specific applications of green processes in the pharmaceutical industry have been reviewed by Cue and Zhang (2009) and some more recent examples include:

(1) The use of synthetic enzyme cascade reactions and less expensive starting materials reduces the number of reaction steps and avoids the isolation of reaction intermediates in the synthesis of nor(pseudo)ephedrine (NPE) which acts an adrenergic re-uptake inhibitor (Sehl et al., 2014).
(2) 'Environmentally friendly' syntheses of paracetamol analogues as potential analgesic and antipyretic agents (Reddy et al., 2014).
(3) Use of catalytic metal-free reductions leading to milder and safer reactions that can be undertaken in air for the synthesis of various medically relevant amines and nitrogen heterocycles (as reviewed by Rueping et al., 2014).
(4) The use of ultrasound-assisted synthesis of acetylcholinesterase (AChE) inhibitors lead to shorter reaction times, higher yields, simpler work-up procedures and purification of the product which did not involve chromatography (Khan et al., 2014).

Green chemistry principles were also used in the development of "The Tiered Protocol for Endocrine Disruption (TiPED)" for use by chemists in the design of new pharmaceuticals (Schug et al., 2013).

8.2 Life-Cycle Assessment (LCA)

As well as the direct effects of a pharmaceutical (and its metabolites) on the environment, the indirect impacts of the pharmaceutical from its manufacture to destruction have to be addressed. As with the manufacture of other household goods and personal care products, the industrial production of pharmaceuticals uses solvents, energy, and water. In addition, the yield of compound that is obtained from starting reagents also needs to be assessed to determine if the entire process is environmentally friendly (Jiménez-González et al., 2004; Wernet et al., 2010). Furthermore, upon reaching the sewage treatment facility, the removal processes needed to degrade a pharmaceutical also require energy and the use of added chemicals such as chlorine gas and flocculants. These will all no doubt contribute to environmental emissions in some manner.

One of the tools that is currently used to allow for all these issues and predict whether a particular chemical is environmentally acceptable when used in a medicine or consumer product in its various applications is *Life-Cycle Assessment* (LCA) (Nieuwlaar, 2013; Kralisch et al., 2015). This involves redrawing the general life-cycle for a product shown earlier in Figure 8.1 to highlight the different input and outputs over a product's life span (Figure 8.2).

The goal of such a life-cycle assessment is a "cradle-to-grave" approach which takes into account the entire life-cycle of materials and associated systems (US General Services Administration Sustainable Facilities Tool Life Cycle Assessment (LCA) Overview, downloadable from http://sftool.gov/plan/400/life-cycle-assessment-lca-overview, accessed 8-8-15):

(1) Quantify or otherwise characterize all the inputs and outputs over a product's life span.
(2) Specify the potential environmental impacts of the material flows.
(3) Consider alternative approaches that might lessen these impacts.

There are four main steps involved in undertaking such a life-cycle assessment (ISO 14044 and 14040, EPA 2006). This is undertaken by a comprehensive "cradle-to-grave" analysis of its environmental effects from the time the compound is synthesized to the time it is destroyed.

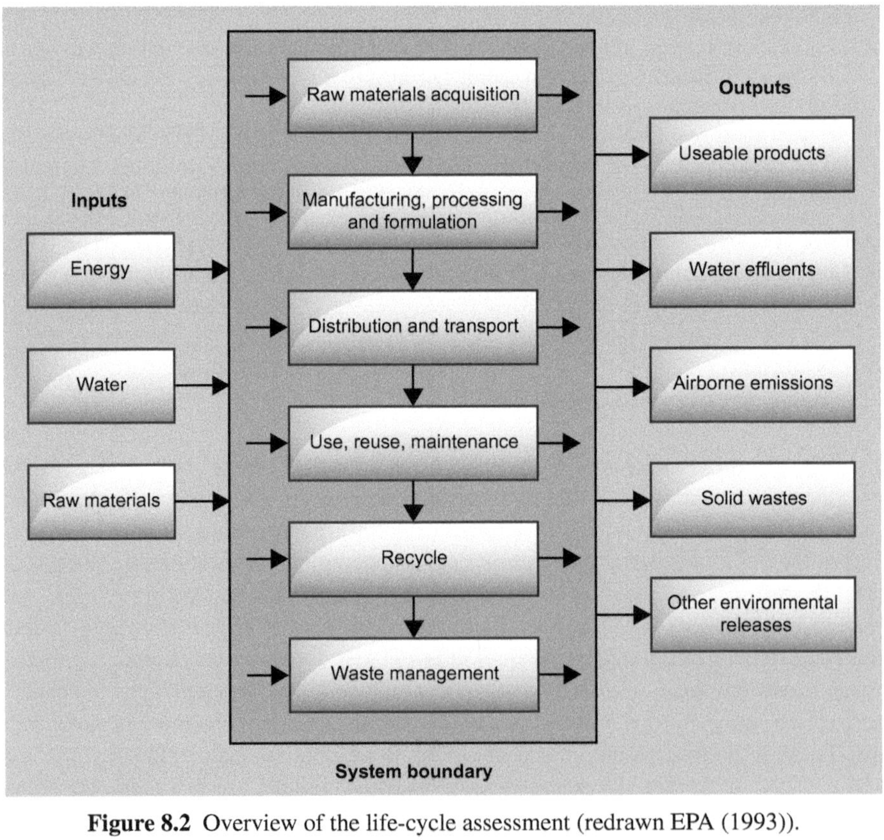

Figure 8.2 Overview of the life-cycle assessment (redrawn EPA (1993)).

Approaches to such environmental management had been practiced in many different ways for years but in 1996 the International Standards Organization (ISO) unified them by establishing the method of LCA with the ISO 14000 series of standards "to promote more effective and efficient environmental management in organizations and to provide useful and usable tools—ones that are cost-effective, system-based, [and] flexible, and reflect the best organizations and the best organizational practices available for gathering, interpreting, and communicating environmentally relevant information" (http://www.iso.org/iso/iso14000; accessed 3-8-15).

In 2006 the EPA published a very comprehensive report "Life Cycle Assessment: Principles and Practice" (EPA/600/R-06/060 downloadable from http://www.epa.gov/nrml/std/lca/lca.html; accessed 6-8-15) which included:

(1) defining the goal and scope of the assessment;
(2) compiling a life-cycle inventory, which is a comprehensive analysis of all the interactions of a product with the environment;
(3) assessing the environmental impact from the data compiled from the life-cycle inventory;
(4) interpreting the overall results of the assessment from data obtained in the life-cycle inventory and the environmental impact assessment.

Benefits of the life-cycle assessment approach include (US General Services Administration Sustainable Facilities Tool Life Cycle Assessment (LCA) Overview, downloadable from http://sftool.gov/plan/400/life-cycle-assessemnt-lca-overview; accessed 8-8-15):

(1) The whole environmental impact is identified.
(2) Environmental effects such as overall energy consumption, water and air emissions can be quantified.
(3) It enables 'like' products to be objectively compared taking into account every aspect.
(4) It goes beyond just the product 'use' phase.
(5) It can help identify inefficient or significant changes in any of the life-cycle phases.
(6) It can lead to a reduction in overall environmental impact and cost.

There are also some challenges to this type of life-cycle assessment including:

(1) Where are the boundaries defined and what is the scope of the model?
(2) What data are available to quantify all material input and outputs at all stages of the defined scope? How reliable is the data?
(3) How can material flows be quantified into environmental impacts?
(4) What are the relative weightings given to environmental impacts and what are the most important ones?

Life-cycle assessments can be performed not only on the environmental risk associated with the use and disposal of the pharmaceutical, but also on the materials and processes needed to produce and destroy the API and its dosage form. A tool for the "cradle-to-cradle" stewardship or "gate-to-gate" life-cycle analysis for a case study of a lyophilized pharmaceutical has been reported by Mata et al. (2012).

Because energy consumption/resource consumption in producing and destroying the pharmaceutical is also considered in a life-cycle assessment, decisions based on life-cycle assessments ensure that an environmental problem is not transferred from one aspect to another (i.e., we can use advanced oxidation processes to degrade pharmaceuticals in sewage, but are we doing more damage to the environment as a result of treating this sewage with advanced oxidation processes than if we just discharge the untreated effluent into the environment?). However, models and assumptions used in these assessments can potentially be simplifications of actual reality, and even though the life-cycle assessments are usually carried out with in the bounds defined by the ISO standards, there is considerable variability in definition of assessment parameters, system boundaries, and the interpretation of results which may limit the utility of the analyses performed (Corominas et al., 2013). A major variation between life-cycle analyses in different studies is the definition of the functional unit, that is the reference quantity as defined in a particular life-cycle assessment where inputs and outputs of two different systems can be measured in a standardized manner. As is apparent in Chapter 3, much debate still exists regarding the most "environmentally friendly" method for the disposal of unused medications even with data generated from a standardized life-cycle assessment (Cook et al., 2012).

A life-cycle assessment for the general use of advanced oxidation processes in the removal of pharmaceutical contaminants has been undertaken (Wenzel et al., 2008) as well as for the process of ozonation in tertiary treatment (Papa et al., 2013). In the former study (Wenzel et al., 2008), it was observed that sand filtration processes were less harmful to the environment than ozonation or membrane bioreactor processes, and sand filtration was estimated to avoid pollution more than induce it. However, it was concluded that the overall pollution induced by operating the tertiary treatment processes resulting from energy consumption was comparable to pollution avoided by removing the contaminants from sewage. Thus, it is uncertain whether using such advanced technologies in sewage treatment is more "environmentally friendly" than just allowing contaminants to be discharged into the environment without treatment. A similar conclusion was reached in the study by Papa et al. (2013) concerning the ozonation of sewage water where it was estimated that the removal of a contaminant by ozonation avoided a similar amount of contamination as that induced by operating the sewage treatment process.

Concerning the environmental impact of particular pharmaceuticals in digested sewage sludge when applied as biosolids onto agricultural land, only six pharmaceuticals and personal care products (galaxolide, tonalide, diazepam, ibuprofen, sulfamethoxazole, and 17α-ethinylestradiol) were considered to have a negative environmental impact on soils. However, a greater negative environmental impact was attributed to heavy metals discharged into the sewage sludge (Hospido et al., 2010). This conclusion is consistent with a more recent life-cycle assessment where the environmental impact of pharmaceuticals in sewage water was considered negligible (Igos et al., 2012) irrespective of whether the sewage water was treated in centralized or decentralized sewage treatment facilities.

Other examples of the application of life-cycle assessment to compare two different wastewater treatment processes include a study by Foley et al. (2010) that showed that microbial fuel and electrolysis cells were more environmentally efficient than conventional anaerobic treatment. A newly developed carbon nanofiber supported catalyst has been shown to have less environmental impact when reducing bromates in waste water compared to a conventional Pd/Al_2O_3 catalyst (Yaseneva et al., 2014).

8.3 Summary

Ultimately, whether a particular pharmaceutical is considered "green" is dependent on its risk assessment as well as the overall life-cycle assessment. Unlike life-cycle assessments, risk assessments are the evaluations of the likelihood (probability) that particular substances are detrimental to the environment or public health (both human and of aquatic species in the case of pharmaceuticals in water). As explained in Chapter 1, risk assessments are usually based on PNECs (predicted no-effect concentrations) and PECs (predicted environmental concentrations). For example, the human health risk assessment of the antiepileptic carbamazepine has been previously studied and despite the environmental persistence of carbamazepine in sewage and surface water, no appreciable risks associated with human exposure to the pharmaceutical could be calculated from such assessment (Cunningham et al., 2010). This result is consistent with the risks discussed for particular pharmaceuticals in Chapter 1.

Inevitably, we do not know the full nature and behavior of many pharmaceuticals in the environment simply because comprehensive information on their behavior, degradation mechanisms, and transformation products in natural and engineered environments are lacking. There is also an increasing number of new pharmaceuticals that we know absolutely nothing about in terms of their presence, persistence, bioaccumulation, and toxicity in the environment (Agüera et al., 2013) although there have been newly developed methodologies to find unknown or suspect pollutants (Hug et al., 2014). As a consequence it is also impossible for the "rules of thumb," life-cycle assessments, or risk assessments to be completely predictive of the "environmental friendliness" of new compounds. Therefore there is currently no conclusive answer as to how "green" are particular pharmaceuticals.

References

Agüera, A., Martínez Bueno, M.J., Fernández-Alba, A.R., 2013. New trends in the analytical determination of emerging contaminants and their transformation products in environmental waters. Environ. Sci. Pollut. Res. 20 (6), 3496–3515.

Anastas, P.T., Warner, J.C., 1998. Green Chemistry: Theory and Practice. Oxford University Press, Oxford, UK.

Bendz, D., Paxeus, N.A., Ginn, T.R., Loge, F.J., 2005. Occurrence and fate of pharmaceutically active compounds in the environment, a case study: Hoje River in Sweden. J. Hazard. Mater. 122 (3), 195–204.

Berkner, S., Thierbach, C., 2013. Biodegradability and transformation of human pharmaceutical active ingredients in environmentally relevant test systems. Environ. Sci. Pollut. Res. 21 (16), 9461–9467.

Boethling, R.S., Sommer, E., DiFiore, D., 2007. Designing small molecules for biodegradability. Chem. Rev. 107 (6), 2207–2227.

Bonvin, F., Omlin, J., Rutler, R., Schweizer, W.B., Alaimo, P.J., Strathmann, T.J., McNeill, K., Kohn, T., 2012. Direct photolysis of human metabolites of the antibiotic sulfamethoxazole: evidence for abiotic back-transformation. Environ. Sci. Technol. 47 (13), 6746–6755.

Chang, H., Hu, J.Y., Wang, L.Z., Shao, B., 2008. Occurrence of sulfonamide antibiotics in sewage treatment plants. Chin. Sci. Bull. 53 (4), 514–520.

Clara, M., Strenn, B., Gans, O., Martinez, E., Kreuzinger, N., Kroiss, H., 2005. Removal of selected pharmaceuticals, fragrances and endocrine disrupting compounds in a membrane bioreactor and conventional wastewater treatment plants. Water Res. 39 (19), 4797–4807.

Cook, S.M., VanDuinen, B.J., Love, N.G., Skerlos, S.J., 2012. Life cycle comparison of environmental emissions from three disposal options for unused pharmaceuticals. Environ. Sci. Technol. 46 (10), 5535–5541.

Corominas, L., Foley, J., Guest, J.S., Hospido, A., Larsen, H.F., Morera, S., Shaw, A., 2013. Life cycle assessment applied to wastewater treatment: state of the art. Water Res. 47 (15), 5480–5492.

Cue, B.W., Zhang, J., 2009. Green process chemistry in the pharmaceutical industry. Green Chem. Lett. Rev. 2 (4), 193–211. http://dx.doi.org/10.1080/17518250903258150.

Cunningham, V.L., Perino, C., D'Aco, V.J., Hartmann, A., Bechter, R., 2010. Human health risk assessment of carbamazepine in surface waters of North America and Europe. Regul. Toxicol. Pharmacol. 56 (3), 343–351.

Daughton, C.G., 2003a. Environmental stewardship of pharmaceuticals: the green pharmacy. In: The Proceedings of the Third International Conference on Pharmaceuticals and Endocrine Disrupting Chemicals in Water. National Ground Water Association, 19–21 March 2003, Minneaplois, Minnesota. (downloadable from the Web: http://www.epa.gov/esd/bios/daughton/ngwa2003.pdf) accessed 31.07.15.

Daughton, C.G., Ruhoy, I.S., 2009. Environmental footprint of pharmaceuticals: the significance of factors beyond direct excretion to sewers. Environ. Toxicol. Chem. 28 (12), 2495–2521.

Eissen, M., Backhaus, D., 2011. Pharmaceuticals in the environment: an educational perspective. Environ. Sci. Pollut. Res. 18 (9), 1555–1566.

EPA 1993. EPA /600/SR-92/245 Report Feb, 1993. Life-Cycle Assessment: Inventory, Guidelines and Principles.

Foley, J.M., Rozendal, R.A., Hertle, C.K., Lant, P.A., Rabaey, K., 2010. Life cycle assessment of high-rate anaerobic treatment, microbial fuel cells, and microbial electrolysis cells. Environ. Sci. Technol. 44, 3629–3637.

Göbel, A., Thomsen, A., McArdell, C.S., Joss, A., Giger, W., 2005. Occurrence and sorption behavior of sulfonamides, macrolides, and trimethoprim in activated sludge treatment. Environ. Sci. Technol. 39 (11), 3981–3989.

Gomes, R.L., Scrimshaw, M.D., Lester, J.N., 2009. Fate of conjugated natural and synthetic steroid estrogens in crude sewage and activated sludge batch studies. Environ. Sci. Technol. 43 (10), 3612–3618.

Hospido, A., Carballa, M., Moreira, M., Omil, F., Lema, J.M., Feijoo, G., 2010. Environmental assessment of anaerobically digested sludge reuse in agriculture: potential impacts of emerging micropollutants. Water Res. 44 (10), 3225–3233.

Hug, C., Ulrich, N., Schulze, T., Brack, W., Krauss, M., 2014. Identification of novel micropollutants in wastewater by a combination of suspect and nontarget screening. Environ. Pollut. 184, 25–32.

Igos, E., Benetto, E., Venditti, S., Kohler, C., Cornelissen, A., Moeller, R., Biwer, A., 2012. Is it better to remove pharmaceuticals in decentralized or conventional wastewater treatment plants? A life cycle assessment comparison. Sci. Total Environ. 438, 533–540.

Jiménez-González, C., Curzons, A., Constable, D.C., Cunningham, V., 2004. Cradle-to-gate life cycle inventory and assessment of pharmaceutical compounds. Int. J. Life Cycle Assess. 9 (2), 114–121.

Khan, A.U., Malik, N., Alam, M., Lee, D.-U., 2014. Ultrasound-assisted synthesis of benzothiazepines and assessment of their in vitro acetylcholinesterase inhibition activity. Green Chem. Lett. Rev. 7 (2), 158–166. http://dx.doi.org/10.1080/17518253.2014.909889.

Khetan, S.K., Collins, T.J., 2007. Human pharmaceuticals in the aquatic environment: a challenge to green chemistry. Chem. Rev. 107, 2319–2364.

Kralisch, D., Ott, D., Gericke, D., 2015. Rules and benefits of lifecycle assessment in green chemical process and synthesis design: a tutorial review. Green Chem. 17, 123–145. http://dx.doi.org/10.1039/C4GC01153H.

Kumar, V., Johnson, A.C., Nakada, N., Yamashita, N., Tanaka, H., 2012. De-conjugation behavior of conjugated estrogens in the raw sewage, activated sludge and river water. J. Hazard. Mater. 227–228, 49–54.

Kummerer, K., 2009. The presence of pharmaceuticals in the environment due to human use—present knowledge and future challenges. J. Environ. Manag. 90 (8), 2354–2366.

Kwon, J.-W., Armbrust, K.L., 2006. Laboratory persistence and fate of fluoxetine in aquatic environments. Environ. Toxicol. Chem. 25 (10), 2561–2568.

Lam, M.W., Young, C.J., Mabury, S.A., 2005. Aqueous photochemical reaction kinetics and transformations of fluoxetine. Environ. Sci. Technol. 39 (2), 513–522.

Larcher, S., Yargeau, V., 2011. Biodegradation of sulfamethoxazole by individual and mixed bacteria. Appl. Microbiol. Biotechnol. 91 (1), 211–218.

Leclercq, M., Mathieu, O., Gomez, E., Casellas, C., Fenet, H., Hillaire-Buys, D., 2009. Presence and fate of carbamazepine, oxcarbazepine, and seven of their metabolites at wastewater treatment plants. Arch. Environ. Contam. Toxicol. 56 (3), 408–415.

Lee, H.-J., Lee, E., Yoon, S.-H., Chang, H.-R., Kim, K., Kwon, J.-H., 2012. Enzymatic and microbial transformation assays for the evaluation of the environmental fate of diclofenac and its metabolites. Chemosphere 87 (8), 969–974.

Lubick, N., 2008. Opening the "green pharmacy". Environ. Sci. Technol. 42 (23), 8620–8621.

Manahan, S.E., 2009. In: Fundamentals of Environmental Chemistry, third ed. CRC Press, Baco Raton, FL.

Maślanka, A., Hubicka, U., Krzek, J., Walczak, M., Izworski, G., 2013. Determination of fluoxetine in the presence of photodegradation products appearing during UVA irradiation in a solid phase by chromatographic-densitometric method, kinetics and identification of photoproducts. Acta Chromatogr. 25 (3), 465–481.

Mata, T.M., Martins, A.A., Neto, B., Martins, M.L., Salcedo, R.L.R., Costa, C.A.V., 2012. LCA tool of sustainability evaluations in the pharmaceutical industry. Chem. Eng. Trans. 26, 261–266.

Miao, X.-S., Yang, J.-J., Metcalfe, C.D., 2005. Carbamazepine and its metabolites in wastewater and in biosolids in a municipal wastewater treatment plant. Environ. Sci. Technol. 39 (19), 7469–7475.

Möhle, E., Kempter, C., Kern, A., Metzger, J.W., 1999. Untersuchungen zum Abbau von Pharmaka in kommunalen Kläranlagen mit HPLC-Electrospray-Massenspektrometrie. Acta Hydrochim. Hydrobiol. 27 (6), 430–436.

Müller, E., Schüssler, W., Horn, H., Lemmer, H., 2013. Aerobic biodegradation of the sulfonamide antibiotic sulfamethoxazole by activated sludge applied as co-substrate and sole carbon and nitrogen source. Chemosphere 92 (8), 969–978.

Nieuwlaar, E., 2013. Life Cycle Assessment and Energy Systems. Reference Module in Earth Systems and Environmental Sciences. Elsevier.

Papa, M., Pedrazzani, R., Bertanza, G., 2013. How green are environmental technologies? A new approach for a global evaluation: the case of WWTP effluents ozonation. Water Res. 47 (11), 3679–3687.

Redshaw, C., Cooke, M., Talbot, H., McGrath, S., Rowland, S., 2008. Low biodegradability of fluoxetine HCl, diazepam and their human metabolites in sewage sludge-amended soil. J. Soils Sediments 8 (4), 217–230.

Reddy, Y.D., Reddy. Ch. V.R., Dubey, P.K., 2014. Green approach for drug design and discovery of paracetamol analogues as potential analgesic and antipyretic agents. Green Chem. Lett. Rev. 7 (1), 24–31. http://dx.doi.org/10.1080/17518253.2014.895426.

Schug, T.T., Abagyan, R., Blumberg, B., Collins, T.J., Crews, D., DeFur, P.L., Dickerson, S.M., Edwards, T.M., Gore, A.C., Guillette, L.J., Hayes, T., Heindel, J.J., Moores, A., Patisaul, H.B., Tal, T.L., Thayer, K.A., Vandenberg, L.N., Warner, J.C., Watson, C.S., vom Saal, F.S., Zoeller, R.T., O'Brien, K.P., Myers, J.P., 2013. Designing endocrine disruption out of the next generation of chemicals. Green Chem. 15, 181–198. http://dx.doi.org/10.1039/C2GC35055F.

Sebastine, I.M., Wakeman, R.J., 2003. Consumption and environmental hazards of pharmaceutical substances in the UK. Process Saf. Environ. Prot. 81 (4), 229–235.

Sehl, T., Hailes, H.C., Ward, J.M., Menyes, U., Martina Pohl, M., Rother, D., 2014. Efficient 2-step biocatalytic strategies for the synthesis of all nor(pseudo)ephedrine isomers. Green Chem. 16, 3341–3348. http://dx.doi.org/10.1039/C4GC00100A.

Shen, L.Q., Beach, E.S., Xiang, Y., Tshudy, D.J., Khanina, N., Horwitz, C.P., Bier, M.E., Collins, T.J., 2011. Rapid, biomimetic degradation in water of the persistent drug sertraline by TAML catalysts and hydrogen peroxide. Environ. Sci. Technol. 45 (18), 7882–7887.

Wenzel, H., Larsen, H.F., Clauson-Kaas, J., Høibye, L., Jacobsen, B.N., 2008. Weighing environmental advantages and disadvantages of advanced wastewater treatment of micro-pollutants using environmental life cycle assessment. Water Sci. Technol. 57 (1), 27–32.

Wernet, G., Conradt, S., Isenring, H., Jiménez-González, C., Hungerbühler, K., 2010. Life cycle assessment of fine chemical production: a case study of pharmaceutical synthesis. Int. J. Life Cycle Assess. 15 (3), 294–303.

Xu, W.H., Zhang, G., Li, X.D., Zou, S.C., Li, P., Hu, Z.H., Li, J., 2007. Occurrence and elimination of antibiotics at four sewage treatment plants in the Pearl River Delta (PRD), South China. Water Res. 41 (19), 4526–4534.

Yaseneva, P., Marti, C.F., Palomares, E., Fan, X., Morgan, T., Perez, P.S., Ronning, M., Huang, F., Yuranova, T., Kiwi-Minsker, L., Derrouiche, S., Lapkin, A.A., 2014. Efficient reduction of bromates using carbon naofibre supported catalysts: Experimental and a comparative life cycle study. Chem. Eng. J. 248, 230–241.

Zheng, W., Zou, Y., Li, X., Machesky, M.L., 2013. Fate of estrogen conjugate 17α-estradiol-3-sulfate in dairy wastewater: comparison of aerobic and anaerobic degradation and metabolite formation. J. Hazard. Mater. 258–259, 109–115.

Summary and Conclusions

9

9.1 The potential environmental risk of pharmaceuticals?

The environmental consequences of pharmaceutical usage have become a global concern. The drivers of this concern are multiple and include the fact that this heterogeneous group of compounds are designed to be biologically active, that they can be potent even at low concentrations, that there is a growing use of these compounds in significant volumes, and importantly that there are advances in analytical techniques that allow the detection of these compounds in the environment even at very low concentrations.

This text has explored the many aspects of the environmental life-cycle of human pharmaceuticals and how this can lead to their ultimate detection, particularly in the natural aquatic environment where they have the potential to adversely affect a range of organisms.

The environmental sources of these pharmaceuticals used by humans were discussed in Chapter 2 and mainly involve the growing dependence on pharmaceutical products for treatments, both curative and preventative. Patient behavior, access to medications, and specific national medication policies can significantly influence the amount and types of medications utilized. The specific usage patterns locally and globally are influenced by patients, prescribers, and funders. It is important to remember that the specific environmental risk is related to the properties of each individual compound, and that not all pharmaceuticals pose the same risk as others.

Human use of these pharmaceuticals leads to their introduction either in the parent form or as a metabolite into sewage water as discussed in Chapter 3. Alternatively, if they are prescribed but then remain unused, they are often disposed of by tipping them down the toilet or sink where they also end up in the sewage. Although there is clearly nothing that can be done to change the human excretion of pharmaceuticals, prescribing practices can be changed to lessen the quantity of unused pharmaceuticals which must be ultimately disposed on in some manner.

There is an growing number of reports from around the globe of the detection of an increasingly wide range of human pharmaceuticals both in wastewater and in natural environmental compartments such as water systems (streams, rivers, lakes, and estuaries) receiving sewage effluent, and in groundwater. This increase in reporting the presence of these pharmaceuticals has in part arisen from improvements in sampling techniques and increased sensitivity in analytical methods as described in Chapter 4. Although still a somewhat time-consuming procedure requiring skilled analytical techniques, the detection and quantitation methods are becoming more commonly used around the world and should lead to an ever-increasing amount of information on the concentrations of many pharmaceuticals in different environmental situations.

The global concern about detectable levels of human pharmaceuticals entering the environment has led to questions regarding the implications of these pharmaceuticals

The Life-Cycle of Pharmaceuticals in the Environment. http://dx.doi.org/10.1016/B978-1-907568-25-1.00009-8

for humans as well as other aquatic biological life forms. As indicated in Chapter 5, toxicity testing, the impact on reproductive function and bioaccumulation of these compounds within aquatic life forms are all of growing interest. However, given that pharmaceuticals must be assessed in both case-by-case and "mixture" settings, and that it is not possible to asses all life forms that may be affected by any given compound or combination, it is difficult to move forward in this area. Additionally, within pharmaceuticals as a group, specific toxicity occurs via a myriad of mechanisms and so this area is subject to considerable uncertainty.

Clearly, one major approach to lowering the concentrations of pharmaceuticals entering the environment is by more efficient wastewater (sewage) treatment as discussed in Chapter 6. Although the concentrations of many pharmaceuticals are significantly reduced even using just primary treatment which involves a crude sedimentation step, some other pharmaceuticals require more advanced and hence costly treatments to reduce their levels beginning with secondary treatment using biological means and then progressing to tertiary methods which can include the use of membrane filtration techniques and advanced redox processes such as ozonation and various photolytic-type treatments. As more is learnt about the environmental levels that can cause unacceptable effects on ecosystems, there will be an increasing demand from the public to implement these more advanced, and hence more costly, wastewater treatment processes.

A very practical way to reduce the concentrations of pharmaceuticals entering the environment is by enacting legislation controlling their prescription, types of disposal methods, levels of wastewater treatment, and maximum levels of discharge into the environment. There is a considerable variation in the types of this legislation between different countries as discussed in Chapter 7.

Green pharmacy incorporating green chemistry principles and related life-cycle assessment as highlighted in Chapter 8 is one of the several potential ways of minimizing the potential environmental risk of the abundant global use of pharmaceuticals as a group and would have particular application for those compounds or groups of compounds that have a high predicted environmental risk.

9.2 Some possible future considerations

The increasingly global concern and hence interest in lessening the effects arising from human pharmaceuticals entering the environment suggests a number of areas that the public, health professionals, wastewater engineers, enforcement agencies, and legislators could usefully address. These might include:

1. Clinicians (or prescribing guidelines) should consider the environmental hazard of a particular pharmaceutical when it is prescribed. This may mean that where two or more drugs are expected to be comparable in terms of efficacy, then those with a lower environmental hazard risk could be considered first-line therapy.
2. Further research could be undertaken into patient behavior regarding collection and use of prescribed medications, and reasons why patients collect medications that they do not intend to use in full.

3. A thorough investigation into methods of medication disposal. Currently, it is not transparent what happens to the medications that are returned to pharmacies for destruction in the different regulatory health areas across a district or country.
4. There needs to be further research into the potential of these unused medications to be "reused" possibly by way of donation. Many of the medications that are presently returned for disposal are in original blister packaging with a long expiry date. While there are concerns regarding tampering and storage, it would be worthwhile to investigate if there is any therapeutic loss of efficacy with these returns.
5. Measurement of the actual levels of the wide range of commonly prescribed pharmaceuticals in each country in a wide range of different environmental situations. This would provide a much more accurate idea of whether there is likely to be any adverse ecotoxicological effects due to the use of pharmaceuticals in that country, rather than having to rely exclusively on the very variable measurements widely reported for other countries and often under very different conditions.
6. Determine ecotoxic parameters for organisms that are unique to a country in response to their exposure to different pharmaceuticals and their metabolites and transformation products. This might identify some uniquely sensitive organisms that are only found in the environment of that country and hence may require special requirements for even lower concentrations of specific pharmaceuticals at the point of discharge than are necessary in other countries.
7. The potential adverse effects that might arise from the widespread use of veterinary pharmaceuticals in different counties have not been addressed in the present text, but their environmental impact may well be as important as that of human pharmaceuticals. This could be particularly important in a country such as New Zealand where a large proportion of the national income is derived from agricultural activities.
8. Further investigations should be undertaken of the enhanced immobilization of pharmaceuticals and their metabolites that occur when treated wastewater effluent is discharged onto land rather than natural waterways in order to further reduce their potential ecotoxicological effects.
9. The effect of the unique environment aspects of some countries, such as the rainfall and temperature patterns and other seasonal effects, may influence the efficiencies of the various existing sewage treatment plants in that country to degrade pharmaceuticals. This variation needs to be investigated and again may necessitate the purchase and installation of additional wastewater treatment equipment in order to meet the required maximum levels for specific pharmaceuticals in the effluent from each plant in the particular country.
10. Further research needs to be undertaken into the ecotoxic properties of the combined mixture of all the pharmaceutical medications used to varying extents in a particular country rather than considering just the ecotoxicity of each compound on its own. It cannot be assumed that the effect of the total mixture will be just the sum of that of the individual components and it may also be mediated by the general background of natural organic matter present in all waste and natural waters.
11. Further research into green pharmacy and the use of more sustainable synthetic pathways for the manufacture of pharmaceuticals and their degradation in wastewater.
12. Further use of life-cycle assessment type models of pharmaceuticals to better understand their complete "cradle-to-grave" cycles and identify gaps in our present knowledge base.

The life-cycle of pharmaceuticals does not start at a prescription and end when a patient takes a medication. From production to disposal or excretion, pharmaceuticals are ultimately discharged in some form into the environment. Pharmaceuticals are heterogeneous groups of xenobiotics whose activity is diverse and concentration-dependant effects are variable. There are many unknowns as to the long-term risks associated with low (subtherapeutic) levels of pharmaceuticals on humans and other species. Water treatment processes are important to manage the currently recognized emergence of environmental pharmaceutical contaminants internationally. However, a whole system approach, specifically the consideration of the entire life-cycle of pharmaceuticals, needs to be conducted to minimize and manage any real or potential environmental risks that are posed by their widespread global use.

Index

Note: Page numbers followed by *f* indicate figures and *t* indicate tables.